T0127881

Science without Boundaries

Interdisciplinarity in Research, Society, and Politics

Willy Østreng

UNIVERSITY PRESS OF AMERICA,® INC.
Lanham • Boulder • New York • Toronto • Plymouth, UK

Copyright © 2010 by
University Press of America,® Inc.
4501 Forbes Boulevard
Suite 200
Lanham, Maryland 20706
UPA Acquisitions Department (301) 459-3366

Estover Road
Plymouth PL6 7PY
United Kingdom

Library of Congress Control Number: 2009932527
ISBN: 978-0-7618-4830-1 (paperback : alk. paper)
eISBN: 978-0-7618-4831-8

To the memory of Stein Kåre Lilloe – a beloved 'Bessa' and never failing friend.

Contents

List of Figures

Abbreviations

ACIA	Arctic Climate Impact Assessment Study
AMSA	Arctic Marine Shipping Assessment
BEAR	Barents Euro Arctic Region
CAS	Centre for Advanced Study at the Norwegian Academy of Science and Letters
CFC	Chlorofluorocarbon
CNIMF	Central Marine Research and Design Institute
CUDOS	Communism, Universalism, Disinterestedness, Originality and (Organized) Scepticism
DEA	Dynamic Environmental Atlas
ECO	Ecological
EIA	Environmental Impact Assessment
EU	European Union
FCD	'Forum of Corridor deliberations'
FNI	Fridtjof Nansen Institute
GIS	Geographical Information System
IBM	International Business Machines
IAG	International Advisory Group
IEC	International Evaluation Committee of INSROP
INSROP	International Northern Sea Route Programme
INSROP-DEA	The Northern Sea Route – Dynamic Environmental Atlas
INSROP-GIS	INSROP Geographical Information System: Software and Database
IPCC	United Nations Intergovernmental Panel on Climate Change
JRC	Joint Research Committee

NATO	North Atlantic Treaty Organisation
NRC	Norwegian Research Council
NSR	Northern Sea Route
NTNU	Norwegian University of Science and Technology
OF	Ocean Futures, Research Institute
RCN	Research Council of Norway
REF	The Natural and Societal Challenges of the Northern Sea Route: A Reference Work
SCS	Steering Committee of Sponsors
SIM	Simulation of Ship Traffic along the Route
SSBN	Strategic Nuclear Submarines
USIT	Centre for Information Technology at the University of Oslo
UAF	University of Alaska Fairbanks
UN	United Nations
USD	United States Dollar
WW II	Second World War

Preface

This book is about the many issues involved in going beyond disciplinary research practices in science, politics and society—addressing the complexities of their interface.

For a long time, interdisciplinarity in research has been burdened by multiple misconceptions. In certain scientific quarters, there has even been harsh opposition. In the mind of its critics, interdisciplinarity is a practice of dubious scientific value and quality, not least due to its assumed lack of adequate methodology. In the light of this opposition, interdisciplinarity has suffered, becoming the least developed and refined approach of science compared with disciplinary research.

At the same time, interdisciplinarity is the practice that the public sector and societal players are increasingly calling upon to deal with complex systemic challenges such as global warming, eco-system management, poverty, governance, extended security, peacemaking etc. This is the case because interdisciplinarity refers to the *interaction of parts* in complex systems, whereas disciplinary studies refer to the *properties of the parts* of the same systems. Against this backdrop, this book addresses the following six interrelated questions:

- How does interdisciplinary research fit into the overall disciplinary organization of the sciences?
- To what extent does interdisciplinary research meet the strict requirements and high scientific standards of the science community in terms of methodology, approach, adjudication, philosophy and theory-building?
- How does the science community adapt to changing circumstances, internally (within the science community) and externally (in relation to government and society)

- How responsive is the science community to societal and political needs?
- To what extent and under what circumstances do governments intervene to influence and change the practice, agenda-setting, conduct and results of research?
- What patterns of interaction, if any, exist between politics, society and science?

Combined, these questions address the *external*, *internal* and *intermediate influences* affecting the course of science at any one time, forging a framework from which interdisciplinarity may or may not emerge as a natural mind-set and approach for researchers.

In Chapter 5, the theories underpinning these questions and influences are tested against the practice and reality of interdisciplinary research. The main testing ground for this reality check is the *International Northern Sea Route Programme* (INSROP)—a six-year collaborative effort initiated in 1993 by Russia, Japan and Norway. Ideally, to enhance the generality, validity and reliability of the conclusions, systematic and extensive comparisons should have been undertaken with cases of a similar format. Two reasons explain why INSROP got prime attention. First, a true search was initiated to trace and identify cases using a similar analytical approach to that used by the INSROP. The outcome was disappointment: Not a single case could be found that featured all or most of the variables applied in the programme, which turned out to be unique in multiple and special ways. Second, as Head of INSROP I got privileged "bonus information" about the inner life and working of the program that most other cases will not disclose to an outsider and match in a comparative analysis. Thus, INSROP took front stage as a source of not-easily-available information. This is not to say that comparisons with other cases were not made at all. Whenever such cases could shed light on one or more of the variables addressed in INSROP they have been included in the discussion, increasing the generality of the conclusions.

This book is primarily a scholarly work serving as a professional trade text for all stakeholders to complex problems, in society and academia. Its primary target group includes sworn disciplinarians and interdisciplinarians, as well as the vast majority of pragmatic academics and societal stakeholders. The book also serves educational purposes as a supplementary text at postgraduate and doctoral levels. The detailed discussion of the various concepts of science (Chapter 1) and the emerging methodology of interdisciplinarity (Chapter 3) will fill an educational void in textbook literature. The same goes for the discussion of post-academic, post-normal and post-positivist science permeating most chapters of the book (Chapters 4, 5 and 6).

In its present form, this manuscript is the story of a planned article which, of necessity, became a book. My original ambition was one of pure moderation, i.e. to provide a condensed case study (15 to 20 pages) from the inside of the INSROP to pass on to others what was learned on job. However, due to the undisciplined flight of fancy I took when peeling away the multiple layers of the complex subject matter of interdisciplinarity, the process momentum grew in tandem with my curiosity. New and important angles of incidences emerged constantly, and they could neither be set aside without being explored in some detail nor be portrayed on a few short pages. Having taken on a life of its own, the manuscript became far more comprehensive.

The script is a product of four interims of writing: *the year 2000*, when I enjoyed a sabbatical at the University of California, Berkeley; *autumn 2001*, when I served as a Visiting Professor at the University of Alaska Fairbanks (UAF), *autumns 2006* and *2008* when my duties as Scientific Director at the Centre for Advanced Study (CAS) in Oslo opened windows of opportunity that allowed me to reach the finish line with a first draft. In the intervening periods, the project took second priority to other research projects and obligations.

In this protracted process, I had the privilege to draw on the profound competence and expertise of highly esteemed colleagues and friends from many research institutions and professions: Professors Harry Scheiber, David Caron and Ernst Haas at the University of California, Berkeley, Professor Edward Miles at the University of Washington, Seattle, Professor Karen Erickson at the University of Alaska Fairbanks, Professor Peter R. Killeen, Arizona State University, Professor Graham Chapman, Lancaster University and Dr. Douglas Brubaker at the Fridtjof Nansen Institute. These individuals contributed critical and constructive comments to individual chapters and/or to the whole of the manuscript. For their time and insightful involvement, I am deeply indebted and grateful.

My appreciation also goes to professors Duncan Snidal at the University of Chicago and William Potter, at the Monterey Institute of International Studies in California who generously set aside time to share with me their rich, lengthy experience of interdisciplinary matters. A most needed dose of additional inspiration was injected into my writings in spring 2000 when my friend, the Norwegian artist Åke Berg, visited Berkeley. For his unconventional, but fruitful way of thinking, I am most thankful.

I am also deeply indebted to the staffs of the research institutions that provided office space, secretarial assistance, inspiration, support and encouragement during the writing and study periods: The *School of Law* (Boalt Hall) and the *Department of Political Studies* at the University of California in Berkeley, the *Department of Political Studies* and the *Northern Studies Program* at the

UAF, the *Centre for Advanced Study* at the *Norwegian Academy of Science and Letters,* the *Department of Sociology and Political Science* at the Norwegian University of Science and Technology (NTNU) in Trondheim, the research institute *Ocean Futures* (OF) in Oslo and the *Thor Heyerdahl Institute* in Larvik, Norway. Special mention should be made of the multiple contributions provided by my colleagues in the *Department of Political Studies at UAF*: professors Jerry McBeath, Jim Gladden, Jonathan Rosenberg, Amy Lovecraft and Karen Erickson. Apart from being excellent discussion partners, they made my stay at this fine university a delightful and stimulating experience, intellectually as well as socially. The staff and fellows at CAS made a similar contribution, not least through the annual series of *interdisciplinary communicative seminars.* The utility and value of these seminars to the book is evident from the number of references made to them. Last but not least, the research staff at *Ocean Futures* acted as a lasting source of encouragement and inspiration. In particular, Drs. Gaudenz Assenza, Arnfinn Jørgensen-Dahl, Steven Sawhill, Øystein B. Thommessen, Johan Kvarving Vik, Lars Lothe, Yngvild Prytz, Karl Magnus Eger and OF's devoted director Jan Magne Markussen gave much needed input that enabled me to stay the course and finish what I began. *Ocean Futures'* unflagging commitment to the idea of interdisciplinarity in research is manifest in its overall objective to apply a holistic approach and provide discipline-straddling analysis to the utilization and management of the oceans and polar regions.

Most of the attractive graphics and illustrations in the book have been designed and produced by my daughter, Cecilie Cottis Østreng Aslagsen who, due to her extraordinary talent and loyalty to the task, has made herself the recipient of countless hugs from a most pleased and proud father. Dag Christian Bjørnsen at Centre for Information Technology at the University of Oslo (USIT) contributed the remainder of the illustrations. He also merits my appreciation. Last but not least, Linda Sivesind has performed wonders with the language in the book, transforming my NorwEnglish into a language of English origin (An adjective here, a comma there, and presto chango, we have a manuscript).

There are also three more women who deserve special thanks and recognition: Sigvor Hamre Thornton, Karen Erickson, and Sonja Cottis Østreng. Without the combined efforts of these women, this book would definitely never have expanded beyond the scope of an article. Sigvor provided accommodation, meals, friendship and moral support throughout my stay at the University of California, whereas Karen Erickson was instrumental in inviting me to the University of Alaska Fairbanks and for getting my stay funded through various awards and grants. Without the cooperative, warm-hearted spirit of my wife Sonja this book would have remained an unfulfilled dream.

She graciously accepted my lengthy absences from home, and handled matters on her own that under normal circumstances would have been shared responsibilities. In keeping my bad conscience at arm's length and providing comfort through countless long-distance phone calls and visits, her contribution to this book is priceless.

This book has received financial support from multiple institutions in Norway and the USA: The Research Council of Norway, the University of Alaska Fairbanks, the American-Scandinavian Foundation in New York and the Centre for Advanced Study in Oslo, Norway. Without the farsightedness of these institutions and their willingness to support a project that extends beyond the realm of disciplines, this book would never have been written.

Any errors or shortcomings related to the content of the book are, of course, the sole responsibility of the author—*mea culpa.*

Oslo, October 2009
Willy Østreng

Chapter One

Interdisciplinary Science and Knowledge: What are They?

"The investigator will always have to make use both of the analytical and of the synthetic method; both of experience and of intellectual construction. Understanding is itself after all a unity . . . arrived at by two processes which appear opposites to the mind."

Johan Hjort, 1920[1]

On the afternoon of 7 May 1959, Dr. Charles Percy Snow entered the prestigious rostrum of Senate House in Cambridge to deliver the annual Rede Lecture. His address on 'The Two Cultures and the Scientific Revolution' unleashed, much to Snow's own surprise, one of the fiercest, most hostile discussions among natural scientists and humanists in the 20th century. What provoked the most was Snow's contention that a gulf of misinterpretations existed between the two ruling knowledge-producing cultures of modern science, the literary intellectuals (humanists) and the natural scientists. C.P. Snow, by training a scientist, by vocation a writer, held that the two cultures had a curious distorted image of each other based on profound mutual suspicion and incomprehension, and that their attitudes were so different that, even on an emotional level, they could not find much common ground. In his opinion, the literary intellectuals believed the scientists to be unaware of Man's condition, whereas the latter assumed the former to be totally lacking in foresight and "anxious to restrict both art and thought to the existential moment."[2] These imageries made the feelings of the one culture, the anti-feelings of the other, producing hostility rather than fellowship between them. What was worse: The prospects that this hostility gap would ever be bridged seemed bleak indeed and history, in Snow's mind, provided no source of inspiration. In the late 1920s, the two cultures managed a kind

1

of frozen smile across the gulf. Then the politeness disappeared, and they just made faces.

C.P. Snow regarded this situation as being all destructive—a sheer loss to the creativity of science. To close the gap between the two cultures was, in his mind, a necessity in the most abstract intellectual sense, as well as in the most practical sense. He was convinced that since the two senses had grown apart, no society was going to be able to think with wisdom. Synthesis of insights, not fragmentation, was what Snow called for in his most controversial lecture.[3]

Snow's intervention is still a matter of academic debate. Although there is a growing awareness in many research circles that there is more need than ever to address holistic processes and problems[4], there are still scientists just making 'faces across the gulf'.' One inescapable fact that strains the topicality of the matter is that the number of scientific cultures has multiplied since the time of C.P. Snow. The two main cultures have grown to become four in that the social and technological sciences have taken on their own distinct cultures, deviating from those of the humanities and natural sciences. More importantly, the four parent cultures have split into numerous specialized branches, as disciplines and sub-disciplines have multiplied many times over in recent decades (see Chapters 2 and 3). The overall aim of this book is to take stock of the variety of scientific cultures and discuss ways of achieving knowledge integration by bridging between them.

This book takes issue with those "making faces across the gulf" and who oppose the scientific value and quality of interdisciplinary research, claiming that it lacks a suitable methodology, scientific approach and theoretical foundation. The counter-argument put forward here is that interdisciplinarity is already anchored in an emerging methodology of its own and that it finds support in post-positivist philosophy, post-academic and post-normal science. At the same time, the book is a political analysis of the interface of science, society and politics and how this interface affects interdisciplinarity as a scientific practice and a measure of societal and political utility.

In science, interdisciplinarity is an ambivalent term. For practical problems it is to a certain extent considered valid and unavoidable, but for theoretical purposes it is "handled with great caution and even with suspicion."[5] Thus, the notion and practice of interdisciplinarity cannot be discussed without reference to science in general. To understand 'science without boundaries' one has to comprehend 'science with boundaries,' since the former depends on the latter. What has been labelled the 'complexity vision' of nature and society is changing our research agenda and has resulted in the formulation of a series of new scientific concepts and perspectives that affect our understanding of what science is and is not.[6] In this vein, the overall question to be discussed in this chapter is: *How does the contested practice of interdisciplinarity fit in with these new concepts and the disciplinary structure of the science community?* The remainder of the chapter will therefore be devoted to discussing this overarching issue by posing a number of questions that will serve as navigational reference points throughout the book: What is science?, What is complex system science?, What is scientific knowledge?, What is disciplinarity?, How does disciplinarity become interdisciplinarity?, What is interdisciplinarity?, and What is interdisciplinary knowledge?

WHAT IS SCIENCE?

It is not easy to describe what scientists are doing and what science is. John Ziman, a noted philosopher of science, claims that attempts to answer such a question are almost as presumptuous as trying to state the meaning of life itself.[7] Yet the question is more puzzling than mysterious.

The concept of science is derived from the Latin word, *scientia,* which means knowledge, and *scientificus*, which means making or producing knowledge. Hence, science is a systematic human attempt to learn about and understand the working of those parts of reality that are explorable by scientific means.[8] However, linguistically, philosophically and historically, the concept of 'science' is exclusively associated with research undertaken to unravel the secrets of the Universe[9]—the physical world, whereas systematic studies within the humanities and to a certain extent also within the social sciences find expression in the concept of 'scholarship.' In the same vein, the notion of 'research' is most often attached to science whereas the concepts of 'study' and 'academic' are used to depict scholarship. These and other linguistic differences have long been sustained in the English-speaking world and legitimized by classical positivism that makes a distinction between *scientific objective studies* based on observation and quantifiable methods and *scholarly studies* of a more subjective nature based on qualitative approaches (see Chapter 3). In the post-positivist era now looming on the horizon, there are certain indications of change affecting the traditional linguistics of systematic studies underscoring the similarities rather than the differences between scholarship and science. As stated by Stefan Collini: "The activities conventionally referred to as 'science' do not, . . . all proceed by experimental methods, do not all cast their findings in quantifiable form, do not all pursue falsification, do not all work on 'nature' rather than human beings; nor are they alone in seeking to produce general laws, replicable results, and cumulative knowledge........it has become more and more widely accepted that different forms of intellectual enquiry properly furnish us with a variety of knowledge and understanding, no one of which constitute the model to which all the others should seek to conform."[10] The most common criteria of what constitutes 'science' also apply to work in the historical-philosophical fields, such as falsifiability, corroborability, verifiability and empirical testability. The procedure of formulating hypotheses and testing them empirically also operates with the interpretation of texts and, by extension, with the reconstruction of historical processes, which is always based on textual interpretations. In philology, "data take the form of the physical properties of manuscripts, linguistic rules and textual and ideological coherence. . . . On this background it is reassuring to conclude that scholarship nonetheless seems to possess and intrinsic self-regulatory ability, which in the long run allows truth to prevail over sensation."[11]

On these grounds, Wolf Lepenies is using the term 'science' in the German sense of Wissenschaft, i.e, *any systematic body of enquiry*.[12] The concepts of 'academic research' and 'post-academic science' are increasingly being used as a collective notion to depict systematic study in all fields and disciplines. Thus, the concept of research is in the process of widening to cover systematic studies of humanists, social scientists and natural scientists alike. This finds expression and reflection in concepts such as *political science, behavioural science, social science, linguistic science, etc.* Scholarship has become identical with the notion of 'soft science,' whereas the concept of 'hard science' is used to depict the quantifiable natural sciences. In this connection, 'science' has become the common denominator of all fields and disciplines. The difference is one of degree rather than one of kind.[13]

Adhering to the 'old' linguistics is also being complicated by the constantly growing number of *multidisciplinary disciplines,* also called *hybrid disciplines,* i.e. conglomerates of disciplines sharing a common focus and/or object of interest or study (see Chapter 4). Many of these hybrids build across the assumed gorge between the natural and behavioural sciences by spinning off from their parent disciplines and finding new expressions in fresh settings called multidisciplinary disciplines. Some of these fresh cluster disciplines embrace both the hard and the soft sciences, making it a bit odd to apply different conceptions of systematic studies to different parts of a unified discipline. As a mix of cognitive psychology, artificial intelligence, linguistics, anthropology, genetics and philosophy, cognitive science is one of several examples of the hybridization of research (see Chapter 2). Geography that aims at building bridges between the social and natural sciences is another (see Chapter 3). In addressing problems such as global warming, ecosystem management of live wild stocks, the formation and depletion of the ozone layer, poverty etc., involving a mixed bag of social and natural scientists, the investigations of the man-made causes of these problems will be labelled scholarship, whereas the research on the natural effects is called science. What, then, to call the research undertaken in the trading zones between the two fields, where social and natural variables intermingle and blend? To reflect the new realities of practical research, the old dichotomy between scholarship and science increasingly seems an anachronism.

In the spirit of post-positivism, the concepts of *academic research* or *academic science* are gradually becoming the collective umbrella covering both scholarship and science (see Chapters 2, 3 and 5). In line with this post-positivist trend, any reference to science in this book is commensurate with the concept of academic science covering the humanities, as well as the social, natural and technological sciences.

Science in this emerging meaning of the term is not alone in seeking knowledge. Apart from scientific knowledge, there is *philosophical knowledge*, *mystic/mythic/religious knowledge* and *practical knowledge* (see Figure 1.1). Philosophical knowledge is arrived at through thought, i.e. *rational speculation guided by stringent logical premises.* Mystic/mythical/religious knowledge applies *feelings, emotions, sensations* and *beliefs* to explore reality, even beyond what is observable, whereas practical knowledge applies

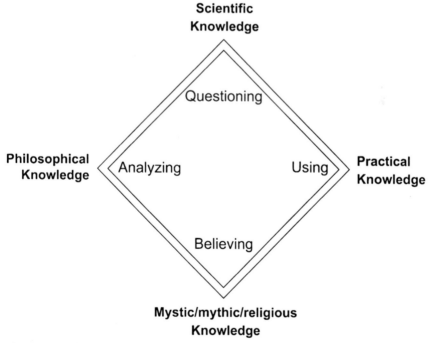

Figure 1.1.   The Four Modes of Knowledge Seeking and Their Bases.

common sense, repetitive experience and *science-based insight* (medical doctors, lawyers, engineers, teachers etc.) as its primary means of operation. Practical knowledge which covers a wide range of societal trades, crafts and professions is also called stakeholder knowledge applied to resolve real life problems. The purpose of this clarification is to discuss how scientific knowledge differs from and complies with other forms of truth-seeking to determine what scientific knowledge is and is not.

Since antiquity, philosophers have been interested in science because it represents the most rigorous attempts by humans to acquire knowledge and to distinguish it from the alleged insights of mysticism, religion and intuition. Science is said to have four facets in that it can be defined by its *method, knowledge, application* and *occupation.* Commonly used, it tends to be an amalgamation of all four.[14] Boiled down to its very essence, the four-faceted science is a conscious artefact and practice of man to produce verifiable and reliable knowledge. In this way, scientific endeavours connote a process or procedure for making inquiries about the world, that is the natural world and the social world and connections between the two, and for evaluating the hypotheses these inquiries engender.[15] This manifests itself in four ways: in a *disciplinary fashion* to breed specialization in *depth* within a restricted niche of reality; in an *interdisciplinary fashion* to foster specialization in *breadth* between two or more disciplines/niches of reality; in an *applied fashion* to resolve practical problems, and in a *basic fashion* to produce new knowledge 'just for the fun of it.' In any event, they concur in applying measures of scientific discovery to explicate their respective objects of study. Scientific methods refer to those measures that increase knowledge by the systematic use of theory, observation, measurements, interpretation and logic—methods that subject potential knowledge to testing.[16]

In its own technical language, science is thus reliable and verifiable information about the physical and human worlds and their interrelations, unmasked by the application of consensual research procedures and methods. As such, science is rigorous, methodical, academic, logical and practical, and has developed a more or less coherent set of ideas. Science speaks first and foremost to the mind, and is different from religion and poetry in that they also speak to emotions. Science stands in the region where the intellectual, the psychological and the sociological axes intersect.[17] It has well-documented historical origins, with a definable scope and content and with recognizable practitioners and exponents. These characteristics imply that the knowledge produced by scientific methods and rigour is the closest man can get to revealing what observable, verifiable reality is all about in both the human and the natural worlds, and to observing the interconnectedness between them.

The driving force of science is to see further, deeper and wider than previous generations could do. Science is thus a tireless collective effort to acquire new, different and improved knowledge. It is distinct from the insights one may get from religion, mysticism and intuition, which have no rules, no stringent methods, nor any construed logic to be applied.[18] Mythology has to do with belief, faith and uncertainty, with incomplete observations and instinctive feelings. 'Non-scientific' concepts, such as religion, harbour no objective yardstick or criteria to help one decide what to believe or not. The ideas that 'win' are contingent on a collection of more arbitrary and subjective criteria. "We must therefore expect religions to include more unprovable statements than science."[19] If religious believes are defined as certainty about something one does not know, scientific facts are defined as confidence in something one knows from the determined application of scientific methods and procedures (see below). Where scientific methods have the ability to accommodate changes in the light of new discoveries, religion is static, founded on dogma and received wisdom, and purporting to represent immutable truth. Religion is also backward-looking to reveal truth, while science looks forward to new vistas, horizons and discoveries.[20] In the words of Galileo (1564–1642): "The Bible shows the way to go to Heaven, not the way the heavens go."[21] Ideologies, belief systems, mysticism, politics, etc. may gain convincing influence among great masses of people simply because they correspond or seem to correspond to the prevailing interests of the people, whereas scientific ideas will spread only because they are true (see the concept of public knowledge below).[22]

Mythology and mysticism serve to fill the gap until scientific knowledge is generated, and sometimes they prove diametrically opposed to real knowledge, so they can never be a lasting alternative. At best, the scientific function of mythology is to act as a stimulant to foster research and scientific inquiry, i.e., to pique curiosity and promote ingenuity. Myths thrive best in the absence of scientific facts. This makes science a method in itself (see Figure 1.1).

In the Greek language, a philosopher is a friend of knowledge, a lover or cultivator of wisdom. Philosophy is thus the mother of the sciences. Already in antiquity, different branches of philosophy were spun off and became specialized scientific fields in their own right. First, the natural sciences and mathematics, in particular geometry, departed from philosophy. Centuries later, sociology and psychology achieved the status of separate scientific disciplines. This process is on-going. Thus, philosophical knowledge is much closer to scientific knowledge than are mysticism/mythicism/religion. There are three reasons for this. First, philosophers seek clarification by way of rational and logic reasoning, which is also the approach of science. Second, a special branch of philosophy, the philosophy of science, claims to be the meta-theoretical framework that provides authoritative guidance about which criteria constitute real science (see Chapter 4). From this perspective, philosophy 'decides' what is science and how various scientific activities should be conducted to qualify as a science. Third, philosophy is academic in approach and part of the scientific community. Meanwhile, philosophy and science are different in that the prime method of philosophy is rational thinking alone, whereas in addition to rational thinking, the methods of science are observation, experiment and measurement.

Science sets limits to philosophy in the sense that one should be suspicious about philosophical positions which are contradictory or incoherent in the light of established scientific knowledge.[23] However, in instances where philosophy extends beyond what is known by science, the speculative inquiry of philosophers may be inspirational to science and, eventually, in due time be subject to testing using scientific methods. There are historical examples of philosophers opposed to established 'scientific' knowledge that were later proven right (or wrong) by fresh scientific inquiries. The Greek philosopher Aristarchos of Samos, who lived in the third century B.C., was one such person who dared to suggest that the Earth moved around the sun at the time when the Earth-centred universe was an established scientific truth.[24] Philosophers like David Hume, Ernst Mach and Immanuel Kant had a great impact on Albert Einstein's thinking and work. The ideas of the empiricist Mach were instrumental to Einstein's formulation of his relativity theory, which did away with the unobserved ether. Einstein also appreciated David Hume's thesis that certain concepts cannot be deduced from Man's perception of experience by logical method. On these grounds, Einstein contended that the relativity theory had proven Mach and Hume right and Kant wrong.[25]

Although philosophy reflects on different aspects of reality and claims to produce extra cognition and insight supplementing that of science, philosophy is no lasting alternative to empirical science. At best, it is more of a supplement, an auxiliary 'science,' adding conceptual clarification, speculative insight, inspiration and vision to the process and conduct of verifiable science (see Figure 1.1). Their partly blurred demarcations are also demonstrated by physics and cosmology, venturing into areas where experimental verification is not possible. One example is the idea of a 'multiverse,' the existence of a huge number of 'pocket universes' like the vast expanding universe domain we see around us, which is part of a much larger physical existence. The second is the 'string theory,' a proposed theory of fundamental physics that unites gravity with quantum physics (see Chapter 3). Together, the two theories have been considered interesting in that they unite quite disparate elements of physics and cosmology in a useful synthesis. Meanwhile, as there is no chance of observationally verifying their main predictions, the question posed by fellow scientists is whether they are scientific proposals or simply philosophical ones.[26] The physicist Lisa Randall—concerned with making predictions about observable phenomena—does not accept such a dichotomy. Rather, she thinks of the two approaches as interconnected and mutually supportive: "Recent advances suggest that extra dimensions, not yet experienced and not yet entirely understood, might nonetheless resolve some of the most basic mysteries of our universe. Extra dimensions could have implications for the world we see, and ideas about them might ultimately reveal connections that we miss in three-dimensional space."[27]

The notion of *symmetry* is another example of the philosophy-science interaction. Symmetry has been an intrinsic part of philosophy, physics and human culture (not least art) ever since antiquity. Plato's conjectured theory on the elementary constituents of the world—earth, water, air and fire—were early on associated with the regular polyhedra, i.e. the cube, octahedron, icosahedron and tetrahedron. Some 1000 years later, the German astronomer Johannes Kepler picked up on Plato's thinking and proposed a model of the solar system based on these polyhedra, now known as *Platonic solids*.[28] In modern physics, symmetry is deeply rooted in physical laws, such as Maxwell's laws of electrodynamics and Einstein's theory of relativity.[29] Thus, modern science ". . . owe a great deal to Plato, who believed that symmetry is the key to understanding the universe. Modern physicists agree."[30] It stems from these examples that the trade of philosophy deals with universal problems in a speculative manner that sometimes may be inside and sometimes outside the reach and scope of verifiable science. To put it differently: when a problem can be defined in concrete and precise terms, it leaves the sphere of philosophy and becomes researchable by the application of scientific methods. Thus, philosophical speculations cannot always be answered by the sciences, but they can still be of scientific utility in terms of guidance and inspiration.

Practical knowledge relates to the know-how gained from *common sense, repetitive experience* and/or *science-based education* in resolving the challenges of everyday life. One obvious example of this type of knowledge is the expertise gained through the practice of various handicrafts. Practical knowledge is insight passed on from generation to generation by repetitive behaviour and observations, and developed further through the introduction of new tools, schooling and continuous adjustments. By this definition, applied research also contributes to the pool of practical knowledge. As discussed below, in post-normal science, scientific knowledge merges with practical knowledge in resolving societal problems of immediate concern. Yet there is an important difference between the two. While the progress of scientific

knowledge depends on a high level of abstraction, theorization, conceptualization and generalization, the progress of practical knowledge can do without the abstractions (see below and Chapters 2 and 6).

These differences in modes of knowledge-seeking do not imply that drawing the boundary between science and non-science is an easy or straightforward task. For instance, the practice of acupuncture has a 'scientific' rationale in China, but in the west it remains a curious empirical technique at best.
Parapsychology and creationism are examples that continue to be attacked
and stigmatized as pseudo-scientific, and the pretensions of their practitioners are often ridiculed. Astrology, once indistinguishable from astronomy, and
homeopathy, which for a while after its inception held real promise of becoming the orthodoxy in medicine, remains firmly saddled with the label of
pseudo-science.[31] There are even scientists who claim that the similarities between modern physics and Eastern mysticism are striking and part of a changing world view, applying to science and society alike.[32] Others are convinced
that issue areas such as art, ethics and religion should and could be integral
parts of science,[33] Last but not least, proponents of post-normal science claim
that the boundary between science and non-science should be transgressed to
enhance the ability of science to resolve practical problems (see below and
Chapters 2 and 6).[34]

In other words, the boundaries between what is and is not science are fluid,
temporary and constantly being challenged by scientists themselves. What
constitutes science at any given time in history is not necessarily identical to
what constitutes science at another time. As pointed out by Barry Barnes et
al., there is indeed a case for saying that scientist's evaluations of fields such
as parapsychology are often impossible to justify even in terms of their own
preferred standards.[35] Here, history is an 'expert witness' (see Chapter 2). The
concept of science is elusive and time-dependent. At any time, there are various criteria which are generally regarded as legitimate bases for demarcating
science. In line with the relativity of the notion, John Merton finds the concept of science to be deceptively inclusive, referring to a variety of distinct
but interrelated items in which the ethos of science, i.e. the set of cultural values governing scientific activities, should be part of the definition (see below).[36]

The Ethos of Science

Science is guided by certain strict values and norms—the ethos or ethics of
science—which are held to be binding on the individual scientists. The claim
is that science has succeeded because scientists comprise a community that is
defined and maintained by adherence to a shared ethic, which serves as a fun-

damental corrective within the science community.[37] These norms have not been codified in any document, but can be inferred from the moral consensus of scientists in their writings and moral indignation directed towards contraventions of the ethos. Thus, the ethos is supposed to have been internalized by the majority of scientists. Four sets of partly overlapping institutional imperatives are suggested by John Merton to make up the ethos — *universalism, communism, disinterestedness* and *organized scepticism.*

Universalism finds expression in the canon that claims of truth are to be subjected to pre-established impersonal criteria consonant with observation and previously confirmed public knowledge (see below). A scientific claim is not to depend on the personal or social attributes of its protagonists. Race, nationality, religion, class, etc. are as such irrelevant: Objectivity precludes particularism. That is to say that universalism is rooted in the impersonal nature of science. If and when society opposes universalism, the ethos of science is subject to serious strain, underscoring that ethnocentrism is not compatible with universalism. In this respect, the ethos of society is on a collision course with the ethos of science.

Communism implies that the findings of science are a product of social collaboration and, as such, assigned to the community at large. That is to say, scientific products do not enter into the exclusive possession of the discoverer and his or her heirs, nor do they bestow upon them special rights of use and disposition. In the world of scientific communism, apart from in relation to patents, scientists' claims to intellectual property are limited to that of scientific recognition and esteem (see below). In principal, scientific results are the common property of mankind. However, as discussed in Chapters 2 and 6, the norm of communism will have different implications in basic, applied and transdisciplinary research. This is so because secrecy, which may be part of applied research, is the antithesis of this norm. Full, open communication of scientific results is a prerequisite for promoting the advancement of knowledge. The results of applied research are to a certain extent the "property" of the party commissioning it. As discussed later in this book, there are ways to alleviate this dilemma, and even to sort it out (see Chapters 2, 5 and 6).

Disinterestedness refers to the norm which implies that the scientist is to have no material, social, political, cultural or other interests to defend or secure in his or her research. The researcher is to be detached from interests of all kinds. This is not to say that disinterestedness equates with altruism or that interested actions equate with egoism. The demand for disinterestedness has a firm basis in the public and verifiable nature of science. The virtual absence of fraud in the annals of science, which, according to Merton, appears exceptional when compared with the records of other spheres of activity, has sometimes been attributed to the personal qualities of scientists, i.e. in the internalized norm of disinterestedness (see below and Chapter 2).

Organized skepticism is interrelated with the other elements of the ethos. To a large extent, this norm is a methodological device, a way of behaving or thinking in which the investigator does not "preserve the cleavage between the sacred and the profane, between that which requires uncritical respect and that which can be objectively analyzed."[38] Organized scepticism is applied not only vis-à-vis other scientists and their results (see below), but also in respect of societal values and interests. This norm implies that no one is immune to scientific scrutiny, which may threaten the existing distribution of societal power and bring science into conflict with the society it is supposed to serve (see Chapter 2).

John Ziman adds *originality* to Merton's list, and uses the acronym *CU-DOS* as a reference to the whole set of five norms. By originality, Ziman means the quest to add new insight to existing knowledge and thereby change it in refinement, depth and breadth—to break away from the puzzle-solving of 'normal science' (see chapters 2 and 5).

In summing up this discussion, science is an internalized value system shared by all disciplines and fields of science, an approach that differs from the other modes of knowledge seeking, a methodology and a cultivation of rationality intended to produce verifiable and reliable knowledge within the confines of the human mind and intellect. This combination of elements produces a certain kind of knowledge: Scientific knowledge.

WHAT IS COMPLEX SYSTEM SCIENCE?

Complex systems are composed of interconnected parts. Thus, in the science of complexity, researchers are supposed to address all relevant variables of a particular system and its ambiance through fusions of unrelated but relevant disciplines. In the book *Frontiers of Complexity*, Peter Coveney and Roger Highfield define complex science as a new way of thinking about the collective behaviour of many basic and interacting units that lead to coherent collective phenomena existing at higher levels than those of the individual units, and *where the whole is more than the sum of its parts*.[39] Breaking complex systems down into their individual parts is only a first approximation of the truth. While it may afford many useful and stage-setting insights, it always behoves scientists to put the pieces together again.[40] Analysis begs synthesis. However, their assumption that the whole is *more* than the sum of its parts calls for a qualification.

Gottlob Frege argues that the whole is a function of the meaning of the parts and their modes of combination. This observation suggests that the impact of the 'sum of the parts' differs between *linear* and *non-linear complex systems*. In linear systems, all components/parts are known and understood down to the smallest detail, causing predictability in terms of its working and outcome, whereas in *non-linear systems,* small deviations in the pattern of interaction between the individual parts may cause unexpected outcomes (chaos theory). Thus, in linear systems the whole is more or less equal to the sum of its parts, whereas in non-linear systems, the whole may either be much more or much less than the sum of its parts (see Chapter 3).[41] In consequence, the whole in non-linear systems is not necessarily *more* than the sum of its components, as assumed by Coveney and Highfield, but *different* from it. As was discovered in the 1920s, the atomic world is full of murkiness and chaos. A particle like an atom does not appear to follow any meaningful, well-defined trajectory so it is hard to pin down to a specific motion. One moment it is in one place, the next in another.[42] Uncertainty is an important feature of non-linear systems, whereas linear systems are predictable and to a certain extent fixed. The key challenge of both systems is to comprehend how complexity arises from simplicity. Let us illustrate the point.

When examining levels of greater and greater complexity in nature, the Nobel Laureate Steven Weinberg sees phenomena emerging that have no counterpart at the simpler levels, least of all at the level of the elementary particles. He sees nothing like intelligence at the level of individual living cells, and nothing like life at the level of atoms and molecules.[43] In the case of living systems, no one would deny that an organism is a collection of atoms. The mistake is to assume that it is nothing but a collection of atoms. Such a claim is in Paul Davies' mind as ridiculous as asserting that a Beethoven symphony is nothing but a collection of notes or that a Dickens novel is nothing but a collection of words.[44] In these examples, the whole interacts with its parts in 'modes of combinations' that add new and sometimes unexpected insights to our understanding of the system. The whole of a particular system is value-added.

The scientific practice designed to unmask the interplay of components in systemic complexes has been given many names: '*a science of new connections,*' '*a science of synthesis,*' '*a science of consilience,*' '*big science,*' '*bold science,*' '*holistic science,*' '*complex science,*' '*symbiotic science,*' '*interdisciplinary science,*' etc. All those designations cover the same reality: scientists look for relationships among clusters of interdependent variables with the aim of addressing their intersections and areas of overlap to reveal insights that individual disciplines cannot provide separately and in isolation from each other.[45] Their focus is on unfamiliar linkages, and they pursue the biggest questions, i.e. those that seem "slightly unsavoury in their breadth and depth."[46]

To unlock the workings of complex systems there is a need for interactive interconnected synthesizers—for interdisciplinary researchers preoccupied with bringing together the bits and pieces of disciplinary research. As put by Jonathan Weiner: If (the living planet) Gaia exists, "she cannot be studied in pieces, with Professor X wondering if the sky is alive, Professor X wondering if mountains are alive, Professor Z wondering if Professor X is alive. If Gaia exists, then all the elements of the planet are connected and working together like the organs of our own bodies."[47] In most areas of knowledge, science possesses piecemeal (disciplinary) knowledge produced by *reductionism*, but lacks the overall synthetic understanding of the interactions produced through the application of the *holistic approach*.[48]

Reductionism entails a reference to the classical Newtonian belief that the dynamics of any complex system can be understood from studying the properties of its parts. Once you know the parts you can derive, at least in principle, the dynamics of the whole. In order to understand a complex system, it should be broken down into its components and each piece should be studied individually by way of disciplinary methods (see Figure 1.2 and Chapters 2 and 3). Here, the focus is on the *properties of the components*.

In holism,[49] the relationship between the parts and the whole is more symmetrical than in reductionism. The assumption underpinning this approach is that the properties of the parts contribute to our understanding of the whole,

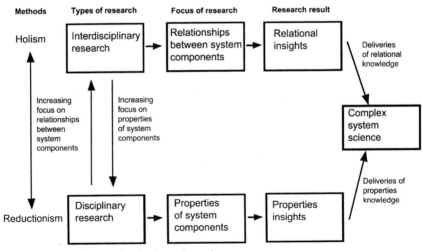

Figure 1.2.   **Holism-Reductionism Dimensions.**

but at the same time these properties can only be fully understood through the dynamics of the whole.[50] Here, the focus is on the *relationships between the components*, i.e. their interconnectedness, interdependencies and interactions (see Figure 1.2 and Chapters 2 and 3). In other words, the study of complexity offers an opportunity to stand back and consider the global interactions of fundamental units, creating integration that crosses the borders of scientific disciplines—to see a grand vision of nature.[51]

Complex systems come in many shapes, designs and sizes. They range from the smallest of systems like cell biology and the functioning of the human brain, to the largest of systems like climate formation, biodiversity, habitat management of fish stocks, ecology, etc. Their similarities highlight their interrelated functioning, their wholeness and multi-faceted coherency.

WHAT IS SCIENTIFIC KNOWLEDGE?

Epistemology, or the theory of knowledge, is driven by two main questions: What is knowledge? and What can we know? If we think we can know something, as nearly everyone does, a third philosophical question arises: How do we know what we know?[52] These questions will be addressed in the following paragraphs.

Conventional scientific knowledge was traditionally understood as the accumulated product of the work of independent rational individuals, making impartial observations in a systematic manner by means of the scientific methods, theories and procedures available at any given time. These individuals produced the eternal objective truth by building their research cumulatively on what their predecessors had done before them. Nature consisted of building blocks to be exposed by scientific methods. In this way, scientific knowledge was supposed to progress through history in a conflict-free harmonious manner by 'standing on the shoulders of (previous) giants (see Chapter 4).'

Over the past 20 years, sociologists, science historians and philosophers have redefined scientific research as a typical knowledge-based social activity, involving the subjective element of the researcher, and the relativity and uncertainty of scientific knowledge. Here, the classic version of scientific knowledge is increasingly being seen as an unduly idealized version of science (see Chapter 4).[53] At present, scientific knowledge is being described by the term *public knowledge,* which hosts subjective as well as objective elements.

To become public knowledge, the results of research have to be generally accepted by the scientific community to be the truth about a specific matter.

Scientific knowledge, which starts with experience, does not become public knowledge until its facts and theories have survived a period of critical study and testing by other competent and distinguished researchers and have, as a consequence, become accepted by the consensus of rational opinion over the widest possible field.[54] Accordingly, science is also communication and discussion to convince other qualified researchers that the results arrived at are both correct and scientifically important.[55] One important instrument in producing public knowledge thus goes through the procedure of critique, reply and rejoinder (see Chapter 3).

But science is something more than communication and discussion. Science is also a social phenomenon "with its own reward system, its revealing snobberies, its interesting patterns of alliance and authority."[56] Sheila Jasanoff observes that the validity of scientific 'facts' and 'theories' depends on tacitly negotiated, interpretive conventions that are shared among communities of researchers. Socially constructed, science thus "retains its authoritative status as long as its underlying premises are not scrutinized too closely or with hostile intent."[57] In this perspective, scientific truths are widely quoted social agreements about what is real, arrived at through a distinctively scientific process of negotiation.[58] The process of gaining universal acceptance of scientific finds is cumbersome and protracted. It moves through periods or stages where the body of scientific results is rated from 'interesting' to 'suggestive,' to 'persuasive, to 'compelling' and finally, until objections cease, to becoming "obvious.'[59] In most cases, there is actually no way to skip any of the steps towards conviction and consensus. The scientific peers "are always there, sceptical, undisciplined, inattentive, uninterested; they form the social group that (researchers) cannot do without."[60] In this way, every researcher needs his peers to bring about the gradual transformation from findings being 'interesting' to becoming 'obvious.' For lively controversy over scientific publications to become public knowledge, the transition is progressive, continuous and often time-consuming. Scientific enterprise is thus corporate, as is knowledge; it represents the consensus of corporate scientists, of the epistemic community who largely determine how we know and what we know.[61] On this backdrop, Richard P. Feynman defines science and scientific knowledge to be the organized scepticism in the reliability of expert opinion.[62]

According to Bruno Latour, science is both chaotic and warlike. It has been noted that the main task of theorists is to fortify their own position against criticism while lobbying plenty of 'critical bombs' into the territory of the opposition.[63] Every researcher is, in his or her own mind, against everyone else, and if a scientist accepts a claim of knowledge to be public, it is often accepted out of expediency, and will be set aside immediately when circumstances allow.[64] In this way, public knowledge is constantly being challenged, tested and criticized. Harmony is not the standard state of science, conflict and disagreements are. Science exists in a societal context. This is not to claim, however, that science is nothing but another societal activity. For an activity to be rated scientific, it has to be governed by strict norms of rationality guided by methods, theories and the conventions/ethos of science.[65]

Thus, public knowledge should not be confused with the publication of scientific results, which are often the subject of disagreement and fierce discussions between scientists (see Chapter 3). For example, failure to acknowledge the process of transformation is why people get frustrated by medical journalism; first the public is informed that researchers have concluded that a certain lifestyle is hazardous to the human health. Later, another team contradicts and disputes the validity of the same conclusion. The resulting public confusion reflects what can happen when scientific results become public knowledge. In this process, no one, neither scientists nor the public, can know what will constitute public knowledge in future.

Some of the most revolutionary scientific discoveries in history were those that lingered for a long time before they gained the acceptance of fellow scientists. Albert Einstein, for example, developed his theory of gravitation in the years 1907–1915 and presented it to the world in a series of papers in 1915–1916. By the mid-1920s, his revolutionary theory had become generally accepted as the correct theory of gravitation, a position it has retained ever since.[66] This experience was duplicated by Professor Alexei Abrikosov who, in 2003, was awarded the Nobel Prize in Physics for his discovery of the type II superconductor. When he first publicized his findings, he was quite simply ignored by fellow physicians in Russia because his conclusions ran contrary to 'accepted' scientific views at the time. To ease the situation, he therefore decided to shelf his discovery for four years, followed by ten more years before the scientific community began to believe in what he had found.[67] Rosalyn Yalow, who won the Nobel Prize for medicine in 1977, had a similar experience when her groundbreaking results were refused by the editors of Science and the Journal of Clinical Investigations until the controversial concept 'insulin antibody' was eliminated from the text. On the basis of this experience she concluded that "The truly imaginative are not being judged by their peers. They have none."[68] There is much truth in this seemingly arrogant statement. It finds among other things, support in the observation that the peer review process, instigated to raise the quality of publications, not only eliminates the worst researchers, but also the best. Basically, experts are experts on what has been, not on what is new.

In a nutshell, the peer review process creates, nurtures and rewards followers rather than innovators.[69] In Thomas Kuhn's scheme, the peer review instrument is a vehicle to safeguard the puzzle-solving of 'normal science,' not 'revolutionary science' or 'creative boundary breaking science' (see Chapters 2, 3 and 4). Several other Nobel Prize laureates have faced severe opposition from the science community when publishing the results of their unconventional research.[70] Such experiences illustrate the harsh reality of challenging and altering existing public knowledge.[71]

In building scientific knowledge, scientists are expected to believe some things in preference to others because of the superior evidence in their favour. The success of scientific results will increase in line with the extent to which they can correctly account for and predict the results of experiments and observations.[72] Steven Weinberg argues that it is not really so important to pinpoint the moment when a researcher became 75 per cent, or 90 per cent or 99 per cent convinced of a theory or a scientific result. The important thing, according to him, is not the decision that a theory is true, but the decision that it is worth taking seriously, i.e. "worth teaching to graduate students, worth writing textbooks about, above all, worth incorporating into one's own research."[73] The concept of a 'moment of worthiness' is illustrated by Michael Zimmerman who claims that if serious action had been taken back in 1974 when the link between the emission of chlorofluorocarbons (CFCs) and the depletion of the ozone layer was first demonstrated by the chemists Rowland and Molina, most of the current problem would have been averted.[74] At that time, the Rowland/Molina hypothesis was rated 'interesting,' but it dragged on, not becoming 'compelling' until the late 1990s. Only rarely will scientific results achieve the status of public knowledge at the time of their publication.

Public knowledge also changes over time. What is deemed by peers to be public knowledge at one point in time may be supplanted by other knowledge later on. The Earth used to be flat according to the public knowledge that existed at the time of Columbus, but few people hold that position today, and none among them claim to be a scientist. All knowledge is corrigible. History teaches us to be humble about any claims to knowledge.[75] It is widely recognized that all scientific concepts are limited and approximate. Science can never provide any complete and definitive understanding. Scientists do not deal with truth in the sense of a precise correspondence between the description and the described phenomena; they deal with limited and approximate descriptions of reality.[76] It is this limited and approximate description that constitutes the scientific truth of a particular phenomenon. Science advances through tentative answers to a series of more and more subtle questions which reap deeper and deeper into the essence of natural phenomena. In this sequence, scientists can improve the reliability, validity, certainty and honesty of their conclusions by paying attention to the rules of scientific inference.[77] The truth of one era may thus turn out to be the misconception of another. It follows from this that the most reliable knowledge available, public knowledge, is relative, culture-specific, corrigible and time-dependent, but at the same time, less so than some of the knowledge (religious) acquired by non-scientific means.

The uncertainty and corrigibility of scientific finds is why research is a never-ending process, and why science cannot do without disciplinarity and interdisciplinarity as complimentary means of discovery. Against this backdrop, some science philosophers conclude that it seems that no one has yet been able to come up with a theory of truth which is neither false nor viciously circular. Scientific truth has proved resistant to philosophical elucidation.[78]

WHAT IS DISCIPLINARITY?

The concept of interdisciplinarity presupposes the existence of disciplines. As such, the disciplines are part of the definition of interdisciplinarity. We will therefore consider the meaning of disciplines before pressing on to define the interrelationship between them.

The Latin term disciplina dates back to antiquity and refers to the instruction of disciples. Initially it referred only to teaching and learning in the domain of the liberal arts, whereas research in terms of experiments or any other form of empirical exploration was not added until the period from 1750 to 1850. The term 'discipline' currently includes both the production of new knowledge through research and its transmission in an educational and organizational context.[79]

Normally, a discipline has a distinct subject matter, a research agenda, a curriculum, an associated theoretical framework and a common approach to study using appropriate techniques for understanding and discovering new knowledge.[80] In this definition, disciplines are relatively fixed, stable and delimited entities of researchers, working and remaining within the academic and intellectual bounds considered theoretically legitimate among themselves.[818] As such, they are recognizable communities of scholars which develop their own conventions to govern the conduct and adjudication of research. They are relatively stable epistemic societies which rely on technical language and particular methods of analysis, and develop standards of evaluation specifically suited to their methodology and objects of analysis.[82] The term 'discipline' signifies a specific set of tools, methods, procedures, concepts and theories that organize experience coherently into a particular 'world view.' Disciplines have their own priorities and restrictions. In this way, a discipline exerts a powerful pull over its members, who collectively comprise a community of shared efforts. This community primarily writes for one another, developing vocabularies that provide efficient shorthand but that simultaneously erect barriers to those outside the discipline. More importantly, the discipline comes to identify a set of questions central to the discipline's inquiry.[83] In Johan Heilbron's mind, modern disciplines are best conceived as units for teaching, research and professional organization.[84]

In the Weber tradition of ideal types, a widely held conception is that the disciplines are distinctly different from one another and that the boundaries between them have been established by the intellectual consensus among representatives of adjacent disciplines. Disciplines define their own niches of reality as topics of exclusive research: Physicians are supposed to concentrate on nature, biologists on biology, psychologists on minds, theologians on God, political scientists on politics, etc. Each disciplinarian is supposed to conduct his or her research *vertically* and *in depth,* and to stay within the restricted bounds of his or her respective field. In this way, the disciplines stand for stability and uniformity.[85] Individual disciplines are specific cultures in their own right. Along with this, the disciplinary organization of research is based on a division of labour. Each discipline is obligated to seek the 'truth' within its respective field on behalf of the scientific community as a whole. To many, this implies that certain topics and issues are considered the 'property' of designated disciplines. This is what is referred to as a *mono-disciplinary* structure (see Figure 1.3.A).

However, as pointed out by Tim Unwin, there is nothing absolute or sacred about disciplinary boundaries. All disciplines have been created and argued about, and there is no single criterion upon which such boundaries can be agreed. Thus the boundaries are arbitrary, elusive, constantly disputed and to a large extent artificial. The fact of the matter is that disciplines are defined in various ways. Following is a discussion of four different attempts to define the term 'disciplines':

First, it has been argued quite simply that a discipline is *the collective activities of its practitioners.* The scope and range of any discipline is thus dependent on the mood and proclivity of its practitioners at any given time, and the boundaries are by definition dynamic and fluid. "Such a definition emphasizes that disciplines are social phenomena that reflect the institutional and political frameworks within which they emerged."[86]

The second attempt made to identify individual disciplines has been through reference to particular *objects of study or subject matters.* For instance, political science is defined by its exclusive focus on the subject matter of politics, at the same time as it is a 'multidisciplinary discipline' in that it employs the scientific approaches and insights of multiple disciplines such as sociology, history, international law, geography, economy, etc., to address politics (see Chapter 4). This definitional approach rests on the assumption that some specific order exists in the world of phenomena, within which practitioners of any discipline simply need to identify their niche. In general, this is the system of definition and justification that is most often resorted to in the process of delimitation between disciplines and sciences.

Third, disciplines have been defined and described in terms of *their methodology or techniques* (see Chapter 4). Here, attempts are made to delimit disciplinary boundaries with reference to a unique set of technical tools. "Disciplines thus expand through the creation of new types of techniques, or through the poaching and development of methods from other disciplines" (see the concepts of *juxtapositional depth and breadth* and *interdisciplinary exchange and adjustment* in Chapter 3).[87]

The fourth definition involves focusing on the sorts of question that disciplines ask, and the ways in which these questions are framed (see the concept of extended mono-disciplinarity, Chapter 3).[88] Questions range from the narrow to the broad, and are hard to capture within the confines of a single discipline. At the outset, questions and answers are discipline-neutral and not the property of any one discipline. The corollary of this is that disciplines defined by the criteria of the questions they pose largely eliminate the significance of boundaries (see below). Even more importantly, there are no automatic contradictions and conflicts of scientific interest between disciplinary and interdisciplinary research. The one takes part on the premises of the other as a cross-border activity. Let us illustrate the point: By erecting a boundary around the cognitive issue of what can and cannot legitimately be said about a fossil record, geologists are bound to attract the attention of biologists. Fossils are, to be sure, rock formations, but they are also acknowledged to be records of plants and animal forms.[89] It is possible to learn more about fossils by integrating the knowledge of both disciplines than by respecting the cognitive restrictions inherent in each of them.

It follows from the above that disciplines are not homogeneous units, and that the stability and orderliness often associated with mono-disciplinary research are overstated and idealized. Bruno Latour argues that scientific disciplines are born free, but everywhere they are in chains to be broken.[90] Actually, disciplines are in a permanent state of flux: new disciplines emerge, established disciplines decline and fade away, or they may grow and disintegrate into several independent specialities. Consequently, what is disciplinary today may be interdisciplinary tomorrow and vice versa.[91] Against this backdrop, some regard the periphery of disciplines to be more like *transition zones* for overlapping than as distinct boundary lines.[92] From this perspective, disciplines have been defined as conglomerates of several subfields with multiple kinds of links to other disciplines and their subfields. In the sociology of science, a discipline is therefore defined as a 'cluster of specialities.'[93] From this vantage point, a discipline is a multidimensional network in which it is difficult to identify a pure core that is independent from other disciplines. Thus, the disciplines are to a large extent connected horizontally to each other through their respective specialties. This is to say that disciplinary specialties, which provide depth to research, are the backbone of interdisciplinarity, which provide breadth to research. The very concept of interdisciplinarity presupposes the existence of disciplines, which are the bureaucratic unit around which education, research and teaching are organized and implemented in most western universities. On this background, a discipline is best defined by its actual use and integrative ability as *an organizational and professional unit of researchers for teaching and research containing a network of professional links to other disciplines through its respective specialties* (see Interfield theories, Chapter 3).[94] In this way, disciplines constitute entities of knowledge integration and organization. This realization is a primary topic in the discourse of interdisciplinarity (see below).[95]

According to Barry Barnes, et al., scientific boundaries are defined and maintained by social groups concerned with protecting and promoting their cognitive authority, intellectual hegemony, professional hegemony, and whatever political and economic power they might be able to command by attaining these things.[96] Thus, the delimitations between disciplines are perceived as boundaries, transition zones and separate multidimensional conglomerates that provide different incentives for integration. These delimitations both inhibit and encourage interdisciplinarity, which can only take place if they are crossed or done away with.

Steven Greenblatt and Giles Gunn argue that understanding boundaries requires determining a number of factors. For example, what do the boundaries enclose and exclude? Are the lines drawn in bold, unbroken strokes or a series of intermittent, irregular dashes? What are the multiple functions and stances from which they are accepted or redrawn? What are the obstacles faced by those who challenge, nullify, or abolish them? What are the nature and import of the activities that occur at points of intersection?[97] The ultimate challenge of interdisciplinarity is thus to understand the nature of boundaries and to find ways of crossing or doing away with them.

CROSSING DISCIPLINARY BOUNDARIES

Transcending disciplinary boundaries may happen in two ways: either by boundary-breaking or by boundary-bridging.[98] The attention of the community of disciplinary researchers are often more focused on boundary-breaking than on boundary-bridging.[99]

Boundaries are broken when practitioners of one discipline turn to a different discipline for a new way of construing their own discipline or seek to bring part of the territory of another discipline under their own wings. Such boundary-breaking practices, long declared to represent subversive imperialism, have now become the accepted mode of behaviour of disciplinary researchers. We witness trespassing on the turf of others when psychologists extend their field of the human psyche to include humanistic disciplines such as art, religion and moral, for example, or when economists circumvent their primary interest in economics to study the impact of social sciences such as education, family, race, politics and fertility, or when political scientists study the political implications of novel technology, culture, etc. When geographers openly admit that they have drawn extensively on ideas derived from political science, sociology and economics, they are simply describing the current state of affairs.[100] The fact is that the same topics are being addressed by many disciplines, and that similar methodologies are applied across starkly contrasting fields. This is what we call *extended mono-disciplinarity* (see Chapter 3). Thus internal conflicts among disciplines are prevalent and constantly ongoing, leading to the establishment of distinct sub-fields within and between established mono-disciplines such as *political-economy, political-sociology, social-psychology, geo-politics, political-geography, socio-biology, economic-ecology* etc. These boundary-breaking entities, also called *hybrid disciplines* or *multidisciplinary disciplines,* carry a unique integrative potential for promoting synthesis of knowledge (see Figure 1.3.B). They are mergers spawned by interdisciplinarity. Interdisciplinary conflicts should not therefore, according to Bruno Latour, be regarded as brakes on the development of science, but rather as their motors.[101]

Disciplinary discontent has the ability to reshape the landscape of science and to reshuffle researchers across old demarcation lines. They are manifestations proving that disciplinary boundaries are arbitrary. They are based on human territoriality and interests, rather than on reality-based functionality. Thus the boundaries between various disciplines, sub-disciplines and units of topical specialization (see Chapter 3) are contingent on accomplishments originating in specific situations and are liable to revision as these situations change.[102] This is so because most research does not fit clearly into one category or the other. The best often combines features of each.[103] From this perspective, boundary breaking is a measure of *forced interdisciplinarity,* i.e. a conquering of 'foreign' territory by peaceful means.

Boundary-bridging is the voluntary and benign version of interdisciplinarity. It happens when representatives of one discipline draw upon the work of related disciplines to solve problems as these problems are defined within their own discipline. These crossovers usually happen in the *border zones* or *interface* where two or more disciplines meet. These zones are the venue of fields, the fuzzy areas where parts of disciplines mingle, blend, change and multiply. This is where hybridization happens (see Chapter 3), and where interdisciplinary cooperation starts. Boundary-bridging preserves disciplinary structure but does away with its hindrances. It provides peaceful coexistence between disciplines, i.e. *benign interdisciplinarity*.

Due to different modes for crossing borders, what has been recognized as a discipline by one university may not enjoy the same status in another. The boundaries that distinguish disciplines are therefore blurred, some more than others. They change over time, and are perceived differently by different people. Disciplinary boundaries are not perpetual, and can be crossed, confused, consolidated and collapsed. They can also be revised, reconceived, redesigned and replaced.[104]

At present, boundary crossings of either category have become widely accepted modes of behaviour among researchers working in multiple disciplines. The science community is gradually coming to acknowledge that there are no such thing as private (mono-disciplinary) ownership of specific domains of research.[105] For centuries, religion was the exclusive domain of theologians. Today, many disciplines, including sociology, psychology, political science, anthropology, philosophy, medical science and even biology (see Chapter 4), claim spirituality as part of their research portfolio and responsibility. In real life situations—the joint focus of all sciences and disciplines— the parameters are interdependent, connected and integrated, and not partitioned and secluded as suggested by the mono-disciplinary structure. It follows from this that topics of research are, in principle, *discipline-neutral* and partly *interchangeable* between scientific domains (see Chapter 2). Their composite nature invites and attracts contributions from multiple disciplines. What is required to bridge C.P. Snow's 'gulf of misinterpretation' is the *curiosity* to explore what is on the other side of the 'fence,' the *courage* to jump over it, and the *willingness* to make deliberate and patient investments in terms of time and effort after touch down (see Chapter 3).

Up until the present, depth of research has prevailed over breadth and synthesis (see Chapter 2). This is so because the disciplines, at least in the western world, are permanent entities organized in university departments and practiced on an everyday basis. Interdisciplinary research, on the other hand, is generally more project-oriented, haphazard, occasional and time-restricted. This implies that the accumulation of interdisciplinary knowledge depends on the pace of progress of disciplinary research. Interdisciplinarity is created and recreated every time a new interdisciplinary team is constituted to address an overarching problem. In this way, interdisciplinary knowledge does not grow in the same manner, to the same degree or at the same pace as disciplinary knowledge. Interdisciplinarity is bound to follow

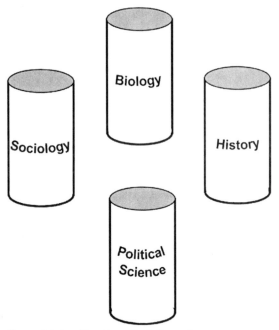

Figure 1.3.A.   The Monodisciplinary Structure.

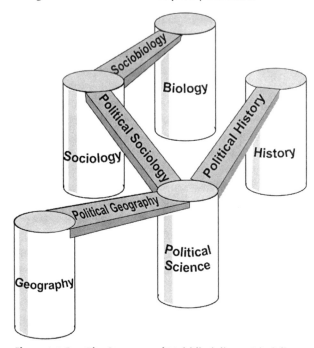

Figure 1.3.B.   The Structure of Multidisciplinary Disciplines.

in the footsteps of disciplinarity, which leads the way. Interdisciplinary knowledge is therefore contingent on the progress of disciplinary knowledge. Disciplinarity is the first step in interdisciplinarity.

WHAT IS INTERDISCIPLINARITY?

The overall objective of interdisciplinarity is to break or bridge the "walls" separating communities of knowledge, i.e. to harvest on each others insights and competence to create an holistic understanding of a given topic, challenge or problem.[106] Two types of interdisciplinarity stand out:

1. *Academic interdisciplinarity*, which concerns the transgression of boundaries between scientific disciplines, and
2. *Transdisciplinarity* which opts to overcome not only the boundaries of scientific disciplines but also those erected between academe and stakeholder expertise in society.[107]

Academic Interdisciplinarity

The objective of academic interdisciplinarity—which can be broken down into different modes of crossdisciplinarity (see Chapter 3)—is to integrate the specialized contributions of two or more disciplines to deal with a complex problem.[108] The craft of this type of interdisciplinarity is to create wholeness out of pieces, to see how the individual contributions of disciplines affect, connect, relate, integrate and interact in composite reality. The quest is to find unity in diversity, to explain how order can emerge from a mass of evolving agents, whether they be atoms, cells or organisms.[109] In so doing, the uni-dimensionality of individual disciplines is pitted against the multi-dimensionality of multiple disciplines, that is, the mono-faceted compared with multi-faceted, the specialist view compared with the generalist view, the specialized answer compared with the compound answer, narrowness against broadness, depth against wholeness. As such, academic interdisciplinarity is a bridge-building/breaking activity occurring when practitioners of one or more disciplines attempt to draw upon the work of other disciplines to solve problems as these problems have been defined either within their own discipline or jointly between disciplines (see Chapter 2). Interdisciplinarity is a means of solving problems and answering questions that cannot be satisfactorily addressed using the single methods and approaches of individual disciplines.[110]

Since the scope of integration is a major factor, a distinction should also be made between *narrow interdisciplinarity* and *broad interdisciplinarity*. The former involves disciplines with more or less the same paradigms and methods, theories and/or concepts. They are *neighbouring disciplines* being adjacent to each other historically and conceptually. Examples include history, sociology and political science, chemistry and pharmacy, and mathematics and information processing sciences. In broad interdisciplinarity, the disciplines differ and are far apart from each other when it comes to concepts, theories and methods. The differences between them complicate integration. Examples include philology and clinical pathology, law and medicine, public health and environmental engineering (see Chapter 4).[111]

When reading the book Frontiers of c*omplexity,* Baruch Blumberg, the Nobel Laureate in medicine, came to realize that he had been practicing complexity as a tool of disciplinary understanding for decades without recognizing the exalted company he had been keeping.[112] It dawned on him that the component description of a system does not contradict the holistic description; the two points of view are complementary, each valid at its own level. They are two different supplementary depictions of a single system.[113] Thus, semantically as well as methodologically, interdisciplinarity presupposes the existence and contributions of disciplines (see Figure 1.3.B).

In 1990, Julia Thompson Klein wrote that to picture the relationship between disciplinarity and interdisciplinarity as a double impasse, and as a fixed choice between one and the other, is to oversimplify the creative interplay that has produced changes in the nature of both disciplinarity and interdisciplinarity.[114] This acknowledgement has given rise to the concept of *disciplined interdisciplinarity* that moves outward from the mastery of disciplinary tools.[115] This is to say that interdisciplinary work depends much more on highly specialized learning and skills, than on universal knowledge.[116] Thus, interdisciplinarity marks no end to disciplines. They actually breed on each other. Disciplinarity is so powerful that it constitutes the 'first principle' of science. Disciplinary specialists are the foundations upon which all else is constructed.[117] This being said, the two practices of research have differences in agendas, and the disciplinary pull can be a barrier to communication when efforts at collaboration are called for in interdisciplinary research.[118] The important thing in this chapter does not involve dealing with the plentiful hurdles to be overcome in interdisciplinarity (see chapters 2 and 3), but to demonstrate that there are no ideological, theoretical or practical contradictions between reductionism and holism in scientific synthesis. In this way, the disciplines form the backbone of interdisciplinarity(see Chapter 2). The two practices are supplementary, not contradictory—they are brothers in arms.

Seemingly contrary to this conclusion is the historical fact that disciplinary research has been going on for decades in relative isolation and independence from interdisciplinarity (see Chapter 2). This is just another way of saying that interdisciplinarity depends on disciplinarity, and not the other way around. Such a conclusion, however, disregards the duality of the objectives of science. If science is to fulfil both its purposes, specialization and synthesis, the disciplines have no other alternative but to converge and engage in interdisciplinarity. From this perspective, it is the objectives of science that call for interdependence between the two practices, not the practices as secluded occurrences. Academic interdisciplinarity as defined above is a code word for diversity and adaptability.[119] Diversity is reflected in the multiplicity of disciplines employed, while adaptability refers to the ability to accommodate the disciplines so that they will co-exist, cooperate and co-produce in the trading zones of domains.

These zones have attracted many names that were indicative of their function: border interdisciplinarity, interdisciplinarity of neighbouring disciplines, zone of interdependence, borderland interdependence, tangential points and intersection of disciplines (see Figure 1.3.B).[120] The problem related to the border zones is that we have many detailed and sophisticated theories about what happens within the various domains, but we have little theory about what happens in the intersection between domains.[121] This is the more surprising since the border zones, from a cognitive point of view are one of the major sources of mental creativity. As pointed out by John Ziman, original ideas rarely come entirely out of the blue. "They are typically novel combinations of existing ideas. To make the connection, one has to cross boundaries between supposedly distinct paradigms—that is, between disciplines."[122] This is restating C.P. Snow's viewpoint: "The clashing point of two subjects, two disciplines, two cultures–of two galaxies, so far as that goes–ought to produce creative chances. In the history of mental activity that has been where some of the breakthroughs came."[123] The creativity occurring at the margins of disciplines has rightly been labeled 'creative marginality.'[124] The assumed creativity resting in the margins of disciplines was empirically tested in a recent Finnish survey. When asked about the reasons for selecting an interdisciplinary approach for their projects, 73 percent of the researchers responded that they were looking for the "production of new and broad knowledge," 54 percent said that "new approaches are interesting and hold potential," whereas 47% said they were looking for "synergies that relate to the sharing of knowledge, skills or resources."[125] In line with this, studies conclude that interdisciplinary research is instrumental in creating new knowledge, regardless of the formal labeling of research into 'disciplinary,' 'multidisciplinary,' or 'interdisciplinary.'[126] At any rate, by not using the interface of border zones, science are letting some of humanities best creative chances go by default (See Chapter 3).[127]

In holism, every problem is defined by the parameters of reality. Reality is the venue of disciplines. It is at the root—the problem as understood by society—where all disciplines meet.[128] For scientific investigation to reflect reality in its composite manner, there is actually no alternative but to consult with the world outside of research institutions. Interdisciplinarity is thus reality based, and is not an artefact of scholarship. At the outset, a problem knows no other boundaries than those set by the problem itself. Its resolution will involve all those disciplines touched by the problem. In interdisciplinarity, reality, as perceived by man, is the point of departure when formulating a research problem. Mono-disciplines seek to resolve problems defined by the confines of disciplines, i.e. those that have been cut and shaped to fit their boundaries. In disciplinary problem-solving, a segment of a holistic problem is cut loose and addressed as if it had an independent existence and no longer related to wholeness. Here, reality is being partitioned among disciplines and addressed in bits and pieces. The holism, diversity and adaptability of crossing boundaries make up an approach that is unique to interdisciplinarity. As pointed out by Bruno Latour, the difference between theory and practice ". . . is a divide that has been *made*. More exactly, it is a unity that has been fractured by the blow of a powerful hammer."[129] In line with this reasoning, some conceive of interdisciplinary research as a means of reuniting action and thought[130] at the junction between pure theoretical research, which emphasizes the pursuit of knowledge, "and that of informed action, which emphasizes usefulness, efficiency, and practical results."[131] This makes also for an extension of the concept of academic interdisciplinarity to become transdisciplinarity.

Transdisciplinarity

Academic interdisciplinarity is part of transdisciplinarity in the sense that both practices cross disciplinary boundaries. They differ, however, in that the latter also traverses the boundaries between science and stakeholder expertise. The transdisciplinary approach drives forward "a debate between research and society at large," involving deliberations about facts, practices and values.[132]

Transcending the boundary between the academe and stakeholder expertise affects the nature and characteristics of scientific knowledge as we know it. The feature that imbues traditional public knowledge with the status of being scientific is its *generality*. As observed, the less circumstantial and conditional a piece of empirical knowledge is, the higher its scientific value. The relevance of pure scientific knowledge depends on its actual or potential contribution to theory and theory-building, that is, to its ability to reveal the fundamental laws of nature and society–to its nomothetic characteristics.[133] The *specificities* of real life cases are to a certain extent ignored in the emphasis of science on the generalities of knowledge. Moreover, the assumptions underpinning traditional scientific knowledge are that it is value-neutral and that statistical methods possess the ability to tame scientific uncertainties (see Chapter 4). The currently prevailing cognition is that certainty and objectivity in science are nothing but unattainable dreams. On these grounds, transdisciplinarians challenge the traditional concept of knowledge in three respects:

First, they place new emphasis on the *specificities* of knowledge, i.e. on *real world problems* as expressed in *individual cases*, containing generalities

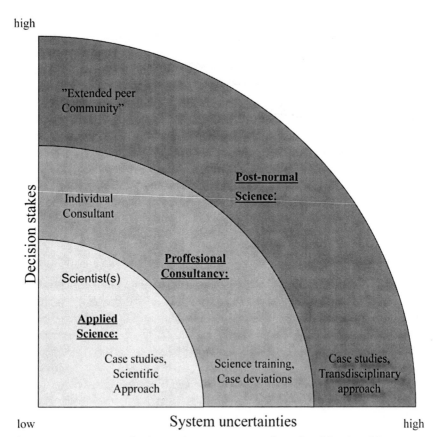

Figure 1.4.   Post-Normal Science Diagram. *Source: Adopted and developed from: Funtowicz and Ravetz: "Values and Uncertainties" in Gertrud Hirsch Hadorn et al. (eds): Handbook of Transdisciplinary Research, Op. ci.t, p.362. (Printed with the permission of the publisher).*

as well as specificities. In pure science, the specificities of an object of study are often overlooked or played down in significance. Only ideographic sciences such as geology and history take a keen interest in the specifics of a study object (see Chapter 4).

Second, to reduce the generality syndrome, knowledge production should involve *a heterogeneous panel of players*, i.e. a wide range of stakeholder expertise representing a raft of conflicting needs, interests and values. In a community of heterogeneous players, scientists are just one category among many knowledge contributors, but they are required to take part because non-scientific players "wish to know as precisely as possible, what the state of the nomothetic art is and how experienced the experts are in handling similar cases."[134] In this 'extended peer community,' the nomothetic contributions of scientists are supplemented and partially offset by the ideographic specificities of stakeholder insights. Everyone has something to learn from everyone else through dialogue and exchange of knowledge.[135] The uniqueness of this approach lies in the partnership between members of different disciplines and stakeholders.[136]

Third, the knowledge production process requires a procedural orientation which leads to shared definitions of problems, conflict resolutions and managerial implementations, i.e. the research process should be based on *deliberative strategic planning*.

On these grounds, transdisciplinarians opt for a new understanding of science and scientific knowledge, that is, a melange of scientific generalities and stakeholder specificities. To enhance the utility of science to society, it is necessary to share efforts and exchange knowledge across boundaries other than purely scientific ones (see Chapter 6). Thus, transdisciplinary research strives for concrete problem-solving in the social and political context through co-operation between those who *know how* and those who *know that*.[137] That leaves a lacuna which may be filled by what has recently been denoted *post-normal science*.

Post-normal science concentrates on questions such as 'What about?' and 'What if?' The issues of 'safety' and 'risk' are examples of this approach. It addresses problems of complexity in which 'systemic uncertainties' and 'decision stakes' are high (see Figure 1.4). Post-normal science is also defined in relation to 'applied science' and 'professional consultancy.' The former is characterised by low systemic uncertainties and decision stakes, while puzzle-solving provides sufficiently effective input into the field of science of relevance to policy. This is what Thomas Kuhn labels 'normal science,' implying that all the routine work of monitoring people and the environment fits into this concept (see Chapters 2 and 4). Professional consultancy takes the intermediate position between the two former positions, in that it is clearly

based on scientific training while it also has to cope with the unexpected outcomes of real life challenges. The work of a surgeon is a good example. On
the one hand, surgeons have to know how the body functions from a general
scientific point of view, and on the other, they are trained to handle the unexpected physical reactions of individual patients to the hardship of surgery.
Similarly, transdisciplinary research and post-normal science are complementary avenues of approach to what is regarded as a brand new understanding of science and its ability to contribute to the solution of persistent, complex societal problems (see Chapters 4 and 6).[138]

The transdisciplinarity practised to promote post-normal science affects the
ethos of science, in particular the requirement of 'disinterestedness' and to a
certain extent also 'organised scepticism'(see above). Stakeholder insights,
values and interests are parts of the resolution of post-normal situations, affecting both the well-being and self-interest of stakeholders. Thus, the traditional ethos (ethics) only partly applies to transdisciplinarity. Some scientists
will therefore find this practice offensive and unacceptable to the high standards of pure science (see Chapter 6). It is also difficult to make a clear-cut
distinction between applied science and post-normal science. Their similarities are striking. Both are case-oriented, addressing problems of a practical
nature for the purpose of servicing the complex needs of society and its players. From this vantage point, applied science and post-normal science are
close cousins of applied science, representing a confluence of practical and
scientific knowledge (see Chapter 2).

The fact that transdisciplinarity is a mix of scientific and non-scientific
knowledge may lead some to dispute its alleged status as a science in its own
right. As clearly acknowledged by its practitioners: "The birth of science is
based on a strict dissociation of scientific knowledge from the various aspects of practical knowledge. The ideal of scientific knowledge as it was
shaped in antiquity is still influential today, although the conception of science and the relationship between science and the life-world has undergone
many changes."[139] They also acknowledge that the partnership between
stakeholders and scientists makes it necessary for the time being to sacrifice
the academic standards of knowledge production and quality control.[140]
These aspects pose many questions of significance: How can knowledge of
different origins, produced and collected using methods and tools of different validity and reliability, be used as a unified measure for resolving societal problems? How does one decide what constitutes stakeholder knowledge,
beliefs and prejudices, and who will take the final decisions in an extended
peer community? What makes the specificities of stakeholder knowledge
more interesting and reliable in problem solving than the specificities ex-

tracted from scientific case studies? What happens in the interface between generalised and specific knowledge, and how do the two forms of knowledge coalesce?

Undoubtedly, there are many important questions to be discussed and answered before the true nature of transdisciplinarity as a science can be established. "Whether it will ever reach the format of an established discipline with sound paradigmatic foundations is questionable, since transdisciplinarity crosses boundaries between forms of knowledge; taking into account the diversity, complexity, uncertainty and values of issues."[141] However, what can be said for sure about transdisciplinarity is that:

- it has a strong component of being scientific in terms of academic interdisciplinarity;
- it has been tested successfully in resolving societal problems in many individual cases;[142]
- it is affiliated with the concept of post-normal science, signalling that it deviates somewhat from what is labelled normal science;
- transdisciplinarians depict their own practice "as a form of research," not as ordinary research, and their activity as a "new science," not as a traditional science;
- it is acknowledged that there is a need to develop the state of the art and to define quality criteria that are appropriate for this kind of research through learning from experience.[143]

In the light of these five bullet points and irrespective of all the tricky unanswered questions posed above, transdisciplinarity will be treated throughout this book as a scientific practice on equal terms to that of academic interdisciplinarity. The convincing argument put forward by Lee Smolin made this an easy choice: "Science requires a delicate balance between conformity and variety.....if science is to move forward, the scientific community must support a variety of approaches to any one problem."[144]

WHAT IS INTERDISCIPLINARY KNOWLEDGE?

According to the above discussion, the purpose of academic interdisciplinarity is *to produce scientific syntheses between two or more disciplines for the sake of scientific progress,* whereas the purpose of transdisciplinarity *is to produce solutions to societal problems through the fusion of knowledge provided by stakeholders and knowledge provided by scientists.* The ultimate aim of the former is to become public knowledge, whereas the latter's prime concern is to enhance the utility of science to society. Owing to the different numbers of disciplines, variables and types of knowledge involved, both versions of interdisciplinarity may take many forms (see Chapter 3).

In its most ambitious guise, all conceivable disciplines, variables and know-how will be used in an all-embracing effort to build synthesised knowledge. In this form, the focus of research is on total reality. This equals the positivist *unity of sciences thesis,* where all disciplines are reducible to the discipline of physics, and/or *consilience theory,* where all disciplines are supposed to be reducible to biology and psychology, and/or *a theory of everything* as discussed by the American philosopher Ken Wilbur (see Chapter 4). In Wilbur's mind, a theory of everything "attempts to include matter, body, mind, soul, and spirit as they appear in self, culture, and nature. A vision that attempts to be comprehensive, balanced and inclusive. A vision that therefore embraces science, art, morals; that equally includes disciplines from physics to spirituality, biology to aesthetics, sociology to contemplative prayer; that shows up in integral politics, integral medicine, integral business, integral spirituality."[145] A theory of everything is by definition a theory that comprises the specificities of individual cases and the generalities of all cases. In principle, such theories represent the ultimate level of synthesis and integration, providing an all-embracing background for the solution to and understanding of all kinds of problems encountered in reality.

In its least ambitious form, interdisciplinarity is intended to synthesize knowledge, either to make it public or to resolve practical problems, involving just two disciplines and their selected variables. In between these two extremes, all possible combinations and numbers of disciplines, variables, parameters and types of insights can be applied to address the composition of a study area/case. Their focus is on issues of various complexities and sizes.

Interdisciplinarity is 'cooperative' in two senses: Either indirectly, by one researcher/stakeholder drawing on multiple sources of practical and/or scientific knowledge for the purpose of converging them into a whole, or *directly*, by several disciplinary researchers/stakeholders pooling resources across all relevant boundaries for the sake of synthesizing their respective segments. Either way, their results are 'corporate.' They are also 'mutually interdependent' in that synthesis cannot take place without the contributions of all the involved experts. In this way, interdisciplinarity is based on the principle of an 'integrated division of labor.' This practice of research is thus founded on *cooperation, corporation, mutual interdependence, mutual respect* and an *integrated division of labor.*

NOTES

1. Hjort (20), pp. 127 and 129.
2. Snow (98), p. 5.
3. Snow (98), pp. 17–50.
4. Robson and Shove (04), p. 375.
5. De Mey (92), p.140.
6. Vatn (05), p. 10.
7. Ziman (98), p. 48.
8. Schumm (98), p. 10.
9. Clugston (04), p. 588.
10. Collini (98), pp.xlv–xlvi.
11. Thomassen (97), pp. 112–113.
12. Lepenies (89), p. 64.
13. Østreng (09), pp. 11–14.
14. Caldwell (82), p. 15.
15. Schumm (98), p. 2.
16. Caldwell (82), p. 16.
17. Ziman (98), p. 52.
18. There are exceptions to the rule, see Capra (00). Here the author concludes that modern physics gradually has come to confirm eastern mysticism.
19. Coveney and Highfield (95), p. 335.
20. Davies (83), p. 220.
21. Cited from Gribbin (05), p. 6.
22. Heisenberg (99), p. 194.
23. Alnes (03), p. 79.
24. The example is cited from Gribbin (05), pp. 5–6.
25. Dongen (07), p. 752. For a more thorough discussion of Einstein's relations to philosophy see: Isaacson (07).
26. Ellis (05), pp. 739–780.
27. Randall (06), p. 3.
28. Nugueira (08), p. 29.
29. Randall (06), pp. 190–193.
30. Nugueira (08), p. 31.
31. These examples are taken from: Barnes, Bloor and Henry (96), pp. 140–168.
32. Capra (00), p. 324.
33. Wilson (98), pp. 229–291.
34. Funtowicz and Ravetz (08), pp. 361–369.
35. Barnes, Bloor and Henry (96), pp. 141–142.
36. Merton (74), p. 268.
37. Smolin (07), p. 301.
38. Merton (74), p. 278.
39. Coveney and Highfield (95), p. 7.
40. Coveney and Highfield (95), p. 234.

41. Gribbin (05), p. 49.
42. Davies (83), pp. 101–102.
43. Weinberg (93), p. 39.
44. Davies (83), p. 63.
45. Anton (00), pp. 169–178.
46. Anton (00), p. 171.
47. Weiner (90), p. x.
48. Vatn (05), pp. 48–53.
49. There are many versions of the holistic approach. One is *mystical* as used in alternative medicine for the union of mind and body, another is anthropological, signifying a systemic account of how social institutions fit together in social practice and ideology. The version used in this book is methodical as presented above.
50. Capra (00), pp. 320–330.
51. Blumberg (95), p. xii.
52. Greco and Sosa (01), p. 1
53. Barnes, Bloor and Henry (96), p. 110.
54. Ziman (98), pp.48–53.
55. Sørensen (99), p. 76.
56. Weinberg (93), p. 185.
57. Jasanoff (92), p. 162.
58. Weinberg (93), p. 186.
59. Wilson (98), p. 64.
60. Latour (99) p. 95.
61. Cetina (99), p. 1.
62. Feynman (69).
63. Hubbard, Kitchin, Bartley and Fuller (02), p. 233.
64. Latour (87).
65. Fjelland (99), p. 23.
66. Weinberg (93), pp. 90–91.
67. *CAS Newsletter*, no. 1, March 2004, 12th year, p. 6.
68. Cited from Sandstrøm, Friberg, Hyenstrand, et al. (05), p. 68.
69. Gadagkar (06), pp. 177–180.
70. Campanario (95), pp. 304–325 and Campanario (06), 302–310.
71. Smolin (07), pp. 308–331.
72. Coveney and Highfield (95), p. 335.
73. Weinberg (93), p. 103.
74. Zimmerman (95), p. 90.
75. Livingstone (05), p. 3.
76. Capra (99), p. 334.
77. King, Keohane and Verba (94), p. 7.
78. Hansen (03), pp. 91–92.
79. Heilbron (04), pp. 26–27.
80. Matthews and Herbert (04), pp. 370–371.
81. Berge and Powell (97).

82. Salter and Hearn (96), p. 20.
83. Caron, Chapin III, Donoghue, et al. (94), pp. 341–342.
84. Heilbron (04), p. 30.
85. Ziman (00), p. 211.
86. Unwin (92), 5–6.
87. Unwin (92), p. 6.
88. Unwin (92), pp. 5–8.
89. Barnes, Bloor and Henry (96), p. 164.
90. Latour (99), p. 296.
91. Granberg (75), p. 3.
92. Matthews and Herbert (04), p. 383.
93. Dogan (01) p. 14852
94. Østreng (08). p. 12.
95. Bruun, Hukkinen, Huutoniemi et al. (05), p. 27.
96. Barnes, Bloor and Henry (96), p. 168.
 97. Greenblatt and Gunn (92).
98. Abrahamsen (87), pp. 355–388.
99. Bechtel (98), p. 111.
100. Hubbard, Kitchin, Bartley et al. (02), p. 202.
101. Latour (99), p. 102.
102. Barnes, Bloor and Henry (96), p. 140.
103. King, Keohane and Verba (94), 1994, p. 4.
104. Greenblatt and Gunn (92), pp. 1–11.
105. Bechtel (88), p. 2.
106. For a fuller definition see Chapter 3.
107. Funtowicz and Ravetz (08), p. 361
108. Bechtel (88), p. 71.
109. Coveney and Highfield (95), p. 8.
110. Thompson Klein (90), p. 196.
111. Bruun, Hukkinen, Huutoniemi et al. (05), p.29.
112. Blumberg (95), p. IX.
113. Davies (83), p. 62.
114. Thompson Klein (90), p. 103.
115. Cited from Thompson Klein (90), p. 106.
116. Thompson Klein (96), p. 157.
117. Clark (83), p. 35.
118. Caron, Chapin III, Donoghue, et al. (94), p. 342.
119. Ziman (00), p. 211.
120. Thompson Klein (96), p. 70.
121. Cartwright (98), p. 237.
122. Ziman (00), p. 212.
123. Snow (98), p. 16.
124. Dogan and Phare (90).
125. Bruun, Hukkinen, Huutoniemi et al (05), p. 104.

126. Bruun, Hukkinen, Huutoniemi et al. (05), p. 10. See also Sandstrøm, Friberg, Hyenstrand et al. (05).

127. Snow (98), p. 16.

128. Caron, Chapin III, Donoghue, et al (94), p. 343.

129. Latour (99), p. 267.

130. Thompson Klein (90), p. 97.

131. Thompson Klein (90), p. 122.

132. Wiesmann, Biber-Klemm, Gossenbacher-Mansuy, et al (08), p. 435.

133. Krohn (08), pp. 369-370.

134. Krohn (08), p. 372.

135. Funtowicz and Ravetz (08), p. 363.

136. Hadorn, Hoffmann-Riem, Biber-Klemm, et al (08), p. vii.

137. De Mey (92), pp. 235-236.

138. Funtowicz and Ravetz (08), p. 362.

139. Hadorn, Biber-Klemm, Grossenbacher-Mansuy, et al. (08), p. 19.

140. Hoffmann-Riem, Biber-Klemm, Grossenbacher-Mansuy, et al. (08), p. 5.

141. Wiesmann, Biber-Klemm, Grossenbacher-Mansuy, et al (08), p. 434.

142. Hirsch Hadorn, Hoffmann-Riem, Biber-Klemm, et al. (08), pp. 43-426.

143. Hirsch Hadorn, Hoffmann-Riem, Biber-Klemm, et al. (08), pp. 43-426.

144. Smolin (07), p. xxii.

145. Wilber (00), p. xii.

Chapter Two

Politics and Science: Partners in Arms?

"The world henceforth will be run by synthesizers, people able to put to-gether the right information at the right time, think critically about it, and make important choices wisely."[1]

Edward O. Wilson, 1998

The scientific community is an integral part of society and the structure of governance. The scientific community has been established and funded by governments, private organizations and/or industry to produce knowledge of practical applicability. It has become a locomotive for modernization; a means to improve the standard of living in the affluent part of the world and a necessary measure that may help developing countries escape their present misery. As such, science is defined by its patronage as an indispensable factor in nation building, development, preservation, change and adjustment. At the same time, scientific progress can lead to the over-exploitation of natural resources and serious environmental problems. Thus the application of scientific knowledge has become an ambiguous good. It represents both a blessing and a danger, providing satisfaction and suffering, excitement and apprehension. The technical civilization "has gone far beyond any control through human forces."[2]

To produce this insight, western democracies have granted their researchers freedom of research, which can be deconstructed into three interrelated freedoms:

• *the freedom of choice*, i.e. the right of an individual researcher to define whatever topic he or she would like to investigate,

- *the freedom of independent conduct*, i.e. the right of a scientist to conduct the research process without external interference of any kind,
- *the freedom of publication*, i.e. right of a researcher to publicize his or her results without censorship or reprisals of any sort.

Thus freedom of research is conditioned by the utility it provides, and the utility of research is conditioned by its freedom. The one cannot be exercised to the exclusion of the other. The two aspects are interdependent and linked together like two sides of the same coin. In other words, science has become a political vehicle in the service of society without being a political hostage. Science is, at one and the same time, *conditionally independent* in that it can follow its own course, and *conditionally dependent* in that it must be attentive to the course requested by its external providers.

It is more correct to say, however, that there are contradictions inherent in and inconsistencies between the two components: The freedom of research principle is undoubtedly constrained by the obligations, and the obligations are challenged by the freedom. As has been pointed out, the more closely science is involved in politics and practical matters, the more pervasive its dependence on social goals, values and wishes. On the other hand, excessive autonomy threatens isolation and can lead to a narrow-minded search for socially irrelevant truths.[3] The truce between the sides can be observed only if they are practiced in a balanced way so as to keep both sides reasonably satisfied. The balancing point is elusive, and depends on the intermingling pattern of societal, political and scientific interests at any given time and place. It is a fixation point in constant movement. The changing dynamics of this elusiveness determines the working parameters for all types of research, be it disciplinary, interdisciplinary, basic or applied. The pattern of intermingling will be determined by the moral and ethical standards of behaviour, not to mention science itself, which have evolved and will continue to evolve in the light of political, social and economic circumstances: "These factors provide the 'selection pressure' and determine in large measure what is or is not suitable" and possible.[4] The function of the freedom of research principle is thus to guard against *improper and undue interference* by and pressure from non-scientific actors in the research process, not a *carte blanch* for researchers to exempt themselves from societal obligations and societal experts (see below). The ethos of science is part of the freedom of research principle.

The overall purpose of this chapter is to discuss the political preconditions under which interdisciplinary research may take place, and to address the ability of the scientific community to deliver what societies need and governments seek. This poses three interrelated questions:

- To what extent do politics matter to interdisciplinary research?
- To what extent does interdisciplinarity meet the social obligations of science?
- To what extent does the freedom of research principle affect politics and interdisciplinary research?

We will answer these questions in four steps;

I. by discussing the varying interaction and dependency patterns of politics and science in different categories of research;

II. by comparing the practice of the science community with the scientific competence asked for by governments;

III. by addressing the political and scientific challenges of complex systems; and

IV. by discussing some measures of compliance between politics and science in managing complex system challenges.

THE INTERACTION AND INTERDEPENDENCY PATTERN OF POLITICS AND SCIENCE

Attempts to apply reliable knowledge for public purposes appear to be as old as government itself. But not until the mid-17th century, did governments begin to employ public knowledge, tested sufficiently to be called *science*, in a policy-oriented way. The establishment of the French Academy of Sciences in 1662 made official a relationship between government and science that has helped transform the world into what it has become today. And as the concerns of society and government became more numerous and complex, and the scope and reliability of science increased, the government-science connection grew in importance.[5] Society's efforts "to speak back to science"[6] have actually been on a steady rise throughout the past half-century. Some characterize this connection as being no less than symbiotic.[7] The links between politics and science differ in the three categories of research: *Practical-instrumental research*, *symbolic-instrumental research* and *knowledge-instrumental research*.[8]

Practical-Instrumental Research

aims at solving current problems for immediate application, be they military, economic, humanitarian, idealistic or the like. This is applied research, disciplinary as well as interdisciplinary, both versions. The clients have two objectives in mind: In addition to finding solutions to the problems impeding the achievement of their interests, they also wish to gain a comparative advantage in relation to their competitors/opponents. In the United States, universities have long been defined as a "major weapon(s) in America's battle for global competitiveness," and for "the bridging of disciplinary, academic, industrial and government boundaries."[9] The metaphor of a partnership between academia and society has been replaced by the metaphor of a purchase order.[10] According to Edward O. Wilson, science, like art, follows patronage.[11] Science has become omnipresent–a mandatory 'detour' for public and industrial decisions to be made.[12]

To keep their edge, clients of practical-instrumental research will have a definite inclination to keep research results secret. Freedom of publication is therefore often conditioned by client interests. Clients are also the initiator of this kind of research in most instances. They know what their interests are, and act accordingly. This also deprives the scientist of his freedom of choice. Here, conditions external to the research process itself are, by implication, governed by the client. However, there are exceptions to the rule.

Occasionally, clients accept *carte blanche* responsibility for funding practical-instrumental projects in total compliance with the freedom of research principle. No curbing or restrictions apply at all. This is not automatically to say that the clients do not perform to the best of their own self interests. Oil companies operating in foreign countries have been known to fund research projects of little or no direct utility for their primary activities with the aim of maintaining a good business relationship with the host government and/or to improve their public image. Here, it is in the client's own interest to show a utilitarian profile by adhering to all the freedoms of research. Their interest is symbolic-instrumental (see below).

In the main, the clients of practical-instrumental research are one of three types: organizations, industry or governments, all of which have different functions and thereby different needs in society. In general, organizations and industries are issue-specific entities, restricting their activities to their own niches or segments of reality. They by and large focus on issues and seek issue-specific answers to resolve issue-specific challenges. When asking for scientific input, they basically avail themselves of the disciplines of science; niche activities require niche expertise.[13] The agendas and responsibilities of governments concentrate on the functioning of society as a whole. Their main, if not exclusive, focus is on the interaction of segments (sectors). Governments are concerned with the complexities of interrelated issues and seek complex system answers to resolve complex system challenges. In this respect, governments need more interdisciplinary answers provided by the synthesizers of knowledge.

Although it is quite possible to make a distinction between interdisciplinary and applied research, the two approaches are often closely intertwined. Interdisciplinary research—both versions—is to a large extent related to the history of governments seeking problem-focused and mission-oriented research. Thus applied research and interdisciplinary research have a common denominator: *the incentive to carry them out is defined by a problem existing independent of either of them.*[14]

The only freedom of research not to be tampered with in practical-instrumental research or, for that matter, in any kind of research, is the freedom of independent conduct. If this freedom is violated, the client's own interests will be the first to suffer. A client deprived of the possibility to base his interest-specific decisions on the 'truth' may fail. His own interests will be at stake. As Peter R. Killeen so elegantly put it: "Good science is portable: Its techniques work independently of one's political persuasion. When this ceases to be the case, science dies and communities are blinded. The introduction of political biases into scientific process–Lysenkoism–can bankrupt a society."[15] Here, true scientists and rational clients have converging interests.

Interest-based science may, however, be vulnerable to corruption. There are rare, but sad examples of researchers bowing to the pecuniary power of the client, adjusting and/or suppressing conclusions to have 'science' prove a particular case in the interest of their client. There are also cases in which data are fabricated. In the event, we are no longer talking about research, but about the distortion of facts and plain fraud. There are even examples of researchers that have made academic careers based on plagiarism, in which "a life in academic 'crime' does seem to have paid."[16] There is, however, reason to believe that outright scientific fraud is relatively rare. The reason is quite simple: In science, it is extremely difficult to get away with fraud. The prerequisites for the acceptance of any scientific results before they become public knowledge are repeatability and independent verification by peers. Here the 'brutality' of peers in scrutinizing each others' results becomes a measure that runs counter to scientific fraud. In science, a person who shows his true colours as a cheater will loose all credibility and remain a cheater in the collective mind of the epistemic community and in the public eye. Practical-instrumental research is based on the premise that clients and scientists alike are committed to seeking the truth on the basis of their own rational self interest.

Practical-instrumental research is the practice of science that most effectively and immediately fulfils the utility obligation of the science community to society as a whole. But it is also the branch that may accept severe restrictions on its freedom of choice and publication. As has been rightly noted, the closer science draws to the political arena, the narrower the distance seems to grow between the processes of science and politics.[17] This may be why some researchers feel that applied research is the kind of research that one would be too listless to carry out if it were not for the fact that someone paid them to do it.[18] Practical-instrumental research is the science activity that is subjected to the most severe criticism from parts of the academic community that dislike the uses to which science and technology are put by political and economic forces: "such as military hardware, surveillance, industrial pollution and destruction of the environment."[19]

In general the holistic reality basis underlying interdisciplinarity pierces the walls between science and politics more effectively than disciplines do. Modern societies are permeated and ruled by expert systems, i.e. by knowledge and expertise. This proximity is reflected in the proliferation of concepts such as 'technological society,' 'information society', 'knowledge society' and 'experimental society.'[20] As such, holism presents a fresh and disturbing challenge as regards how to define the relationship between interdisciplinary science and politics.

The closeness between politics and interdisciplinary research also involves the reality of the freedom of research principle, and the obligation of science to provide useful and applicable knowledge to its benefactors in society and government. No compromises may, however, apply to the freedom of independent conduct. Practical-instrumental research is thus undertaken on the premise that two of the three freedoms of research may, under certain circumstances, to some extent be put aside and slightly compromised in relation to the strict standards of the ideal types in the Weber tradition.

Symbolic-Instrumental Research

primarily serves client interests. It comes in two versions–one opting for *territorial influence* and one aiming at *influencing issue areas*. The former is initiated to demonstrate that the government in question possesses scientific capacity capable, should the need arise, of being used as a basis for claiming influence in territorial matters in non-scientific fields. Here the client aims to ensure the presence/activity of researchers in a region, e.g. Antarctica, outer space, the abyssal plain, or wherever and whenever he wishes to assert himself and his national interests.[21] Such a presence will signal two things: first, the state's interest in and affiliation with the area in question, and second, the government's political determination to play an active part in the development of the area in future.

The second version pertains to industrial firms, organizations and governments currying favour with a government to achieve benefits in areas other than research. One example would be an international oil company initiating research favoured by a government in order to improve its own position with a view to the allocation of oil and gas concessions. In such case, the results of research would not be of primary interest to the company. The only reason for initiating the research would be to humour the government into a good and favourable mood towards the company. Governments which use research as a measure to achieve political gains are another example (See Chapter 5). In instances like these, research becomes symbolic–a measure for achieving favours in fields other than science. It is applied as a vehicle to achieve influence.

At the symbolic level, the scientists are pawns in the game run by the clients, and as such, are full-time employees in fulfilling a societal obligation given high priority by the client. In some cases, scientists are not aware of their overriding function in relation to government or society. This, in turn, also means that as long as the content and organization of the research represents no threat to clients' interests, researchers will be granted freedom of choice, conduct and publication. This freedom will be undermined, however, if scientific choices were to make it difficult to maintain primary interests, or if the client–rightly or wrongly–perceives research to harm his interests. Thus, symbolic-instrumental research provides both conditionally good and conditionally poor grounds for scientific freedom. Symbolic-instrumental research provides freedom within the bounds of client interests. As long as the scientific activity does not interfere negatively with client objectives, all three freedoms will apply. Where they collide, the opposite may occur. In this kind of research, scientists are well advised to know their way around the minefields of client interests. As amply demonstrated in Antarctica, symbolic-instrumental research may benefit from involving *science politicians*, i.e. researchers who, in addition to carrying out active research themselves, are knowledgeable about or have even involved in the previous decision-making process of the client to initiate research in the first place.[22] Knowledge about the client's situation will never harm research activities, although it may to a certain extent curtail the freedom of research. In this kind of research, the real objective of the client is not always explicit but may, in principle, be 'hidden' from the implementers. Being aware of the secrets, i.e. the hidden agenda, is a means of keeping to the straight and narrow path of science (See Chapter 5).

In principle, symbolic-instrumental research can either be conducted in a disciplinary or an interdisciplinary manner. However, to avoid hurting client interests, disciplinary research seems the less risky. This is because disciplinary research is in-depth research conducted within the confines of a restricted niche, while interdisciplinary research stretches out across disciplinary boundaries, integrating knowledge on a broad front. The broader the perspective, the greater the likelihood for touching the circles of client interests that are not to be touched in order to avoid any conflict of interests (See Chapter 5).

Knowledge-Instrumental Research

is the very *raison d'être* of basic research. Its sole purpose is experimental or theoretical work undertaken primarily to acquire new knowledge without any particular application in view. Knowledge-instrumental research is curiosity-driven. Or as John Ziman put it: "basic research is what you are doing when you don't know what you are doing it for."[23] It is the research that one undertakes for the "sheer joy of understanding beautiful phenomena."[24] Knowledge-instrumental research is often also called 'pure science', 'real science', 'fundamental science', 'strategic science' etc. to distinguish it from applied research, and it is estimated to account for at least 10% of scientific activity.[25] It is the knowledge production engine of the science community *par excellence* and it makes no attempts to solve immediate problems (practical-instrumental research) or to support the hidden interests of clients (symbolic-instrumental research). It is the research that is not driven by bureaucratic, political or commercial directives, but rather by the inner needs of science and scholarship.[26] Indeed, basic research is a high-risk activity: it can fail, but it can also lead to breakthroughs. The greatest scientific breakthroughs in history have been produced by representatives of the knowledge-instrumental community. Its motive force is scientific curiosity, and its goal is to increase knowledge and insight for the sake of scientific progress.

In the 1830s, when William Whewell coined the term 'scientist' as a substitute for 'natural philosopher', he saw a scientist as someone engaged in a unique social role who required protection and had an autonomous existence from the rest of society.[27] He regarded science in knowledge-instrumental terms as a pure, autonomous activity, separate from technology, politics and industry, i.e. as a realm of independent existence. Scientists' only obligation was to find the 'truth.' In certain quarters, this ideal ultimately developed into academic snobbism in that practical-instrumental research was regarded as an occupation for second-rate minds. At the start of his career, C. P. Snow took precisely that line himself, stating that he long prided himself on the fact that the science he was doing could not under any conceivable circumstances, have any practical use. He concluded: "The more firmly one could make that claim, the more superior one felt."[28] As demonstrated in Chapter 1, he was later forced to pocket his pride and superiority.

As conceived by Whewell's followers, knowledge-instrumental research usually underlines that science and scientists have a kind of objectivity which is not necessarily compatible with the subjectivity of politics; no man and no nation has a monopoly on science and knowledge. Science is the common heritage of mankind. As such, science transcends the demarcation lines of politics and unites forces across national borders. It allows for total freedom of research and is controlled by nobody but the scientists themselves.

For the past 150 years, knowledge-instrumental research has basically been disciplinary in approach. Depth has taken precedence over breadth. The implication of this is not that knowledge-instrumental research cannot be interdisciplinary. As pointed out by Edward O. Wilson:" Scientists have broken down many kinds of systems. They think they know most of the elements and forces. The next task is to reassemble them. . . . That in simplest terms is the great challenge of scientific holism."[29] Resisting linking discoveries by causal explanations is to diminish their credibility and wave aside the synthetic scientific method, "demonstrably the most powerful instrument hitherto created by the human mind,"[30] he continues. In other words, what is lacking in knowledge-instrumental research is the pooling together of disciplinary knowledge to create an interdisciplinary basic understanding of how complex systems work. Wholeness is missing, and there is no scientific reason why basic research should be reserved for the disciplines alone. The current state of affairs is nothing but an omission of the sciences, not a necessity of science. If interdisciplinary research is conducted solely to meet the practical needs of governments, then complex system challenges–the ultimate objective of science–are left to be defined by politicians rather than by the knowledge-producing community. For basic scientists to reclaim scientific agenda-setting from the politicians, there is a need for high quality interdisciplinary basic science.

It has been noted that the notion of pure science cannot be defined in terms of policy because policy is all about future action.[31] This does not, however, imply that knowledge-instrumental research in its present form is of no utility to society or the political system, or that it has not been assigned any obligation to serve society. The notion of a science isolated from the rest of society is, according to Bruno Latour, as meaningless as the idea of a system of arteries disconnected from the system of veins.[32] The main bridge between science and society is *educational.*[33] When governments support the disciplinary version of knowledge-instrumental research, they do so because it may prove useful, at least in the long haul. No rational society is inclined to use taxpayer's money to fund research that has no other purpose than to calm or satisfy the curiosity of an individual scientist. Basic research is actually founded on an unwritten contract between the individual scientist and society, stating that scientific progress of no immediate utility at one point in time, may, eventually prove applicable for resolving the challenges of practical life. According to Steven Weinberg, this expectation has generally proved correct.[34] On this ground, the President of the Royal Society, Lord Rees of Ludow, distinguishes between two kinds of science, *applied science* and *"not-yet-applied-science."*[35] Suffice it to take one extreme example: the struggle to define and understand the smallest, indivisible building block of matter and the universe: *the concept of the atom.* It started out as a project of pure philosophical speculation in ancient Greece (Democritus), moved on through the era of modern or classical science (starting with Galileo and furthered by Newton) and ended up with Oppenheimer in 1945. The process took more than 2000 years before the idea was of practical-instrumental utility in terms of weapons of mass destruction and nuclear energy. The enormous impact of nuclear science on the political and economic structure of the international system is beyond dispute. The possession of nuclear weapons altered the international balance of power and imposed lasting changes on the World order. It still does. Thus, the political influence of basic physics on the general political situation is greater than ever before.[36] Ultimately, knowledge-instrumental research usually ends up being of use to society. In most instances, it takes time and patience before it happens. The most sophisticated governments acknowledge that scientific progress through basic research is the lifeblood of applied research and that the former is needed to maintain the quality of the latter. "Basic research is the first step in the innovation process," the former Norwegian Minister of Education and Research, Kristin Clemet pointed out in 2004.[37] The results of basic science thus form the very foundation of fundamental knowledge to which applied scientists have no alternative but to resort in order to resolve practical problems of some urgency. From this perspective, basic science is the first step in making science applicable. As stated, investing in basic research becomes a down payment on products and processes that will fuel economic growth and productivity.[38] Philippe Busquin, the EU Commissioner for Research, took stock of this when stating that instead of asking about the future of basic research in Europe, one should ask about the future of Europe without basic research. The implication being that in EU's endeavours to become the world's most competitive and dynamic knowledge-based economy, basic research has been awarded a place of pride and utility.[39] The assumption is that the nations leading in research, not least basic research, have an edge when it comes to commercialization.[40]

Some contend that the dichotomy between pure and applied science does not actually exist. Jacob Bronowski is one of them, denying any sharp boundary between pure and applied science on the grounds that science is full of useful inventions relating directly to social needs.[41] Here he is on solid ground. History shows that basic research sometimes quickly turns into concrete results and even commercial products without anyone having anticipated or planned for it at the outset. For several decades during the 20th century, the defining principle of western science was based on the principle of separating pure science from applied research, whereas Soviet scientists refused to recognize the distinction between the two. In the aftermath of the Sputnik shock in the late 1950s, U.S. science planners considered it to be in the national interest to redefine their position and to stop making such a strong distinction between pure and applied science, whereas their Soviet colleagues used the increased political leverage they enjoyed in the aftermath of their triumph to re-establish the official legitimacy of fundamental, basic science.[42] These historical examples show that the distinction between the two practices is a combination of shifting historical circumstances and group vanity rather than of expediency.

According to John Ziman's conception, research culture has gradually changed into a *post-academic period* in which basic research merges with the technoscience (applied science) and strategic science needs of governments. In this definition *post-academic science* and *post-normal science* are related concepts,[43] and will therefore be used interchangeably henceforth. The time has long passed when basic research was conducted only in universities and was undertaken solely for its own sake. Well known examples of this are the research conducted in applied research centres such as IBM, Microsoft, Bell Laboratories and the Max Planck Institute, and the market-driven research conducted at universities. In Norway, nearly 50 per cent of all university research is applied. The differences between universities and private applied research institutions and between applied and basic science are getting more and more blurred. According to Ziman, this development calls for a new scientific concept: *academic science,* where the distinction between all types and practices of research are becoming increasingly irrelevant. The most recent step in reshaping the conception of science came with the introduction of the transdisciplinary approach making stakeholder and scientific knowledge equal partners in problem solving.[44] This is not to say, however, that the concept of basic research has lost all meaning. Rather, the need is to redefine basic research within the context of academic and post-normal science as an activity with no *direct* links to any given application, with multiple institutional homes, and with the objective of furthering knowledge and stimulating societal discourse.[45] In Ziman's redefinition, the *indirect* utility of basic research is given pride of place.

Ziman also points out that apart from providing training so that experts can handle the realm of technoscience, basic research provides *world pictures* and *scientific attitudes* of immediate use to society. The world pictures provided by science serve to create awareness of areas of public concern. Scientific concepts such as *ecology, biodiversity, global warming, corporate state, democratic deficiency, environmental security,* etc. have been adopted by the political system as benchmarks for defining the content of political discourse, nationally as well as internationally. Thus, important concepts arising from academic science engender political debate, give rise to political agendas, direct the attention and energy of governments, and affect political decisions. Academic science is thus crucial to the way societies work, and even to their very survival.

Ziman goes on to argue that scientific attitudes are also important to understand how pluralistic societies function and sustain themselves. The scientific tenet that dogmas are for doubting (see Chapter 1), theories for testing, assumed facts for disconfirming and established authorities for deflating, reinforces the necessity of maintaining the public debate. Since scientists are constantly fighting among themselves about what constitutes public knowledge at any given time, it seems even more important that the doubts expressed in science also find their expression in the public debate. Disagreements and compromises are at the very heart of the way pluralistic societies work. In this way, pluralistic societies receive important impetus from the critical culture of academic science, and in so doing lubricate and sustain the very foundation of societies' functioning and existence (see Chapter 1).[46] Thus, as expressed through academic science, the basic elements of research indirectly entail societal utility in terms of raising the awareness of public concerns, shaping the public debate and supporting the workings of a pluralistic society.

The traditional distinction between disciplinary and interdisciplinary research is an ideal informed by tradition and the prestige of high-level theory, i.e. "an abstraction that assumes that disciplines are theory-centred and society's problems will fall inevitably outside the scope of pure disciplinary study."[47] In this perspective, applied research becomes an extended branch of basic research (see Figure 2.2). Historically speaking, interdisciplinary research started out as applied science and spread to the non-applied sciences (see below and Chapter 4).[48] Applied interdisciplinary research is therefore not, as sometimes contended, an illegitimate child of basic science. It is rather a sibling in that the two types of research are related and belong to the same family. They require active triangulation of depth, breadth, and synthesis.[49]

The practices of research, i.e. *basic, applied, disciplinary* and *interdisciplinary research*, are intertwined, and the science community cannot afford to interpret the freedom of research principle to the exclusion or at the expense of its utility obligation to society. This most vividly comes to expression in transdisciplinary research.

The Public and Knowledge-Instrumental Research

As is illustrated in Figure 2.1, basic research is to a certain extent a carrier of scientific input into all the other practices of science. They not only breed, but also feed on each other, and the spider in the web consists of knowledge-instrumental particulars, i.e. basic disciplinary research. At present, knowledge-instrumental particulars deliver input to practical-instrumental synthesizing, although the delivery route should ideally have gone through knowledge-instrumental synthesis.

However, in the public consciousness, considerable uncertainties exist as to the exact utility of knowledge-instrumental research, which can also be highly resource-intensive. Authorities will therefore tend to keep expenditures down,

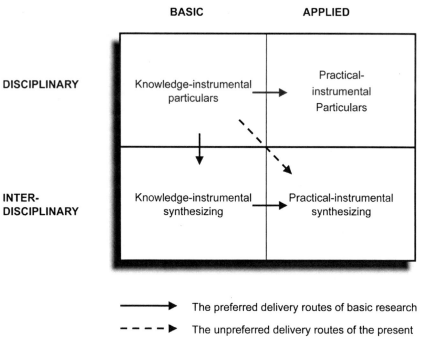

Figure 2.1.   The Interconnectedness of and Delivery Routes between Practices of Research.

although without falling out with their own researchers, whose presence is necessary to accomplish the political and societal objectives of practical — and symbolic-instrumental research. In this perspective, knowledge-instrumental research is needed to keep scientists content until such time as they will be called upon by governments to accomplish their political function associated with practical- and symbolic-instrumental research.

However, the lack of public acknowledgement of the immediate utility of basic research makes knowledge-instrumental research easy prey for social criticism and even political hostility. Widespread ignorance, which also applies to parts of the science community, of the interconnectedness of research practices can easily be exploited for political purposes. In most non-democratic and a fair number of third world countries, freedom of research is hampered and tampered with by governments with impunity, as they refuse to accept the risk of researchers coming up with unexpected and unacceptable results that could possibly be of value to the political opposition and its quest for power. Authoritarian societies to both the right and the left have been unable consistently to reconcile science with ideology.[50] In some of these countries, the social sciences in particular are defined as instruments of politics, i.e. as politically destabilizing factors. Accordingly, science is strictly controlled, directed and governed by the authorities. The distinctions between various categories of sciences are regarded either as irrelevant or as too risky to accept. Knowledge-instrumental research is rarely purported to be different from the other categories of research and is rated, in practical-instrumental terms, by the authorities with a view to its function and orientation.

Less drastic, but equally hurtful to science are measures such as ridicule, cuts in funding and deliberate neglect, all of which undermine the foundation on which knowledge-instrumental research rests. Even the most liberal, science-dependent of all democracies, the USA, has skeletons in her closet.[51] The anti-American trials staged by Senator McCarthy in the 1950s which victimized many academics on false pretenses were simply one of many tragic incidents. Michael Zimmerman comments, "when political considerations become paramount, free inquiry and public welfare often take a back seat."[52] In this perspective, the freedom of research principle is more of an ideal than a reality, and governments are likely to have their way because the state, in many countries, is the single most important provider of funding for universities[53] — the time-honoured designated Mecca of basic research. Thus, knowledge-instrumental research is not, as is often believed, immune to political interference. On the contrary, it is probably easy prey for populist politics.

The types and practices of research are intertwined in an interconnected Circle of Science, where the utility production is the collective result of all scientific efforts (see Figure 2.2). In 1993, the Norwegian government took stock of this acknowledgement and established the Norwegian Research Council (NRC) by merging five separate research councils that had been founded on the divisions between basic and applied research and between the social and the natural/technological sciences. The idea was that the effort should move towards a match between the organizational set-up and the reality of scientific practice and innovation. Since research took place across disciplinary boundaries and ministerial demarcations, having several research councils would only prohibit synthetic thinking.[54] What has been denominated the 'new NRC' aims at tying basic and applied research closer together and facilitating interplay across disciplinary boundaries.[55] Here the Circle of Science has got an organizational expression very different from the set up usually applied to universities.

To preserve the credibility of science as independent of economic and political interests, all rights attached to the freedom of research principle should ideally be part of research. However, when ideals interface with reality, adjustments are often made in the assumptions that apply to the latter. Compromises

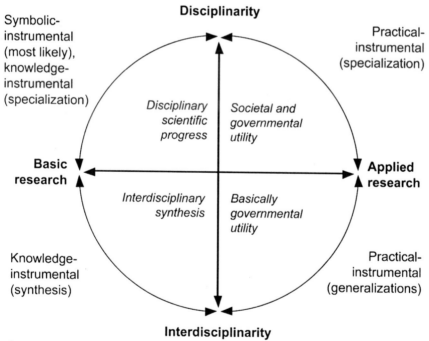

Figure 2.2.   The Circle of Science: The Interconnections, Interaction Pattern and Utility Production of the Four Practices and Three Categories of Research.

have to be struck (see Chapter 5). This begs the question of which of the three rights is most susceptible to bending, without violating the validity and reliability of the sciences?

The one right that can never be hampered or tampered with is the right of independent conduct; and the credibility and utility of the sciences hinges on that right. More importantly, the utility of research cannot be separated from its validity and reliability. Practical applicability and utility depend totally on the existence of the freedom of independent conduct. This right is non-negotiable and immune to compromise. Next in rank is the right of publication, which, when implemented, makes the results of research and the practice of conduct available to the scrutiny of peers and thus to the development of public knowledge (see Chapter 1). As such, clients and scientists alike will, in many cases, benefit from integrating this correctly into their cooperation. However, in actual practice, clients may have reasons for postponing publication for some time (see below). As long as we are talking about postponement of publication, however long, the important thing is to be conducive to compromise and to comply with scientific standards. The right of choice is the most prone to compromise. This right is exercised in full within the context of knowledge-instrumental research, and it is conditioned in symbolic-instrumental research and less so in practical–instrumental research. The question of who was first to conceive the specific idea of applied research, i.e. the client or the scientist, is far less important than the fact that the latter ultimately accepted implementation on scientific grounds.

Generally speaking, academic freedom is protected, although not without the science community granting concessions and showing sensitivity to the interests of clients and/or sponsors. This applies to all categories of research, but it is most pronounced in practical-instrumental undertakings, in particular in transdisciplinary research. The science community has a sufficiently strong political basis in the freedom of research principle to speak their mind, to influence political decisions that affect their own working conditions, and to advise and inform their governments. In democratic societies, values often take on a life of their own, outside of the reach and control of contemporary politics. These values have become part of consensual politics that unite the population across party lines and become strong threads in the fabric of society. The freedom of research principle has achieved such high status because it is the ultimate expression of freedom of thought and speech, and thus become a hallmark of civilized, democratic society. No expression of the freedom of speech and thought can be stronger than that attributed to some of the best minds of society. In societal upheavals, universities often take on the leadership for articulating discontent and organizing protests. They are the 'value-conscience' of most political systems, also those that are anti-democratic.

Having the political system grant freedom of expression to science is tantamount to providing 'arms' to a potential future opponent. The freedom of research principle is therefore the best guarantee for sustaining freedom of speech and thought of the public and vice versa. Freedom of expression has become the common legacy of democratic citizenship. As such, governments which once advocated that the principle be internalized as a common value of their respective societies, have, because of their former success, restricted their own leeway to ignore and/or combat it later on. Here, the politics of the past stand tall against any politics of the present that would discard the politics of the past. The leeway of governments has shrunk, and the science community has a weapon to be actively applied in alliance with the populace. From this perspective, the science community is not deprived of political power vis-à-vis its own government.

In line with this reasoning, Werner Heisenberg agrees that the political influence of science has grown stronger since World War II. This is not primarily due to the freedom of research principle, but rather to the existence of nuclear weapons, which, in his opinion, has burdened the scientist with a double responsibility. On the one hand, scientists can choose to take part in the administration of their countries by applying science for the benefit of the community, and they will then have to shoulder responsibility for the decisions made. On the other hand, they may voluntarily withdraw from any political participation, but will still be responsible for the decisions made, not least the wrong ones that could eventually have been prevented if they had chosen differently.[56] Either way, scientists play a political role. Politics is part of science and science has become part of politics, not only as a measure in nation building and modernization, but also as a bearer of democratic values. Governments have to devote attention to the interests of science and science to the interests of governments. There must be a balance between the obligation of utility and freedom of research. In what way and how this will happen, will at all times be decided by the dynamic intermingling pattern of science and politics. The search for the fixation point is a never-ending quest. 'Independent science' is, and will always be, part of politics, which by definition is biased and value-laden. The form taken by the intermingling is important not only to science, but also to politics. In this process, the interests of the one are, in Bruno Latour's mind, translated into the interests of the other, i.e. "when their goals are frustrated, actors take detours through the goals of others, resulting in a general drift, the language of one actor being substituted for the language of the other."[57]

Michael Zimmerman prefers the intermingling to be replaced by a division of labour. In his view, government officials ought to look at the 'big picture' and set the direction for national research efforts. Rather than worrying about individual research projects, political leaders and representatives should be thinking about research initiatives that will best serve the country. "This is a much more difficult task than finding fault with a study of Peruvian brothels but one that is too important to be left to the scientists alone. We need real leadership if we are to make meaningful progress on such critical topics as issues of women's health care, AIDS, and environmental protection and restoration."[58] However, the campaign for a 'better understanding of science' is not balanced by an effort to achieve better scientific understanding of the public.[59] In this respect, the science community still has a long way to go.

To what extent is there compliance between science production and utility, and in what way, if any, do democratic governments use their influence to provide for the research they request? Or to put it the other way around: Does the implementation of the freedom of research principle satisfy all the needs of government and society?

THE PRACTICE OF SCIENCE AND THE NEEDS OF GOVERNMENTS

In the 20th century, disciplinary science has been so successful that outsiders sometimes picture it as a kind of monolithic corporate organization like IBM or Microsoft. This imagery has nourished the belief that scientists look alike, act alike, think alike, and speak the same jargon. No imagery could be more wrong. In the reflective mind of Jonathan Weiner, science has, contrary to this image, developed into the very opposite—a Tower of Babel, where few speaks the language of the others, and few seem eager to learn more languages than their own disciplinary mother tongue. The language confusion makes handling the success of science a problem. The accumulation of scientific knowledge has been so overwhelming that no one can hold it within "the horizon of a single mind. There are chemical oceanographers and there are physical oceanographers. There are stratospheric chemists and there are tropospheric chemists. These people do not mix. They mix as little as the sea and the air they study, or as the various layers of the atmosphere."[60]

Many people are of the opinion that specialization has got out of hand. "A petrologist studies rocks; a pedologist studies soils. The first one sieves the soil and throws away the rocks. The second one picks up the rocks and brushes off the soil. Out in the field, they bump into each other like Laurel and Hardy, by accident, when they are both backing up."[61] This long-drawn sigh from Weiner pinpoints the crux of scientific development–the rapid and to some extent uncontrolled specialization taking place in all scientific disciplines. The truth of the matter is that the average scientist no longer manages to keep abreast of the knowledge accumulation in his or her discipline. The burden of disciplinary knowledge has been growing very rapidly for quite some time. Concerns about the danger of overspecialization emerged quite early, dating back to the late Middle Ages.[62] In 1892, Professor Karl Pearson at the University of London gave this account of the situation: "Scarcely any specialist of to-day is really master of all the work which has been done in his own comparatively small field. Facts and their classifications have been accumulating at such a rate, that nobody seems to have leisure to recognize the relations of sub-groups to the whole."[63] Since this utterance, the pace of the process has picked up exponentially. To cope, for decades the scientific community has deliberately divided disciplines into smaller and more manageable entities, and the smaller disciplines into sub-disciplines which have been further broken down into units of topical specialization which are becoming ever lighter in weight, thinner in scope and deeper in penetration (see Chapter 3).

Although the disciplinary organization of science persists, the professional frame of reference and the identity of scientists are no longer primarily within the traditional disciplines, but rather with smaller units of topical specialization. Tony Becher argues that what are variously described as sub-disciplines and/or specialisms form their own counter-cultures which may press against the overall culture of the discipline of which they form part, and may thus seem to threaten its unity.[64] Disciplines are composed of clusters of specialities, i.e. units of topical specialization that form the micro-environments where research and communication take place (see Chapter 1).[65] The traditional organization of the sciences into disciplines as we used to know them seems to choke under the burden of increasing disciplinary knowledge.

The days are long gone when an individual genius like Isaac Newton could be on the cutting edge of physics, mathematics, optics, theology, alchemy, and all the other abiding interests in what was then called—not science, but simply natural philosophy.[66] Historically speaking, the specialization of science has moved through three distinct stages: The first phase, the philosophical one, was truly pre-disciplinary in orientation. Aristotle was a pioneer in physics and biology. Plato made physics, cosmology and mathematics core elements in his philosophical thinking. To a large extent, biology and physics were the epitome of ancient philosophy.[67] In short, the many philosophical luminaries who followed in the footsteps of Greek thinkers were scientists in search of an organized conception of reality. Their quest went far beyond the special sciences as they are defined today. Although the earliest documented use of the term interdisciplinarity in research did not appear until the 1920s in the social sciences,[68] the old Greek philosophers approached and studied reality in broad interrelated terms.

The second phase was inaugurated by Galileo. The focus of science was on the 'simple questions' of natural philosophy, "looking to explain why apples fall to the ground and why the Sun rises in the east, "[69] ignoring for the time being the complexities addressed by ancient thinkers.[70] Specialization became the new credo. The speedy progress and success of specialist science made the burden of disciplinary knowledge grow into an increasingly heavy burden on the individual scientists. In the late 1890s, Karl Pearson complained that it was "as if individual workers in both Europe and America were bringing their stones to one great building and piling them on and cementing them together without regard to any general plan or to their individual neighbour's work."[71] The credibility of researchers as scientific experts was increasingly linked to their specialities.

In the third phase, the disciplinary burden of knowledge became so overwhelming to the scientific community that it saw no alternative but to continue the partitioning of science into smaller sub-disciplinary units of topical specialization. The search for smallness was the driving impulse.[72] This process went on unaffected until the middle of the 20th century at which time most of the simple questions had been answered.[73]

Over the past 500 years, the focus of science has shrunk from addressing the big picture of the universe to examining the smallest particles, e.g. molecules, atoms, quarks, nanomaterials, etc. According to Mattei Dogan, the history of scientific advancement is the history of concatenated specialization.[74] The number of specialities has become staggering. By 1987, there were 8 530 definable fields of knowledge in the sciences. By 1990, roughly 8000 research topics in natural science alone were being sustained by specialized networks.[75] In sociology alone, recent estimates suggest that there are some 50 specialities. The same applies to many other disciplines.[76] The ongoing fragmentation of knowledge is not a reflection of the real world, but, in Edward O. Wilson's mind, an artefact of scholarship.[77]

The process of specialization has evolved and been fuelled by multiple forces. First, the training and education of researchers does not orient them to "the wide contours of the world." In science, grants and honours are awarded for discoveries, not for scholarship and wisdom[78] Success as a scholar often means spending an entire career in a constricted area of formal study. Second, fragmentation is also the rational choice of individual scientists who want to take control of their professional lives, that is, to overcome the burden of disciplinary knowledge. Fragmentation and specialization may entail a note of resignation, representing a survival kit for disciplinary scientists. Although specialities are considered parts of a larger discipline, the modern scientist does not usually attempt to master his discipline in its entirety. Instead he limits his attention to one field, or perhaps two.[79] Thus, parts of each formal discipline gradually become unknown in their entirety even to those who profess to further other parts of the discipline. In the case of sociology, the aspiration to be a general sociologist was deemed a realistic goal in the 1930s and 1940s. This was the case because the common body of core concepts and the body of accumulated research were small enough to make it possible for an individual researcher to make significant contributions and speak authoritatively about the field in general. Fifty years later, it is "difficult to imagine the genius necessary for such accomplishments . . ."[80] Third, the changing topography of the subdivision of disciplines did not only spring up and grow naturally; the disciplines, boundaries, standards, and priorities also evolved from processes involving differences of scientific opinion and straightforward conflicts between individual scientists.[81] Fourth, it also reflects the enormous success of the sciences in applying the method of reduction. In general, scientists have been so intrigued by the discoveries promoted by this method that instead of reverting back to see how their discoveries fit in with the totality, they have continued to dig deeper and deeper into their specialities to continuously reduce the size of their units of topical specialization.

In the past 50 years, compensatory reactions towards the specialization of the sciences have come from within the science community as well as from the political system. The feeling is that the units of topical specialization have grown too small, and that there is a pressing need to bring the bits and pieces of reductionism back into 'wholes' again. In this perspective, interdisciplinarity is a new stage or phase in the evolution of science.[82]

THE COMPENSATORY REACTION
TOWARDS INTERDISCIPLINARITY

During the 20th century, there were two interdisciplinary waves within the academe. The first occurred in the social sciences. It began at the close of World War I and lasted until the late 1930s. This movement was a sort of bridge building effort in that quantitative methods were borrowed from the natural sciences and applied for 'soft science' use (see Chapter 4).[83]

The second wave dates to the last 30 years of the 20th century. During that period, a subtle restructuring of knowledge took place, "featuring a growing permeability of disciplinary boundaries, a blurring and mixing of genres, a post-modern return to grand theory and cosmology, with even profound epistemological crises sharing one commonality," i.e. they have all been labelled interdisciplinary.[84] In line with this assumption, the *New York Times* ran an article in April 1986 entitled *Scholarly Disciplines: Breaking Out*, in which it was reported that many professors at leading universities were feeling frustrated by the narrowness of their separate departments and that they wanted to ask bigger questions than those within the purview of their individual disciplines. Their alternative agenda was to examine how other disciplines interpret the human world of language, knowledge and culture: "Out of such yearnings has come a current of change that is affecting the way academics think, write and teach. Some believe it is challenging the basic system of American universities."[85] Groups and committees were formed all over the United States, challenging the departmental system based on disciplines. Some of these groups produced influential journals that explicitly ignored disciplinary boundaries and gave rise to new courses, straddling specialities and disciplines.[86] A multitude of interdisciplinary programmes for research and education popped up, first in the USA, then in Europe, Oceania and elsewhere. Some universities even came to the conclusion that they intend to pioneer the field of academic interdisciplinarity to foster and further the interplay between and cooperation across disciplinary boundaries.[87]

Following World War II, the US government opted to pool insights and methods across disciplinary delimitations in so-called *geographical area studies*. These studies were based on the assumption that some of the politically awakening regions of the world that were of particular interest to the United States had specific similarities in culture, language and history that made them suitable areas for systematic interdisciplinary research. By bridging gaps between disciplines, teams of scholars hoped to work towards unity of regional knowledge (see Chapter 4).[88] Among the many regions assigned such status by the US government were the Middle East, Latin America, Western Europe, Eastern Europe, China and the Soviet Union. Designed to produce knowledge about the contemporary foreign cultures of potential 'enemy peoples,' area studies started out as a minor enterprise in the war efforts.[89] The pivotal motive of this initiative was political. Foreign policy decision-making was to feed on and benefit from research and the best integrated knowledge available to policy makers. Policy formulation was to be rational not emotional, comprehensive not particularistic. The Cold War became the golden age of area studies.

Many of the most prestigious universities in the United States accepted the challenge of putting substance into the concept of area studies. This movement was manifested in new social science courses and programmes, and integrated departments in several universities. Many teams took up teaching for the sake of feeding area specialists into the bureaucratic structure. In return, some universities received substantial public funding. For example, by 1988 there were about six hundred self-declared area studies programmes on American campuses. A survey of the field four years earlier revealed that 80 or so programmes at the top of the range in terms of number and quality, received annual support from the U.S. government. Between 1959 and 1981, these programmes produced 88 000 students with academic degrees in language and area studies. Since then, a stable corps of some 7000 academic area specialists has been scattered throughout higher education, both within and outside organized research centres in the USA.[90]

Recruitment to these groups took place first and foremost from the ranks of the social sciences, but also from the humanities and to a lesser extent from the natural sciences. Area studies became per definition interdisciplinary in academic composition as well as in working practice. These teams created a new culture of science in that they gained an 'interdisciplinary competence' or even 'identity' by taking part in evaluation committees for each others' students and doctoral candidates, by reading each others' books and by publishing their work in the same interdisciplinary journals that disregarded disciplinary demarcations (see the methodological concepts of 'interdisciplinary competence', 'identity', 'synergy' and 'dual competence' in Chapter 3). Area studies thus clearly demonstrated that the delineations between many of the social sciences were arbitrary and partly artificial, and that crossing boundaries between them gave rise to useful novel insight.[91] This initiative soon spread to universities in other parts of the world, including the Soviet Union, Western Europe, Japan, India, Australia and many Latin American countries. In some of these countries, the United States became the number one object of area study in support of the formulation of their external policies.

Interdisciplinarity and Mission-Oriented Research

It is said that societies have problems, universities have disciplines. This is an ironic heart-felt sigh reflecting the lack of coherence between what societies perceive to be problematic and the way universities are organized and work to meet the needs of society.

Society's problems do not automatically come in discipline-shaped blocks. Most often, if not always, they come in the composite shape of reality. This is why a major force in changing the university structure towards more interdisciplinarity was not intellectual, but rather political and financial.[92] Some even claim that interdisciplinarity is not a theoretical concept, but a practical one ". . . that arises from the unsolved problems of society rather than from science itself (see Chapter 1)."[93]

Historically speaking, political needs have been the single most important catalyst for breeding interdisciplinary research. During World War II, the military's need for a new turbo-engine led to cooperation among physicists and chemists, an effort now regarded as part of the early history of the field of solid-state physics. The most renowned mission was the transdisciplinary

Manhattan project to build an atomic bomb, a cooperative effort on the part of science, industry and the US Army. More than 100 000 people worked together with chemists, physicists, mathematicians and different types of engineers to resolve the construction challenges related to making the atomic bomb work. Here the specificities of stakeholder knowledge and the generality of science based knowledge merged into an unusual but successful unity. The 1957 launch of Sputnik by the USSR triggered the establishment of transdisciplinary US engineering centres, which ultimately became 'sociotechnical think tanks' for the US government. Such activities also laid the basis for America's computer revolution. War contracts greatly hastened the first steps on the electronic super highway which lead to a micro-electronic revolution—first in intelligence, cryptography and military computations, and then in business machines, civilian computers and a world-wide web of electronic communications like the Internet. In this process, military funded projects greatly stimulated the rapid growth of US universities and laboratories, a trend that continued into the post-war world.[94] These are examples of organizations/projects/centres fulfilling their social mission, irrespective of how controversial some of these projects were in society at large, and also within and between the sciences and the military-industrial complex. It seems as if the extraordinary times of history provide additional impetus to break down walls that have been artificially erected between sectors of society, whereas ordinary times seem to produce conditions for reconstructing barriers and partitions by clearly defined sector delimitations. To counteract such a pattern, the US National Science Foundation was established to support and promote the interconnectedness of basic and applied research, in times of crisis as well as peace. Against the backdrop of the external and internal reactions to science specialization, a reasonable assumption would be that interdisciplinarity should be treated as a legitimate and welcomed offspring of disciplinarity, rather than as an excrescence of science.

THE ESTEEM AND SCIENTIFIC STATUS
OF INTERDISCIPLINARITY

Interdisciplinarity has become the talk of the town. At present, it is both the beauty and the beast of science—a fuzzy idea that has been around for long, but never taken firm root in the disciplinary organization of science. It is both excluded from and included in the good society of academic life, engendering annoyance in some and fear and hope in others.

Interdisciplinarity, and not least transdisciplinarity, has become the ugly duckling of disciplinarity for some, while for others, it represents the swan of science. Some regard interdisciplinarity as an amorphous belief system introduced by individuals not apt to make a career in the disciplinary organization of science. Others see it as mandatory to fulfil scientific synthesis, the ultimate goal of science. A third group thinks of it as an invention of politics rather than of science, i.e. as a panacea to resolve the complex systemic challenges of society (see Chapters 5 and 6); Some regard it to be a major source of scientific creativity, an unavoidable measure to foster progress and vision in research (see Chapters 4 and 6); Others openly admit they do not know what it is all about, and there are even those who vehemently discard it as an annoying element that detracts public attention and funding away from the disciplines. In other words: Interdisciplinarity is a most contested practice of science. Why is this?

The Problems of Interdisciplinary Research

The point of departure for interdisciplinary research is far from optimal. Most universities the world over train specialists rather than scientists able to work at a tangential point between disciplines and sciences. Scientific merit is achieved through specialist studies. To earn the ultimate honour of research, a full professorship, documented disciplinary excellence is required. In a study on the barriers of multidisciplinary research in the Nordic countries commissioned by the Nordic Council in 1995, the authors conclude that researchers working across established disciplines are in a much more difficult position than those resorting to the disciplines, and that the disciplinary faithful are favoured, formally and informally, in science competitions.[95] They concluded: "Universities appear to be oppressive environments for interdisciplinary research with little understanding or inclination to facilitate extra-departmental, multidisciplinary research.....Careers are surely not 'launched' by participation in interdisciplinary projects; rather those careers may be stifled by pursuits occurring outside the mainstream of disciplines." This is the case because the socialization and licensing of researchers take place within the rather narrow confines of disciplines. As pointed out by Gunnar Skirbekk, in specialist teaching, *de-education* plays an important role in purifying disciplinary orientation, i.e. one gravitates actively away from certain ways of posing questions and seeing different angles of the case.[96] Becoming a disciplinary specialist implies learning about what the discipline is and what it is not. In the words of C.P. Snow: "over an immense range of intellectual experience, a whole group (of scientists have become) tone-deaf. Except that this tone-deafness doesn't come by nature, but by training, or rather the absence of training. As with the tone-deaf, they don't know what they miss.....Yet their own ignorance and their own specialization are just as startling."[97]

Through the process of de-education or tone-deafness, the message is loud and clear: *Keep off, no trespassing or intrusion.* The result of this is that the intersections of disciplines by and large are no one's primary concern or responsibility. They are often left as lacunae in the disciplinary organization of the sciences; the *no man's land* of research. When a phenomena is defined as belonging to a particular discipline, something else is defined as falling outside its realm.[98] What falls outside the bounds of one specialty is not necessarily picked up by another. The risk is that parts of reality may simply not be studied sufficiently as a consequence of the disciplinary organization of the sciences. In consequence, Gerald Holton urges his colleagues to be scientists first and specialists second.[99] In other words, do not lose sight of the big picture as a result of specialization.

The idea that disciplines and sub-disciplines are the only components of science work is not merely a product of de-education, it is also rooted in institutional arrangements. Universities are generally organized into departments, and the politics of individual disciplines are often hostile to interdisciplinarity. Journals are mostly disciplinary in orientation; most research councils are organized into disciplinary entities, as are publishing houses. The criteria for granting research support reflect the organization of universities and research councils, and disciplines are the focal point of graduate degrees.[100] Although, this institutional structure is gradually being modified by the establishment of university centres and cross-disciplinary programmes, journals and conferences, interdisciplinarity commands little academic prestige within the science community itself. Renee Friedman claims that the concept of interdisciplinary centres is actually ironically meant, in that it implies centrality, but many of the centres are peripheral to the main academic enterprise. "Most centres are *at* the university, not *of* the university."[101] In geographical area studies, the professors and other personnel being trained were in enclaves away from the rest of the campus.[102] This may partly explain why the majority of people engaged in interdisciplinary work feel they lack a common identity, and as a result, often find themselves 'homeless, i.e. in a state of social and intellectual marginality.[103] As will be discussed in Chapter 6, this feeling is in the process of changing.

The scientific results of interdisciplinary research are often received with considerable scepticism and labelled charlatanism by the community of sworn disciplinary and sub-disciplinary researchers. The claim is that interdisciplinary research is a fad, lacking substance and good scholarship—a cover for dilettantism.[104] Nor is this characterization completely without merit. To earn a reputation as an expert in several disciplines, and to hold the knowledge burden of multiple disciplines "within the horizon of a single mind" seems in most instances beyond human reach and capacity. Most scientists feel that the burden of disciplinary knowledge is more than over-whelming. The warning issued to moderate the ambitions of interdisciplinary research and to harmonize them with the restrictions applying to Man's cog-nitive capacity, seems more than appropriate.[105] If individual researchers, de-spite the mental restrictions that apply, employ a great number of disciplines in their own research, the accusations of charlatanism may be proven right, and the evidence of their case dismissed. This should not, however, be con-fused with any proposition that individual researchers can no longer engage in interdisciplinary research on their own.[106] Bridging small units of topical specialization across two, or perhaps three, disciplines belonging to the same family of science, for instance, history, political science and sociology, is still within the reach of many scientists. This call for moderation in the ambition of what a single individual can realistically do on his or her own. For grander schemes involving many disciplines across two or more branches of learning, cooperation between the cognitive capacity of multiple minds seem to be the only organizational solution to success.[107] Interdisciplinarity is not a mono-logue, but a discourse between disciplinarians from different specialities. This also applies to individuals working on their own. To succeed, they have to interact virtually with other specialists and to familiarize themselves with the issues, materials, approaches and methodologies of adjacent disciplines.

Basically, bad research depends on the quality of the researcher, not on the practice of research. Crossing borders may actually be a measure of de-edu-cation from disciplinary constraints; those that come from outside of a field often contribute new insight to the enrichment of the invaded discipline. This is simply a matter of not seeing the forest for the trees. In other words, disci-plinary enrichment can result from crossing boundaries, but not necessarily from blind loyalty to a particular world view.

Another challenge confronting interdisciplinary science is the Tower of Babel. Robert O. Keohane confronted this challenge when he published his classical work, *After Hegemony* in 1984, developing the concept of world political economy by crossing the boundaries between politics and economy. To attract a wide readership, he deliberately tried to eliminate professional jargon and to develop terms by using ordinary language. In so doing, he re-alized that the key analytical concepts he applied could easily be misunder-stood. He therefore urged the readers to "be careful not to seize on words and phrases out of context as clues to pigeonholing (his) argument. Is it 'liberal' because I discuss cooperation, or 'mercantilist' because I emphasize the role of power and the impact of hegemony? Am I a 'radical' because I take Marx-ian concepts seriously, or a 'conservative' because I talk about order? The simple-mindedness of such inferences should be obvious."[108] By joining the insights of two disciplines, he developed bridging concepts that made sense in the interface between fields.[109] What is more, his book is a reference work in international relations theory, and a 'must' for students of multinational politics and economy. Thus the only viable way to overcome the pitfalls of the Tower of Babel is to engage in the learning process of interdisciplinary work (see Chapter 3). This is what many disciplinarians decline to do, adding to the longevity of linguistic confusion.

A related, but nonetheless somewhat different problem is that individuals who take part in interdisciplinary projects are often deprived of the possibility to assess the scientific quality of team members coming from other disciplines.[110] This undoubtedly poses a real challenge to the overall quality of the project. If there are doubts about the quality of the work of the others and one lacks the qualifications to judge for oneself, any serious researcher would be hesitant to build his own research on possibly questionable contributions. That would eventually bring scientific progress to a halt and crumble the utility of science. As has been pointed out, the need for faith, verging on *blind faith*, is commonly apparent at the inception of team work when the individual disciplinary players are often complete strangers or mere acquaintances. Under such conditions, a great deal of blind faith on the part of the different team members is a precondition for cooperation. Gradually, the different team members' visions clear and they can begin to conditionalize what was previously blind faith. The process of conditionalization may eventually replace an environment characterized by *mutual faith* by one that is characterized by *mutual trust*.[111] However the phenomenon is denominated, e.g. blind faith, mutual faith or mutual trust, it is necessary to be convinced that the others have the disciplinary stature and competence necessary to make solid disciplinary contributions to the interdisciplinary purpose of the cooperation. At the end of the day, faith and trust, whether they are warranted or not, are part of the multi- and interdisciplinary game. No chain is stronger than its weakest link. As long as one is not in a position to secure the most highly esteemed team members from the highest echelons of the sciences, interdisciplinary research is based on a certain degree of trust. One solution might be to invite only those with an established reputation for high-quality disciplinary expertise from the merited ranks of full professors. Research on career patterns may provide some comfort in this respect as it supports the widely held belief that senior faculty members are the most likely recruits and perhaps the best suited for interdisciplinary activities. They are the ones who can risk taking time out from the disciplinary mainstream without harming their careers, and they are often also the ones in need of new challenges.[112]

It has frequently been claimed that it is difficult to verify the scientific quality of interdisciplinary products and results. In disciplinary reports, this is taken care of by the continuous scrutiny of peers who zealously and perhaps jealously check each others' contributions.[113] In academic interdisciplinary reports, the contributions from the various disciplines/stakeholders may be scattered piecemeal throughout a manuscript, mingling together in a way that can make disciplinary peer reviewing a most time-consuming, difficult, or even impossible endeavour. Ideally, the quality of academic interdisciplinary works should be checked by two types of peers, i.e. disciplinary and interdisciplinary. In most cases, only the expertise of the former has been called upon for quality control. This may result in reviewers demanding more rigour in their own areas of expertise than in the results of the synthetic approach.[114] Excluding the latter can make the results of evaluations biased, skewed, one-sided and, consequently, unsatisfactory. Disciplinary experts are trained to look for disciplinary contributions, not for the *interrelationships between contributions*. Interdisciplinary activities should therefore be judged on how well they achieve their objectives and how well they integrate knowledge. The appropriate approach for judging quality "is not to impose standard assumptions relating to individual disciplines, but to insert into the ideology of excellence, *an interdisciplinary-specific discourse*."[115] This poses the question of whether interdisciplinary competence and methodology are available and sufficiently developed to be used for evaluation purposes. Many would dispute that question. In this book, we take the opposite stand, based on two alternative interrelated approaches: The first stems from the transdisciplinary movement who argues that there is a need for an *extended peer community*, not least to bridge the gap between scientific expertise and public concerns. In this way, scientific evaluations become a dialogue among all the stakeholders in a problem, from scientists themselves to practitioners.[116] The second approach relates to the assumption and documentation presented in this book that interdisciplinary competence and methodology have been developed into an adequate scientific tool in their own right to be applied not only in research by also in scientific adjudication. This competence is available from a multitude of multidisciplinary disciplines, providing measures of unification between the involved disciplines (see Chapters 3 and 4).

The challenges of interdisciplinary work are myriad and compound. The arguments against interdisciplinarity aired by its most ardent critics actually represent a valuable source of improvement, refinement and resolution. Their qualms about whether connections can be made between individual disciplines and even more challenging between science and stakeholders, and whether any one concept or theory can be so general as to include all or families of disciplines, are fundamental questions that remain to be further sorted out. Throughout the history of science, new ideas have always caused scientific disputes and led to polemical publications criticizing the new ideas. Such criticism has often been helpful to their development.[117] In the dialectics of the discourse, the critics should be commended for their contributions to transforming a practice of imperfection into a practice of scientific regard.

Interdisciplinarity still has a long way to go before it can compete with the disciplines, sub-disciplines and units of topical specialization in terms of funding, positions, number of students and, not least, scientific respect and merit. Although, the argument is constantly sounded publicly that knowledge is increasingly interdisciplinary, financial cutbacks in actual practice curb and even eliminate programmes striving to integrate. In terms of number, there were fewer officially declared interdisciplinary programmes in 1990 than in the 1970s.[118] Interdisciplinary activities has often fallen victim, in part, to the academic equivalent of last hired, first fired.[119] The institutional obstacles to such programmes have been formidable. In the late 1980s and early 1990s, most problem-focused research in universities functioned "as an adhocracy."[120] In a survey conducted at that time, in meeting with 100 leaders of interdisciplinary environmental studies programmes at 56 universities, Russell Peterson found most of them felt they were treated as second-class citizens on their campuses. They felt they had lost ground in terms of budgets, degree approval and faculty tenure cases.[121] This situation is about to change (see Chapter 6). Being a devoted interdisciplinarian requires a fair amount of avocation and stubbornness. Or as expressed by Alexei Abrikosov: "If you do something that breaks with conventional thinking, it's difficult to become recognized. But if you break with conventional thinking, the chances are also pretty high that you have made a real discovery."[122] This statement calls for a qualification. It does not purport that most things that break with traditions are important, but that breaking with tradition can be important to scientific advances.

THE POLITICAL AND SCIENTIFIC CHALLENGES OF COMPLEX SYSTEMS

We live in unique times when it comes to the practice of science and politics. On the one hand, the disciplinary scientific community is making more breakthroughs in knowledge production than ever before. Most fields of society are being fuelled by disciplinary scientific progress. Disciplinary research has created miracles, not least with a view to the technological progress of society. In this perspective, scientists and politicians have every reason to embrace each other. Together, they have succeeded in addressing the needs and aspirations of their fellow citizens. The disciplines have no doubt made tangible contributions and fulfilled many of their obligations to society. The problem is that what have become disciplinary successes, have developed into interdisciplinary challenges. In medicine, the breakthroughs have been staggering, but diseased people soon realize that the remedies prescribed to cure their diseases are likely to create new problems in the form of side effects. The cure has become both a blessing and a curse. There is a need to correct for the problems that disciplinary approaches render. The overriding challenge of today is to produce accurate and complete descriptions of complex systems.

Most of the issues vexing humanity daily cannot be resolved without the convergence of knowledge from many sciences and disciplines. For the sake of illustration, we will address the complexity of ecosystems and the preconditions underpinning their political management and scientific understanding.

Ecology, Science and Politics

Rachel Carson's most debated book was *Silent Spring,* published in 1965. It showed beyond reasonable doubt that the application and introduction of science-based technology in society was the number one cause of harm to the environment. Before that, environmental awareness and concern were not prominent at universities or among governments. As has been pointed out, environmental studies entered universities on waves of social capital and interdisciplinary rhetoric.[123] Man's preoccupation with the functioning of ecosystems soon became a prime scientific concern, given that the human impact on the planet never has been greater and the need to understand the impact never more urgent.

An ecosystem is a *specialized community of organisms interacting with each other and with the environment in which they live.* Thus defined, an ecosystem is made up of living and non-living environmental components that form a life support system within an orderly working totality.[124] Organisms and their assemblages are assumed to be the most complex systems known: ". . . by constructing themselves from molecule to cell to organism, (they) surely display whatever deep laws of complexity and emergence lie within our reach."[125] In theory, one ecosystem is identifiable and distinct from others as they vary with regard to composition and the interaction pattern of their life support systems. A tropical ecosystem is certainly different from a polar ecosystem. Meanwhile, it is believed that the various ecosystems interact and depend on each others' functioning. Large-scale actions taken in one ecosystem may therefore be felt in another, despite being very far removed either temporally or spatially from their original locus.[126] In theory, this interaction pattern is believed to string planet Earth together as a functioning whole and will at any given time reflect the soundness of interactions. The new paradigm in science has found its most appropriate formulation in the emerging theory of self-organizing systems in which living organisms, social systems and ecosystems are strung together in a unified conception of life, mind, matter and evolution.[127] Basically, this is the scientific idea underlying Jim Lovelock's concept of Gaia, the living planet, which sees the Earth as a self-regulating system in which conditions suitable for life are maintained by feedback processes involving both living and non-living parts of the planet.[128] These feedback mechanisms are invoked to explain the relative constancy of the climate, the surprisingly moderate levels of salt in the oceans, the constant level of oxygen over the past hundred million years, and why life forms are so diverse. This idea does not imply that the planet itself is 'alive'; the name Gaia, from the Greek Earth-goddess, is just a nifty name, not to be taken literally.[129] In line with the Gaia concept, however, more and more scientists now believe that complex systems are self-regulating and that nature can no longer be regarded as passive as was commonly believed at the time of Newton. The world is unstable and complex and irregularities play a significant role (see Chapter 4). Planet Earth is physically and biologically assumed to be a *global village*.[130] To prove or disprove these assumptions, the paramount challenge of science is to break down and then synthesize the assemblage of organisms that occupy ecosystems. Most studies in ecology focus on just one or two species of organisms at a time, out of the thousands occupying a typical habitat: "Yet they are aware that the fate of each species is determined by the diverse actions of scores of hundreds of species that variously photosynthesize, browse, graze, decompose, hunt, fell prey, and turn soil around the target species. The ecologists know this principle very well, but they still can do little about predicting its precise manifestations in any particular case. . . . (T)he ecologist faces immensurable dynamic relationships among still largely unknown combinations of species."[131] It is important to understand how such complexities emerge and work. Science does not have that understanding. As it turns out, in environmental affairs, the failure of science to synthesize its basic disciplinary findings has left the political system with less effective tools to handle the environmental challenges facing humanity today. Synthesized science may produce a set of science-based options or policies upon which policymakers will and can act.[132]

The Biosphere, Technosphere, Sociosphere and Noosphere

The political challenge of these spheres is aptly couched in the interdisciplinary reasoning of Patricia M. Mische in her thought-provoking article, *Ecological Security and the Need to Reconceptualize Sovereignty*, published in 1989. Her point of departure is that although no one wished it so, humankind is the first species to become a geophysical force in its own right, altering

Earth's climate: "a role previously reserved for tectonics, sun flares, and glacial cycles."[133] The geophysical force are, in her mind, formed at the intersection of *national sovereignty*, permitting governments to do whatever they like within their respective territorial borders, and *technological advances*, allowing for large-scale encroachments in the working of planet Earth. The exercise of state sovereignty no longer complies with the sovereignty of working ecosystems. The two sovereignties are at loggerheads: the sovereignty of nature is no longer congruent with the sovereignty of politics. Therefore, political sovereignty must not be thought of in static terms, but as a dynamic interactive process involving a system of relationships and "flow of energy and information between different spheres of sovereignty."[134] In the relations between the human and non-human worlds, there are four systems affecting the dynamics of sovereignty: the *biosphere,* the *technosphere*, the *sociosphere* and the *noosphere*.

The biosphere is the sphere of life. It is composed of the atmosphere, lithosphere and the hydrosphere, all of which support living organisms. The biosphere is increasingly affected by human activities in terms of large-scale encroachments. The biosphere makes up the total interaction of ecosystems.

The technosphere is the system of structures made by humans and put into the space of the biosphere. It refers to such things as human settlements, cities, villages, factories, roads, dams, irrigation, etc. The applications are designed to satisfy human needs, but they also may have unintended consequences on the functioning of the biosphere. The technosphere has been developed to the point of making large-scale, devastating encroachments on the biosphere possible.

The sociosphere is comprised of the political, cultural and economic institutions humans have developed to manage their relations with each other and with the two other spheres. It is in this sphere decisions are being made to apply the results of the technosphere into the biosphere.

The noosphere refers to the spirit, mind, consciousness and reflective thought which envelopes the biosphere. It is in this sphere that the fundamental thoughts of development are made.

These spheres interact, making up the parameters of the human existence and progress. However, as noted, in this interplay the noosphere depends on the existence of the biosphere, whereas the biosphere will persist without the noosphere. This is so because if you destroy the biosphere, all human minds are also destroyed.[135] The biosphere is part of the noosphere, and not vice versa.

Up to the present, in exercising political sovereignty, the interaction between the human and nonhuman worlds within the lager biosphere and the dynamics involved in the interrelationships of all four spheres have been insufficient and negligent.[136] To sustain the life-supporting ecosystems of the Earth, this practice has to be corrected. To make the necessary changes, Mische suggests that four basic principles or premises be implemented: I. The *sovereignty of the Earth* should precede and supersede human sovereignties, II. *The Earth is indivisible*; the world is politically divided, but ecologically united in interactive ecosystems and bioregions. All bioregions are, to some degree, interdependent. III. *Bioregional alliances should be formed*; most environmental threats cannot be resolved by one nation alone. Thus, states that inhabit or border on the same bioregion can be effective through collective, cooperative actions.[137] IV. *Eco-values for ecological security* have to be developed and put to practice.[138] To truly understand the complexities involved in the overall functioning and malfunctioning of these spheres, the natural and social sciences, along with the humanities, have to contribute to the process of knowledge unification and syntheses. If we compare political reasoning with interdisciplinarity in research, there are striking similarities. The bio- and technospheres refer to the natural sciences, the sociosphere to the social sciences and the noosphere to the humanities. This is also reflected in the scientific concept of an ecosystem, which implies a hybridity of approaches and perspectives. The concept appears in social and geographical sciences, resource management, environmental impact assessment, planning and decision-making, and social science research. Yet the predominant academic tendency has been to represent it to some extent by disciplinary approaches.[139] The interdisciplinary reasoning of Mische is not reflected in the political practices and needs of governments.

The preamble to the *Rio Declaration of 1992* states that humanity stands at a defining moment in history: "We are confronted with a perpetuation of disparities between and within nations, a worsening of poverty, hunger, ill health and illiteracy, and the continuing deterioration of the ecosystems on which we depend for our well-being."[140] To reverse this sad state of affairs, the declaration defines science as one of the means for achieving the overall objective of sustainable development, i.e. development that meets the needs of the present without compromising the ability of future generations to meet their own needs. In this respect, the sciences are called upon to provide information to better enable the formulation and selection of environmental strategies in decision-making processes. The sciences are expected to be responsive to emerging needs, and to take part in the interactive processes between the sciences and policy-making.[141] At the same time, Agenda 21 acknowledges the success of disciplinary research, stating that there is a wealth of data and information to be applied to improve the management of sustainable development. However, finding "the appropriate information at the required time and at *the relevant scale of aggregation is a difficult task*."[142] To meet political needs, the document calls for governments to consider undertaking the *insti-tutional changes* necessary at the national level to achieve the necessary *inte-gration of environmental and developmental information.* In compliance with this line of argument, the Agenda goes on to state that *methods* for assessing the interaction between different environmental, demographic, social and de-velopmental parameters are neither sufficiently developed nor applied. *Indi-cators of sustainable development* are in great demand but they have not been developed sufficiently for effective application. In short, there is a severe lack of interdisciplinary scientific information and tools to come to grips with is-sues related to the political objectives of sustainable development. The over-all goal is to strengthen the capacity to collect and use multisectoral informa-tion in decision-making processes.[143] The Agenda also asks for efforts to bridge the gap between what is needed by decision-making bodies at the lo-cal, regional and global levels, and what is provided by the scientific com-munity at the same levels. One of the key objectives of Agenda 21 is to im-prove and increase the fundamental understanding of the linkages between human and natural environmental systems, i.e. to improve the analytical and predictive tools required to better understand the environmental impacts of development.

Judging from the recommendations of Agenda 21, governments have come a long way towards generating awareness of the need to integrate the physi-cal, economic and social sciences in order better to understand the impacts of economic and social behaviour on the environment and of environmental degradation on local as well as global economies.[144] This has forced the emer-gence of a new development in the cognition of humans. In the context of the environmental crisis, ecologists, businessmen, political philosophers, ac-tivists, etc., are now seriously talking about granting nonhuman species some sort of rights and possibly even legal standing. Bruno Latour argues that po-litical representation of nonhumans seems not only plausible but necessary at this point, although the notion would have seemed ludicrous or indecent not long ago. "We used to deride primitive peoples who imagined that the disor-der in society, a pollution, could threaten the natural order. We no longer laugh so heartily, as we abstain from using aerosols for fear of the sky may fall on our heads. Like the 'primitives', we fear the pollution caused by our negligence—which means of course that neither 'they' nor 'we' have ever been primitive."[145] Here Latour argues that 'indigenous knowledge' should be conceded to in political processes. The concept relates to the experiences ac-cumulated over time and passed on from one generation to another on the special conditions and circumstances under which local people live, act and survive. The concept was first introduced to international politics in the late 1980s through the Rovaniemi Process—an inter-governmental initiative among the Arctic states to develop a strategy for sustaining and protecting the Arctic environment from irreparable human encroachments. Upon the establishment of the Arctic Council in 1996, indigenous knowledge and participation were institutionalised in the working of the new organisation as a permanent element alongside science.[146] For science to come to grips with the complexities of sustainability, stakeholder knowledge appeared to be an important supplement to science. The same came to expression in the Rio Declaration, which called for a strengthening of the capacity to collect and use *multisectoral information* in decision-making processes.[147] These are early political manifestations of the need to apply the transdisciplinary approach as expressed in post-normal science. Transdisciplinarians argue that serious harm has been caused by traditional science "ignoring the uncertainty of scientific knowledge, by neglecting the user's knowledge, and by failing to consider contextual conditions of applications."[148] To fit the problem structure of society, the need is for a 'new social contract for the science'[149] or a 'new commitment of science'[150] to cultivate further both expressions of interdisciplinarity (see Chapter 6).

In a world in which resources are finite, humans face the same constraints as all other species; the environment sets a limit on the number of individuals that can be supported, and when that limit or carrying capacity is exceeded for an appreciable length of time, quality of life begins to deteriorate and then population size begins to decrease. Scientists urge the public and political system to accept that the limits to growth are as basic a law of nature as is gravity.[151] After all, comprehending the complexity of life is the greatest challenge facing modern science today. What humanity stands to gain by succeeding in this endeavour is evident: A proper understanding of the living economy of the planet is the key to safeguarding its future.[152] To illustrate the challenge, a leading biologist contends that economic theory ignores the basic biological reality of limits to growth and that some economists advance the view that limitless growth is not only possible but a goal toward which society should aspire.[153] To avoid such misapprehensions, the recommended remedy is to merge the insights of biology and economy and, for that matter, the insights of all relevant disciplines, to avoid anyone from going forward on the basis of their extra-disciplinary ignorance. It has become public knowledge that long-term, large-scale environmental change and human welfare are closely linked, "such that preconditions of future conditions must be made with explicit consideration of these linkages. Such analysis necessitates new kinds of interdisciplinary research."[154] The wake up call being made by more and more scientists is that the transformation of the earth by humanity is omnipresent, continuing and imperfectly understood. It is a phenomenon that requires unprecedented foresight because choices made today reach far and perhaps irrevocably into our future.[155] Humanity has reached a historical point in time where ecological soundness has become the *high-politics of the collective of world states*. Ecology has become politics — political ecology, and politics has become ecology — ecological politics. Although science may not be able to precisely forecast the behaviour of a complex system, "nonlinear dynamics shows that we can gain some insights into its global behaviour — for instance, through the knowledge of the system's set of attractors. These insights may provide the bedrock of understanding for future decision-making (see Chapter 1)."[156] At any rate, integrative scientific studies of complex living systems are mandatory to support the need for global action on a rational basis.[157] The distinction is between disciplinary knowledge integrated into a coherent matrix but with missing interstitial data identified, as opposed to separate, uncoordinated inputs from the disciplines for the analysis of a policy problem with greater risk of missing data being undisclosed.[158] That being said, the large topics raised by C.P. Snow are not the exclusive property of any one discipline (see Chapter 1) — "indeed, they legitimately claim the attention of any educated citizen, and should not be confined to a set of academic pigeon-holes."[159] Questions about topics such as educational structure, social attitudes, government policy-making and so on are intimately involved and relevant in coping with these complex issues.[160]

Fragmentation and Segmentation

Science and politics alike would benefit from a merging, rather than separation, of disciplines and expertise. Here the needs of the two realms are congruent. Yet their ability to handle complex system challenges, individually as well as collectively, is curbed by fragmentation, which abounds not only in universities, but also in public life. Whereas the sciences are fragmented into disciplinary departments, ministries are the building blocks of governments. Ministries are in turn comprised of a variety of sub-units (departments, directorates, etc.) which are responsible for delivering specialized services to their respective segments of society. Magnus Fladmark pinpoints the problem: "The situation we have today is that when the politicians look for . . . input from universities, they have to knock at many doors. Behind each door there is likely to be a highly focused specialist, rather than a generalist with a strategic and integrative view of culture. We see the mirror image of this when looking at the system of government. When academics go to politicians for advice on priorities and support, they are passed around the houses of a vast array of central departments, quangoes and local authority officers. Each is likely to be responsible for a specialist function, but it will not always be clear how particular functions fit into a larger strategic framework, if indeed such a framework actually exists."[161] The characteristic application of scientific method to public affairs resembles the approach of a reductionist scientist. The results are a specialized use of science pursuant to specialized agency tasks.[162]

Segmentation, fragmentation and specialization are thus characteristic of modern society, not only of the sciences. When crises require an integrated response, the appropriate ministries are constrained by their own organizational structures,[163] and the sciences by their departmental organization. Modern societies seem to be under the "tyranny of small decisions"[164] based on the exclusiveness of the reductionism. From an organizational point of view, no one seems adequately capable and/or fit to handle the challenges of complex issues.

As has been pointed out by a group of American multidisciplinary researchers, while scientists can carefully explore segments of our vast and complex planet, legislators and policymakers must work through opposition and compromise to produce documents that may be ambiguous or policies that may have uncertain effects. Institutionalizing the role of science in the making of policy on problems of complex issues represents an important challenge which tests the capacities of both legislators and scientists to understand the needs of the other and to overcome the obstacles to communication.[15] For the scientist, the question is one of what one knows in the absolute sense. For the policymaker, the issue is one of making choices, and the question is consequently one of relative knowledge. For politicians, politics is the art of the possible, and knowledge becomes one more tool towards that end. But as has been observed, authority in science is testable knowledge subject to change with very little notice, whereas authority in government is based upon cultural norms for which the validity tests of science are largely irrelevant.[166] These differences suggest that policymakers should be more appreciative of the gaps, uncertainties and changeability in scientific knowledge and must learn how to deal with those gaps, while scientists need to understand how to respond better to the needs and mentality of policymakers by presenting results that will relay the uncertainties they have, yet also convey clearly what is understood so that the understandings of the Academy may appropriately influence the choices that need to be made.[167] In this sense, the peril of scientific knowledge is not that it is incomplete or epistemic, because incomplete knowledge is still more than the policymakers would otherwise have had. The danger, in fact, is that it will appear to be more to the policymaker than it really is. Politicians often think about epistemic knowledge as objective and final, but that is not necessarily the case (see Chapters 1 and 4). These differences can only be bridged if the two parties educate each other regarding the differences between their cultures and perceptions of reality. As has been pointed out, contempt for politicians is what creates the widest consensus in academic circles.[168] For their part, politicians probably make the assumed lack of any pragmatic grasp and ability of scientists a measure of derision to unite them across party divides. The negative effects of fragmentation and segmentation that impede adequate coping with complex-system challenges will most likely be aggravated by the two spheres operating in different realities, speaking different tongues and exercising contempt for the trade of each others' professions. As observed by Johan Heilbron, disciplines are to the academic world what nation-states are to the political realm, or firms and corporations to the field of business. He concludes that all these organizational structures are institutional regimes typical of the modern era.[169] The Nobel Laureate Murrey Gell-Mann argues that we must get away from the idea that serious work is restricted to beating to death well-defined problems in a narrow discipline, while broadly integrated thinking is relegated to cocktail parties. "In academic life, in bureaucracies, and elsewhere, we encounter a lack of respect for the task of integration."[170] A sense of broad perspective is absent from the interface between science and government.[171] The present situation requires science to expand its boundaries to include different validation processes, perspectives, and types of knowledge. "In particular, it requires the gap between scientific expertise and public concerns to be bridged."[172] In this respect, governments have a superior responsibility for applying their financial, legal and political incentives to foster change among the organs of government and the subordinate sectors of society.

MEASURES OF SCIENCE AND THE NEEDS OF POLITICS

This paragraph will address three different avenues to make scientific practice meet the needs of governmental and societal politics. The first refers to Thomas Kuhn's historical scheme on the progress of science through shifts of scientific paradigms. The second relates to exogenous changes on the science community and the third stresses a middle way in which new interdisciplinary measures are being introduced to supplement the disciplinary organization of science.

Shifts of Scientific Paradigms

The *first law of philosophy* says that for each philosopher there exists and equal but opposite philosopher.[173] This made, William Bechtel—himself a philosopher of science—to declare that no one should take philosophers as final authorities.[174] Thomas Kuhn followed that sage advice. When he published his book on *The Structure of Scientific Revolutions* in 1962, he delivered a powerful intellectual blow or at least staunch challenge to the predominant philosophy of the time, positivism (see Chapter 4). Contrary to the positivist position that science accumulates knowledge over time, Kuhn argues that science develops through shifts of scientific paradigms triggered by repetitive intellectual revolutions. In Kuhn's perception, positivism is nothing but a scientific paradigm to be replaced in due time by another. Or in our context: Science is perfectly fit to adapt to new and changing circumstances and demands.

Kuhn argues that the discovery of serious shortcomings in the ability of social institutions to handle crises and to deliver what they are supposed to deliver, is a precondition for political as well as scientific revolutions. A scientific revolution in the historical scheme of Kuhn implies a shift in the paradigms of science. Although, the concept of a paradigm was not clearly defined in the first edition of Kuhn's book,[175] the second edition took stock of the initial conceptual vagueness and defined a paradigm as the *constellation of perceptions, procedures, conceptions, models, values and techniques shared by the scientific community.* In this way, a paradigm is the consensual *professional matrix* of science at any one time—the accepted way of gathering or producing scientific knowledge.[176] In Kuhn's historic scheme, academic disciplines move through phases of development characterized by different assumptions about how scientific work should proceed. When a matrix or paradigm changes, *normal science* changes. Normal science is conducted according to the prevailing paradigm. As conceived by Kuhn, normal science does not aim at generating new insights or exploring new phenomena. On the contrary, phenomena that do not fit in with the values of the existing paradigm/matrix are often overlooked and neglected. Normal science restricts itself to dealing with phenomena and theories that have already been chosen and defined by the value composition of the existing paradigm. In Kuhn's conception, normal science is therefore nothing but puzzle-solving in that its overall purpose is to more finely hone existing theories and knowledge. The British physicist Lord Kelvin illustrated Kuhn's concept of normal science convincingly in an address to the British Association for the Advancement of Science in 1900, stating: "There is nothing new to be discovered in physics now. All that remains is more and more precise measurements." As commented by Lisa Randall, Lord Kelvin was famously incorrect: "very soon after he uttered those words, relativity and quantum mechanics revolutionized physics and blossomed into the different areas of physics that people work on today."[177] The young Max Planck encountered this same attitude in 1875 when he started at the University of Munich and was urged by his professor not to study science because there was nothing left to be discovered.[178]

Shifts of paradigms, according to Kuhn, occur when the normal science of a certain era has exhausted its ability to secure continuous scientific progress. One paradigm is then replaced by another, and the shift will provide new values and techniques for science to aspire to for the sake of scientific progress. The old paradigm provides no more guidance, i.e. its values and preconditions have faded. Thus, moving from one paradigm to another calls for a fundamental re-conceptualization of the world on the part of academics: "they need to look at the world in a new way and learn a new language for talking about it."[179]

Landmark years as 1900 for Max Planck's quantum theory and 1905 for Albert Einstein's special theory of relativity show how new ideas and methods only gradually entered the world of physics.[180] The new paradigm then refreshes normal science, which starts a different process of puzzle-solving defined by the values of the new paradigm. In this scheme, the stability of normal science is punctuated by occasional revolutions, showing that science itself is capable of internal reform and change. When the time is ripe, a revolution will take place and it will result in a shift of paradigms. According to Kuhn, the history of science goes through five distinct stages: *immature science*, *normal science*, *crisis science*, *revolutionary science* and *resolution*, which are followed by a return to the normal science of the new paradigm. Thus, a closed loop involves the four last stages in an infinite historical process of science development (see Figure 2.3).[181]

It should be noted, however, that Kuhn developed his historical scheme on the basis of the natural sciences. Notwithstanding, the universal applicability of the notion of a paradigm has been adopted lock, stock and barrel and without much resistance by other fields of research as well. Søren Kjørup comments on behalf of the humanities that the application of the concept is beyond discussion, but that its consequences are different from the natural sciences. Whereas the paradigms of the latter are supposed to replace each other, the paradigms of the former have a tendency to live side by side for an extended period of time. This is due to the greater liberality and tradition of the human sciences for having multiple paradigms flourish or exist in parallel as supplementary approaches.[182] The social sciences

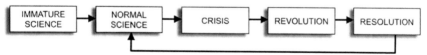

Figure 2.3.   Kuhn's Five Stages in the History of Scientific Disciplines. *Source: William Bechtel: Philosophy of Science. An Overview for Cognitive Sceince, Lawrence Erlbaum Associates, Publishers, Hove and London, 1988, p.52. (Printed with the permission of the publisher).*

seem to copy the humanist pattern. As discussed in Chapter 4, the positivists, post-positivists, post-modernists and post-normal scientists of the present have not (yet) superseded each other, but rather represent different schools of though—competing paradigms—in addressing social phenomena.

This notwithstanding, there are voices within the scientific community that contend that science is approaching a new age of synthesis[183] aimed at fulfilling the ultimate goal of science to re-create the original assembly of the elements of complex systems.[184] Ziauddin Sardar insists that the context of science has changed and brought the uncertainties inherent in complex systems to the fore. Fritjof Capra follows suit, saying that the new paradigm or matrix can be described in various ways, e.g. it can be called a holistic world view, seeing the world as an integrated whole, or it can be called an ecological view based on the concept of *deep ecology*, which refuses to see the world as a collection of isolated objects, but rather as a network of phenomena that are fundamentally interconnected and interdependent. This is a paradigm with a "very profound sense of connectedness, of context, of relationship, of belonging."[185] The concept of post-normal science indicates that a shift of paradigm is about to happen in that a fresh approach and a new type of knowledge is being construed enhancing the ability and effectiveness of the social sciences to engage in societal problem-solving. Some even predict that science is rapidly heading towards a *post-disciplinary era and landscape* (see Chapter 4).[186] The idea that a shift may be imminent is gaining ground within the epistemic community, but the pace of the shift is at best gradual and slow. Some scholars argue that the on-going shift—irrespective of pace—is caused by a troubling feeling among scientists that normal science is running out of creativity, and that what creativity there is, is increasingly occurring in the interface between traditional disciplines (see Chapters 4 and 6).[187] Still, the prevailing paradigm is that of reductionism which acts as a fairly effective roadblock to prevent the new paradigm of synthesis from getting the upper hand.

It is a fact, however, that representatives of both the science community and the political system are acknowledging that the environmental crisis is mounting and that humankind is short of adequate and effective measures to counteract adverse trends. The question then is whether a shift of paradigm from within the science community itself will come in time to meet the urgency of complex system politics. In Kuhn's scheme, shifts of this kind often take a generation to be fulfilled, and the change goes through two distinct steps: Upon the introduction of the new paradigm, just a few researchers convert, and when the most ardent opponents—the old guard—are gone, the whole profession tends to convert fairly swiftly.[188] Many a scientist would say that such a time-span is far too long for complex system-politics to wait until effective action is initiated. Some scientists contend that humankind will "be frying on land and boiling in ocean water" before it can be epistemologically proven that global warming is Man related and triggered.[189]

To enhance the likelihood of reversing the present trend of mounting crises, intermediary measures should be mobilized either by changing the present organization of universities (a fairly unlikely prospect in the near future) or by different institutions being established in the interface between the academe and lawmakers to facilitate communication.[190]

Exogenous Changes

If the shift does not come in time, a change can, alternatively, be stimulated from outside science itself. Governments do indeed have the financial wherewithal to change the direction of science. They have done it before and can do it again. As far back as in the late Middle Ages, when broad minded thinking and research enjoyed prominence in science, governments actually used their power to generate disciplines. The term discipline was mainly applied to three areas: in Paris, to theology and the arts; in Bologna, to the law; and in Salerno, to medicine.[191] These demands for specialization were external to the educational institutions, but the legal and medical faculties at Bologna and Salerno respectively recognized it as being in their best interest to harness the needs of governments. In this respect, the sector's own acclaim of what governments wanted produced an effective outcome for both parties.

Reforms forced upon a sector from the outside without the consent of those affected are doomed to be instruments of little effectiveness. The approval of the implementers is vital to the effectiveness of the reform. The relatively poor experiences of government-induced interdisciplinarity during the past 50 years do not lend themselves to frequent repetitions. If two parties can work out an agreement between themselves on how to implement changes, the likelihood is that the reform will fulfil its own objective. Such a move may, however, be interpreted as an encroachment on the freedom of research principle and as an unreasonable and extra societal obligation put upon the shoulders of researchers. The point of balance between societal obligation and freedom of research may be invoked as tilting in the 'wrong' direction. Democratic values may be launched against the force of big government. A great deal of time may elapse before this tug-of-war situation is eventually settled. Meanwhile, is there a golden mean?

Intermediary Measures

The Gulbenkian Commission argues that there is, and that knowledge-instrumental interdisciplinarity can be fostered without imposing radical changes on the organizational setup of universities.[192] The present division of labour between disciplines should not, in the mind of the Commission, be discontinued. What is needed is to build on what is already there and to make deliberate efforts to strengthen the organization of intellectual activity across disciplines. The multiple interdisciplinary programmes that have seen the light of day up until 1997 do not, according to the Commission, fulfil the function of knowledge-instrumental interdisciplinarity. Those programmes are alien excrescences 'forced' upon the university structure to meet external, not internal needs. They have never become fully integrated with the disciplines to nourish cross-fertilization and synthesis of research. To breed a change, the money providers for science will have to use their financial power to transform the interdisciplinary excrescences to grow inwards into the structure, rather than outwards. The modest measures proposed by the Commission may serve as a mind-bender for coming up with more measures that are adequate to start an inward-integration of interdisciplinarity in universities and research institutions. The measures proposed are these:

- *Expansion of institutions within or associated with universities, which are willing to bring scientists from different disciplines together to work for one year on specific and important themes.* Institutions of this kind are already in existence, but there are far too few of them. The recruitment of team members should take place on a broad basis in terms of gender, geographical area, culture and disciplines, and all participants should be well prepared before starting their collaboration and exchange of scientific viewpoints. The short time available makes initial consensus on basic viewpoints within the team a significant precondition for making cooperation worthwhile.
- *Establishment of integrated, interdisciplinary research programmes within the universities that have specific intellectual objectives and sufficient funding to last for an extended period of time* (for instance, five years, see Chapter 3). These programmes are different from existing research centres, which have often been granted eternal life, and are supposed to come up with their own funding. The ad hoc quality of such programmes should create a mechanism for continuous experimentation. In so doing, one should consider deferring the launch of educational programmes, leaving participants to demonstrate for themselves and others the utility and validity of their approaches.
- *All university professors should be hired to perform research and teach in two departments.* Today, the norm is that a professor is affiliated with one department only, usually the department in which he or she holds his or her highest degree. The proposal is to turn this structure around to create a new

structure where all tenured personnel are affiliated with at least two departments, i.e. the one where the professor got his disciplinary training and the other where he has demonstrated an interest or done relevant work. To avoid any of the departments trying to prevent this from happening, there should be a requirement that at least 25 per cent of the members of a department do not hold a degree in that discipline.

Recently, a Chair of this type—the Thor Heyerdahl professorship—was established at the Norwegian University of the Life Sciences in Ås outside of Oslo. The incumbent will be working at the interface of health, environment and development and is obligated to develop and further refine interdisciplinary methodology and theory of universal scientific application.

- *PhD students should work in two departments.* The situation is the same for post-graduate students as for professors. They are all affiliated with one department only and are supposed to make their dissertations within the discipline offered by their department. The Commission recommends changing the present structure and making it compulsory for doctoral students to belong to one department, but take a certain number of courses or do research under the auspices of another department. In the mind of the Commission, if administered in a liberal and serious manner, such measures will gradually change the present and the future.

- *To develop new procedures for hiring interdisciplinary scholars.* While a steadily growing number of universities and colleges in the United States have adopted formal procedures for hiring and reviewing interdisciplinary faculty members, few have a comprehensive approach to the entire pre-tenure experience. Once hired, interdisciplinary scholars frequently face a set of common difficulties in their research, teaching and administrative roles. To a large extent, they are being evaluated and treated on terms that apply to disciplinarians. However, changes may already be underway. In a fairly recent survey among 19 US universities and colleges, three answered yes to the question if they had codified the process for interdisciplinary hiring, tenure and promotion, whereas four said that such a codification process was under way.[193]

- *To stimulate transdisciplinary research and post-normal science in universities* (see Chapter 6).

Others have suggested the establishment of a new kind of professorial chairs at universities to foster interdisciplinarity. Such chairs should be filled by people from either the formal sciences—logic and mathematics—or from the philosophy of sciences (see Chapter 4), or by researchers that have acquired interdisciplinary competence through practice.[194] In reviewing the experiences gained during a long-term Swedish interdisciplinary study, the evaluation committee for the project suggested the establishment of a chair for a *bridge scientist,* i.e. a researcher entering into cooperation with specialists from other disciplines with the aim of producing synthesized knowledge on the basis of their ability to explain their own unit of topical specialization, including both limitations and merits, to others (see Chapter 3). It is assumed that such a chair would provide invaluable professional and intellectual links between the sciences and their disciplines. According to the Committee, a philosopher of science could be a suitable profession for such a chair (see Chapters 2 and 4).[195] Others have suggested the establishment of a new kind of position: *"pan-university professors without tenure,"* who are willing to risk taking time out of the disciplinary mainstream. Such a position would be subjected to periodic reviews, and incumbents should have the possibility to take disciplinary 'leaves of absence' to update their disciplinary skills.[196] These chairs should be *of* a university, not *at* a university, and their purpose and designation should be to breed interdisciplinarity between departments involving multiple disciplines. These and other measures may be introduced in universities based on the uneasiness within the scientific community itself regarding the present state of specialization, and the political needs and financial resources of governments. As intermediary measures, they may accelerate the process towards a shift of scientific paradigms, furthering interdisciplinarity before a definitive shift of paradigm occurs.

Science takes time, as do shifts of paradigms. The answers to complex system challenges will not come over night. This being said, if "we are to deal effectively with current environmental crises and avert future ones, and if we are to put technology to work on our behalf instead of becoming its victim, then we, as members of democratic society, must soon make a large number of complex decisions."[197] This calls for governments to counteract their own fragmentation and help science do the same. A synthesis of politics and science seems inevitable if we are to cope with complex system challenges. Or, as put by the philosopher Ken Wilber: " the world in general . . . is now at something of a branch point: we can continue the road of scientific materialism, fragmented pluralism, and deconstructive postmodernism; or we can indeed choose a more integral, more embracing, more inclusive path to travel."[198] It remains to be seen whether the two sectors, individually and/or jointly, will manage to engender the changes required.

NOTES

1. Wilson (99), p. 294.
2. Heisenberg (99), p. 189.

3. Roll-Hansen (05), p. 11.

4. Coveney and Highfield (95), p. 333.

5. Caldwell (82), p. 29.

6. Nowotny, Scott and Gibbons (01), p. 1

7. Obasi (92), p. 21.

8. Bohlin (88), p. 19, and Østreng (89), pp. 89–90.

9. Cited from Thompson Klein (96), p. 181.

10. Clarke (95), p. 152.

11. Wilson (99), p. 101.

12. Sørensen (99), p. 92

13. Berge and Powel (97), p. 17.

14. Berge and Powel (97), p. 17.

15. Killeen (03).

16. Martin (08), pp. 60–71.

17. Jasanoff (92), p. 158.

18. Berge and Powell (97), p. 17.

19. Sardar (00), p. 4.

20. Knorr Cetina (99), p. 5.

21. Østreng (89), pp. 88–113.

22. Østreng (89), p. 94.

23. Ziman (00), p. 18.

24. Weinberg (93), p. 59.

25. Ziman (00), p. 17.

26. Wittrock (02), p. 10.

27. Sardar (00), p. 9.

28. Snow (98), p. 32.

29. Wilson (99), p. 93.

30. Wilson (99), p. 207.

31. Ziman (00), p. 19.

32. Latour (99), p. 80.

33. Ziman (00), p. 53.

34. Weinberg (93), p. 282.

35. Ludow (08).

36. Heisenberg (99), p. 27.

37. Clemet (04).

38. Cited from Thompson Klein (96), p. 202.

39. Letnes (04).

40. Bioengineering Systems Research in the United States: An Overview (87), p. 100.

41. Bronowski (65), pp. 7–8.

42. Kojevnikov (04), p. 305.

43. Elzinga (08), p. 349.

44. Hirsch Hadorn, Biber-Klemm,Walter Grossenbacher-Mansuy, et al. (08), pp. 19–39.

45. Commission of the European Community (04), p. 5.
46. Ziman (04).
47. Thompson Klein (90), p. 97.
48. Høyrup (00), pp. 367–372.
49. Thompson Klein (96), p. 212.
50. Caldwell (82), p. 123.
51. Zimmerman (95), pp. 68–69.
52. Zimmerman (95), p. 71.
53. Kjørup (99), p. 46.
54. Hambro (02/1).
55. Hambro (02/2).
56. Heisenberg (99), p. 192.
57. Latour (99), p. 89.
58. Zimmerman (95), pp. 77–78.
59. Ziman (00), p. 54.
60. Weiner (90), p.198.
61. Weiner (90), pp. 198–199.
62. Thompson Klein (90), p. 20.
63. Pearson (37), p. 17.
64. Becher (90), p. 335.
65. Chubin (76), p. 455.
66. Weiner (90), p. 193.
67. Orman Quine (81), p. 191.
68. Bruun, Hukkinen, Huutoniemi, et al. (05), p. 22.
69. Weiner (90), p. 198.
70. Gribbin (05), p. 1.
71. Pearson (37), p. 17.
72. Wilson (99), p. 54.
73. Gribbin (05), p. 1.
74. Dogan (01), p. 14851.
75. Thompson Klein (96), p. 42.
76. Dogan (01), p. 14854.
77. Wilson (99), p. 8.
78. Wilson (99), p. 42.
79. Somit and Tanenhaus (64), p. 49.
80. Turner (), p. 70.
81. Friedman (01), p. 3.
82. Thompson Klein (90), p. 12.
83. Thompson Klein (96), p. 11.
84. Thompson Klein (90), p. 11.
85. *New York Times* (86), p. 14.
86. Wallerstein et al. (97), p. 65.
87. One of many examples in this respect is my own university: The Norwegian University of Science and Technology in Trondheim, who has made interdisciplinarity one of its main research objectives. See NTNU's *Program for tverrvitenskaplig*

forskning (97), Trondheim. When it comes to US universities adopting an interdisciplinary educational and research policy, see the excellent book by Thompson Klein (90), pp. 47–54, 157–181.

88. Thompson Klein (90), p. 25.

89. Thompson Klein (96), p. 107.

90. Thompson Klein (96), p. 109.

91. Wallerstein et al. (97). pp. 38–40.

92. Roy (79), p. 42.

93. Cited from Thompson Klein (90), p. 42.

94. Knudsen (99), p. 192.

95. Wisted and Mathisen (95), p. 16.

96. Skirbekk (94), p. 2.

97. Snow (98), p. 14.

98. Wisted and Mathisen (95), p. 14. The translation is the responsibility of the author.

99. Cited from Thompson Klein (90), p. 138.

100. Kaje (99), p. 4.

101. Friedman (78), p. 63.

102. Thompson Klein (96), p. 107.

103. Thompson Klein (90), p. 13.

104. Bauer (90), pp. 105–119.

105. Wisted and Mathisen (95), p. 13.

106. Skirbekk (94), p. 2.

107. Berge and Powell (97), p. 5.

108. Keohane (84), pp. 11–12.

109. Weiner (90), p. 198.

110. Wisted and Mathisen (95), p. 18.

111. Berge and Powell (97), p. 26. The italics of concepts have been added by this author.

112. Thompson Klein (90), p. 182.

113. Ziman (98), p. 51.

114. Pfirman, Collins, Lowes et al. (05).

115. Thompson Klein (96), p. 211.

116. Hirsch Hadorn et al. (08), and Sardar (00), pp. 64–65.

117. Heisenberg (99), p. 167.

118. Peterson (90), p. 223.

119. Thompson Klein (96), p. 209.

120. Thompson Klein (90), p. 139.

121. Cited from Thompson Klein (96), p. 97.

122. The CAS Newsletter (04/1).

123. Thompson Klein (96), p. 96.

124. Østreng (99/1), p. 8.

125. Wilson (99), p. 96.

126. Zimmerman (95), p. 99.

127. Capra (00), p. 339.

128. Lovelock (79).

129. Gribbin (99), p. 24.

130. Zimmerman (95), pp. 98–99.

131. Wilson (99), p. 92.

132. Caldwell (82), p. 107.

133. Wilson (99), p. 303.

134. Mische (89), p. 396.

135. Wilber (00), p. 98.

136. Mische (89), p. 397.

137. Østreng (99/1), pp. 1–20.

138. For a full account of this reasoning see Mische (89), pp. 389–427.

139. Thompson Klein (96), p. 60.

140. Agenda 21: Programme of Action for Sustainable Development (92), p. 15.

141. Agenda 21: Programme of Action for Sustainable Development (92), pp. 257–263.

142. Agenda 21: Programme of Action for Sustainable Development (92), p. 286. The italics have been added by the author.

143. Agenda 21: Programme of Action for Sustainable Development (92), p. 284.

144. Agenda 21: Programme of Action for Sustainable Development (92) pp. 259 and 260.

145. Latour (99), p. 202.

146. Østreng (99/1), pp. 39–44.

147. The italics have been added by the author.

148. Hoffmann-Riem, Biber-Klemm, Grossenbacher-Mansuy, et al. (08), p. 4.

149. Lubchenco (98), pp. 491–497.

150. Cetto (00).

151. Zimmerman (95), pp. 83–84.

152. Coveney and Highfield (95), p. 11.

153. Zimmerman (95) pp. 83–84.

154. Sax (94), p. 195.

155. Caron, Chapin III, Donoghue, et al. (94), p. 341.

156. Coveney and Highfield (95), p. 331.

157. Coveney and Highfield (95), p. 332.

158. Caldwell (82), p. 18.

159. Collini (98), p. viii.

160. Stefan Collini: "Introduction" in C.P. Snow: *The Two Cultures......*, Op. cit. p. xliii.

161. Fladmark (98), p. 224.

162. Caldwell (82), pp. 75–76.

163. Hershberg (88), p. 14.

164. Caldwell (82), p. 114.

165. Caron, Chapin III, Donoghue, et al. (94), pp. 244–245.

166. Caldwell (82), p. 123.

167. Caron, Chapin III, Donoghue, et al. (94), p. 345.

168. Latour (99), p. 245.

169. Heilbron (04), p. 25.
170. Cited from Coveney and Highfield (95), p. 8.
171. Fladmark (98), p. 227.
172. Sardar (00), p. 64.
173. Boman (09), p. 112.
174. Bechtel (88), p. 119.
175. Mastermann (65).
176. Kuhn (72), p. 175.
177. Randall (06), p. 85
178. Weinberg (93), pp. 13-14.
179. Hubbard, Kitchin, Bartley, et al. (02), p. 23.
180. Friedman (01), p. 40.
181. Bechtel (88), p. 52.
182. Kjørup (99), p. 115.
183. Wilson (99), p. 12.
184. Wilson (99), p. 230.
185. Capra (99), pp. 6 and 326.
186. Hubbard, Kitchin, Bartley et al. (02), pp. 57–62.
187. Pfirman, Collins, Lowes et al. (05), p. 15.
188. Kuhn (72), p. 151.
189. Peet (98), p. 7.
190. Caron, Chapin III, Donoghue, et al. (94), p. 344.
191. Thompson Klein (90), p. 20.
192. Wallerstein (97), pp. 84–87.
193. Pfirman, Collins, Lowes, et al. (05), p. B15 (internet).
194. Brock, Comitas, Sigurd, et al. (86) pp. 98–99.
195. Brock, Comitas, Sigurd, et al. (86) p. 99.
196. Thompson Klein (90), p. 133.
197. Zimmerman (95), p. xii.
198. Wilber (00), p. xiii.

Chapter Three

The Synthetic Entities, Concepts and Methodologies of Academic Interdisciplinarity

<None>*"Contemporary science is like a people without leaders, like an army without officers. We have thousands of men digging up thousands of fragments, but . . . where are those working across instead of in depth, those that shall string the pieces together and reach a result? They are missing. Today you have to be a member of the Holly Clan, walk the designated road, be a specialist.....My goal is first and foremost to challenge the belief in the Clan. We need a new form of science, researchers that walk across, build up—and put together."*[1]

Thor Heyerdahl, 1942

Research that straddles the boundaries between disciplines is known by many names. Among the most common are: interdisciplinarity, transdisciplinarity, multidisciplinarity, pluridisciplinarity, polydisciplinarity, supradisciplinarity and metadisciplinarity. The definitions of these integrative modes are equally multifarious, varying from country to country, institution to institution, from one part of a campus to another and even among members of the same team. However, they all have a common denominator: *the idea of synthesis of knowledge.* Their objective is to answer complex questions, address broad issues and resolve problems beyond the scope of any one discipline, whether on a limited or grand scale (see Chapter 4).

The purpose of this chapter is to identify the entities of knowledge integration, define the various modes of integration, clarify their differences, group them according to their integrative potential, and make them operable on the basis of a variety of methodological tools, skills and attitudes. The focus will be on the integrative distinctions between multi- and interdisciplinarity aiming at sorting out and clarifying conceptual misunderstandings and uncertainties. The assumption underpinning this chapter is that academic interdisciplinary research is a distinct research field in its own right and on its own merits, and that its approach is methodologically unique and supplementary to that of individual disciplines. In particular, this chapter takes issue with the widespread misconception that there are no such things as interdisciplinary methods.[2]

The field of methodology deals with ways of achieving scientific knowledge.[3] Stanley Schumm defines scientific method as a way of doing anything in line with a regular plan,[4] i.e. the application of systematic procedures and means of generating, analyzing and interpreting data. Thus, scientific methods provide measures leading from questions to answers. These measures are of three kinds: *scientific attitudes, skills* and *tools* (see Figure 3.2). In combination, they are supposed to correct/reduce the subjectivity of the scientist on the results of his or her research (see Chapter 4). The distinction made between attitudes, skills and tools may sometimes be blurred and the areas may even overlap. In essence they are, however, supplementary and interrelated. Scientific attitudes refer to the *problem orientation of research* and the *awareness* of the idiosyncrasies, tools, concepts and techniques of other disciplines. An attitude may be broad and inclusive or narrow and exclusive. Skills basically refer to the *training* and *quality of experience* a researcher gains over time to fulfil whatever he or she sets out to do using scientific tools that are *concrete techniques* for unmasking reality by way of statistical analysis, interviews, observations, mathematical measurements, synergetic discourse, etc.

This chapter will therefore also be used to operationalize these measures and to assess their individual and combined abilities to forge disciplinary items of knowledge into interrelated systems. According to their integrative abilities, these measures will then be subsumed and grouped under the different modes of integration applying to academic interdisciplinarity (see Figures 3.1, 3.2 and 3.3). However, first we must determine what the entities of knowledge integration are. More specifically, at what level of knowledge— disciplinary or sub-disciplinary—is convergence most likely to occur?

THE ENTITIES OF INTEGRATION

As observed by the noted science writer John Gribbin: "The fate of specialists in any one area of science is to focus more and more narrowly on their special topic, learning more and more about less and less, until eventually they end up knowing everything about nothing. It was in order to avoid such a fate that, many years ago, I chose to become writer of science rather than a scientific researcher."[5] Gribbin's concern is that unless the present course is corrected, modern science will ultimately leave society with a breed of scientists who are so specialized that they have no one to speak to, no one to discuss with, no one to talk to, no one to learn from and no one to report to. They may end up in a state of secluded professionalism—as isolated islands of knowledge without bridges between them. This situation has been vilified in the discourse as a negative force that *exclusively* promotes fragmentation and specialization in absurdum. There are convincing reasons to doubt the validity of such an assumption.

Bruno Latour takes the opposite stand, arguing that an isolated specialist is a contradiction of terms because no one can specialize without the concurrent 'autonomization' of a small group of peers.[6] In substantiating his point, he refers to the observation that scientists totally on their own doing field research in desolate parts of the world never stop speaking in a *virtual arena of colleagues* with whom they constantly argue *in absentia* as if the "wooded landscape had been transformed into the wooden panelling of a conference room."[7] In a similar vein, physicists are accused of being in the habit of constantly talking to each other at blackboards with no one else present.[8] In other words, a specialist needs other specialists to talk to, to disagree with, to convince, to argue with, to be stimulated by, to quarrel with, to despise and to look up to. This virtual arguing seems to be an important aspect in the dialectic of scientific progress and in forming public knowledge. Consequently, specialists will never specialize in absurdum because that would deprive them of interlocutors. The human psyche will see to it that specialization stops short of seclusion. Humans exhibit human characteristics even if they are specialists.

In compliance with Latour's reasoning, the late Norwegian biologist Johan Hjort claims that the deeper we delve into a problem, the more we feel that it is itself a part of a whole great structure erected by science and thought.[9] The assumption is that specialization makes us aware of wholeness, and that the tinier the units of topical specialization become, the more easily we will be able to identify the trading zones between disciplines and specialities. Specialized studies to unmask the nuances of factual details are what is required to reconstruct wholes from bits and pieces. Jon Elster, who has devoted most of his academic life to interdisciplinary research, claims that the passing of time has made him almost obsessed with facts and details. In his experience, the richness of nuances can only be unlocked by exploring deep into the details—the deeper, the better. "The resulting nuances may then form the basis for fresh theoretical and conceptual propositions and improvements."[10] The basic oneness of the universe is one of the most important revelations of modern physics, according to Fritjof Capra: "It becomes apparent at the atomic level and manifests itself more and more as one penetrates deeper into the matter, down into the realm of subatomic particles."[11] Detailed work in science is, as argued by Barry Barnes et al., never intelligible purely by reference to the esoteric conventions and concerns of the speciality in which it is performed. It always has significance for allied or opposed specialities, and is always liable to evaluation as an element of science in general and an instance of what is conventionally accepted as science in particular.[12] Increasingly, specializations overlap and transcend disciplinary boundaries (see Chapter 1).

One of the salient features of units of topical specialization is their relative lack of stability compared with parent disciplines. Human cells are, for example, in a state of constant flux. They subdivide and recombine, changing shape and disposition in a never-ending process. Some subunits even "exhibit an 'anarchic tendency' to appear more closely allied with counterparts in the heartlands of other disciplines than to subunits of their own disciplines."[13] The assumption is that units of topical specialization of one disciplinary origin may have significant features in common with component entities in other disciplines, imbuing them with the potential to promote greater interdisciplinary understanding. As the focus of research is on ever smaller minutiae, the disciplinary orientation of research is gradually being replaced by a holistic problem orientation, i.e. the effort to understand all the implications and ramifications of a given unit of topical specialization. The smaller the unit, the greater the likelihood that disciplinary boundaries will be transcended to highlight the complexity of a unit of topical specialization. Thus, the emergence of complexity leads to the gradual erosion of boundaries of the special branch.[14] When one finally knows "everything about nothing," one has actually exhausted the possibilities of one's own discipline to learn more. The only way to expand one's own insight is to harvest from the turfs of other disciplines pertinent to one's own speciality. The cross-fertilization between disciplinary units of topical specialization may, in due time, spill over into broader areas of expertise and gradually provide more favourable conditions for interdisciplinary research on a grander scale. As was the case at the time of Galileo, generalizations were possible because the amount of knowledge was comprehensible within 'the horizon of a single mind.' Thus, the Gribbin syndrome may prove to be the only effective medicine for science to foster wholeness from particulars, nominating the units of topical specialization as the synthetic nucleus of interdisciplinarity. Specialization is not the antithesis of complex science or the opposite of interdisciplinarity, but rather its foremost prerequisite. What we are witnessing today is simply one more expression of the dialectics of science, i.e. smallness creates wholeness, wholeness requires smallness and wholeness is different from the sum of the units of topical specialization (smallness) in non-linear systems.[15] The ultimate premise here is that specialities, unlike disciplines, have no inherent boundaries.[16]

There are no established borders to cross, just new insights to be added and gained. The narrowest of specializations reduces the distance between sub-disciplines and necessitates a manifold approach, calling for what has been termed an *aggregative approach*, meaning the assemblage of tools and procedures from multiple specialities. This cognition seems to have been lost in the cascade of regrets and complaints expressed about what has been termed 'overspecialization'.

Specialization has given rise to a number of interactions as disciplinarians approach each others' borders.[17] The corollary of this is that the concept of academic interdisciplinarity is misleading in that it overlooks or disregards the process of specialization. It is the combination of specialities, not the combination of whole disciplines that produces integration. In Mattei Dogan's mind, the concept of interdisciplinarity should therefore be avoided and replaced by the notion of *hybridization of specialities*. Hybridization implies an overlapping of and contact between segments of more than two disciplines, a recombination of knowledge and competence in new specialized fields. Many of the most creative specialities are hybrid specialities and recombination among them arises in the borrowing of concepts, theories and methods (see Chapter 4). Hybridization takes place because specialization leaves gaps between disciplines and specialities fill the gaps between two or more existing fields. In the history of science, a twofold process can be observed: specialization within disciplines, accompanied by their fragmentation, is the first process, whereas the recombination of specialities across disciplinary borders is the second. A *hybrid discipline* or *multidisciplinary discipline* is thus a combination of two or more branches of knowledge.[18] Jean Piaget calls hybridization the *genetic recombination* of science, underlining the artificial nature of disciplinary boundaries.[19] On this account, Dogan concludes that innovation in each discipline depends largely on exchanges with other fields belonging to other disciplines.[20] Socio-metric studies to a large extent support this conclusion in that there is more communication between specialities belonging to different disciplines than between specialities within the same disciplines.[21] Julia Thompson Klein reminds us that an embryologist and a geneticist may be more alike in knowledge, techniques and interests than two chemists: "In this circumstance, it is (still) proper to call collaboration between the geneticist and the embryologist 'interdisciplinary' while classifying the joint work of two chemists who labour to understand each other as 'disciplinary' research."[22]

Dogan is right. The hybridization of specialities is the main feature that facilitates the integration of scientific knowledge. But in the light of this, he denounces the concept of interdisciplinarity as obsolete, irrelevant, replaceable and not fitting the reality of scientific integration. In so doing, he becomes overly formalistic. The crux of interdisciplinarity has always been and still is to provide for integration between recognizable fields of complementary expertise. Whether these fields are known as specialities, areas of topical specialization, sub-disciplines, hybrid disciplines, multidisciplinary disciplines, monodisciplines, stakeholder knowledge or merely disciplines (see Chapter 4) has no bearing on the operational objective of the concept of interdisciplinarity. From this perspective, the notion of hybridization of specialities does not contravene or add anything new to the traditional integrative meaning of interdisciplinarity. What it does do, is to identify specialities as the entity of integration and explain their interrelationships without challenging the fact that the specialities have been developed and cultivated under the auspices of a monodisciplinary structure that is still at work. The focus of the hybridization of specialities should rather be on the transition zones where trading between units of topical specialization takes place.[23] In these zones, specialities either overlap, touch, mix or merge, easing interdisciplinary exchanges and interactions when it comes to concepts, tools, insights and theories. Here, hybridization is both cause and effect and has led to more recombinations and border crossings over the past three decades than in the previous millennium, according to Julia Thompson Klein.[24] On these grounds, the concept of interdisciplinarity will continue to be used throughout this book. The same goes for the term 'disciplines' which, throughout this book, is used interchangeably with sub-disciplines, specialities and units of topical specialization.

CHANGES IN THE MONODISCIPLINARY STRUCTURE

The practice of boundary crossings in search of topics on the turfs of others—long declared subversive imperialism—has now become the accepted mode of behaviour for disciplinary researchers. Gradually, the research community has come to acknowledge that there is no such thing as private (monodisciplinary) property rights attached to specific domains of research. In actual conduct, scientists are allowed to do whatever they like, as long as they do the job properly on consensual scientific premises,[25] i.e. *normal science* in Thomas Kuhn's terms. In principle, the domain of research has become the common property of all sciences and disciplines. This is what we call *extended monodisciplinarity*. Thus, the traditional monodisciplinary organization of research is gradually fading away, and a new structure, based on two distinct pillars is emerging. The first pillar is the fragmentation and hybridization resulting in units of topical specialization and sub-disciplines, and the second is extended monodisciplinarity, where research is moved by 'imperialistic expansion' into the territory of other disciplines (and even into stakeholder insights), either by breaking or bridging across boundaries. The original demarcation lines between monodisciplines are now being attacked on two fronts from within the old structure itself. The one extends the field to include parts of other disciplines, while the other breaks the disciplines down into sub-units of specialization. Thus, the sub-disciplines and the units of topical specialization have become the dominant features of the new structure of research. The monodisciplines have become umbrella concepts, embracing a variety of specialities. A discipline is thus a cluster of specialities subjected to continued hybridization (see definition in Chapter 1 and Figure 3.1). In this vein, the concern about specialization expressed by Gribbin is not a sign of crisis but a symptom of scientific growth and expansion fostering interdisciplinarity.

Although the process of hybridization is self-propelled in the light of the growth of knowledge, the process of synthesis is dependent on the application of scientific methods, i.e. the combination of attitudes, skills and measures, resulting from composite questions and seeking holistic answers.

THE CONCEPTS OF SYNTHESIS AND THEIR
METHODOLOGICAL EXPRESSIONS

Multidisciplinarity, Pluridisciplinarity and Polydisciplinarity. These three concepts cover the same reality and are often used interchangeably. For the sake of simplicity, the concept of multidisciplinarity, currently the most frequently used, will be applied in this book. Multidisciplinarity can be defined in two ways in simple and synergetic terms.

The simple definition regards multidisciplinarity solely as a compounded offshoot of monodisciplinarity, i.e. *as a juxtaposition of individual disciplines that work in concert and/or in parallel on a common research problem.* Here, the disciplines mainly contribute on the basis of their own premises, procedures and methodology, presenting separate reports in their disciplinary terminology joined only by external editorial linkages.[26] Diversity is high, adaptability low. This is the simplest form of multidisciplinarity and the form that, for obvious reasons, is the most accepted by sworn disciplinary researchers. It represents no imminent threat to the structure and approach of monodisciplinarity and is the natural extension of monodisciplinarity on its own terms. In *simple multidisciplinarity,* the relationship between the disciplines may be cumulative and mutual, but not interactive. Parallel investigations and borrowed information are no guarantee of ultimate convergence.[27] The disciplines are neither changed nor enriched, and the lack of a well-defined matrix of interactions means relationships are likely to be limited and transitory (see Figures 3.1 and 3.2).

Simple multidisciplinarity has been characterized as a cafeteria of courses in many disciplines.[28] On these grounds, some allege that multidisciplinarity in research has no genuine meaning.[29] It is nothing but disciplinarity in parallel. This conclusion is arguable at best, and wrong at worst. In any case, it is superficial. It disregards the fact that multidisciplinarity differs from disciplinarity in terms of methodological measures, e.g. *interdisciplinary curiosity, holistic problem-orientation, aggregative approach* (attitudes), *interdisciplinary practice,* (skill), *juxtapositional depth and breadth,* and *interdisciplinary exchange and adjustment* (tools). Let us examine these measures.

Holistic Problem-Orientation

As discussed in Chapter 1, interdisciplinarity is based on holism and interdisciplinary curiosity as to how single discipline-based parameters function in composite organisms and systems, be they societal, humanistic or natural. In holism, or system thinking, the individual components can only be understood in relation to the function they collectively perform to ensure that the whole works. Interdisciplinarity is therefore contextual, i.e. oriented towards the connections and interdependence of system parts. This outlook is the opposite of the analytical approach (disciplinary/reductionism), which is to take a system apart for the sake of studying and understanding its parts as individual components, whereas holism seek to understand how the parts perform jointly within the framework of a complex system. The essential properties of a system culminate in attributes that individual components do not possess alone, i.e. the old adage that the whole is more than or different from the sum of its parts. The focus of system theory is therefore on the interaction of components. As an attitude, holistic-problem orientation thus implies that the contextual approach will add new insights to the analytical (disciplinary) approach. Interdisciplinary research is looking for the additional insights that systems can only provide as whole systems. Simple multidisciplinarity applies a holistic-orientation, although in modest form.

Interdisciplinary Sensitivity, Practice, Competence, Identity and Community

The fact that scientists search for suitable research themes across disciplinary borders is, if not interdisciplinarity in itself, an important prerequisite for promoting it. Without the ability to see from different sides of various boundaries, it is not possible either to redraw the boundaries themselves or, more crucially, to change what can be seen as a result.[30] As pointed out by Gunnar Skirbekk, when scientists in search of research topics traverse the monodisciplinary structure, there is a need to develop and internalize what he calls a *reflexive distance* to one's own conceptual and methodological preconditions. That is, the results reached within one's own units of topical specialization, has to be attuned to what other disciplinary units say about the same phenomenon. In presenting the results, the preconditions for research should be made explicit in order for other disciplines to understand what the results do or do not say. There is a need for *interdisciplinary sensitivity.*

Interdisciplinary sensitivity is a methodological mindset requiring mutual respect and humility on the part of all specialists involved in extended monodisciplinarity. They share a curiosity about what others can bring to the table to expand their own insight and to promote the synthesis of knowledge. To achieve this does not, in Skirbekk's mind, imply that disciplines have to cooperate. Interdisciplinary sensitivity is something which, even at the present stage of topical specialization, has to be developed in order to sustain the traversing search for topics throughout the whole of the new structure. It is a concept that embraces individuals who practice extended monodisciplinary research, as well as cooperative groups involved in interdisciplinary boundary crossings. Skirbekk admits, however, that the attitude of interdisciplinary sensitivity will thrive better in multidisciplinary research communities than in monocultures.[31] Interdisciplinary sensitivity may, in turn, instil an interest in working on broader perspectives than those offered by the narrow confines of the units of topical specialization and monodisciplines. This will gradually result in what has been labelled *interdisciplinary practice*, which implies that the disciplinary bonds are transcended for the sake of interdisciplinarity, but without leaving the disciplinary approach and expertise.[32]

As a scientific skill, disciplinary sensitivity may further be developed into *interdisciplinary competence,* which is long-term experience and training in research cooperation extending beyond the confines of one's own discipline.[33] As has been noted, the ideal interdisciplinarian should combine thorough disciplinary skills with a radical research philosophy that enables him or her to pursue projects or ideas considered unconventional by the discipline.[34] This is interdisciplinary competence. Here, interdisciplinary sensitivity is transformed via interdisciplinary practice into interdisciplinary competence, which in turn may create a community of *interdisciplinary identity* among those possessing the necessary skills.

In this context, the term interdisciplinary identity refers to the way in which disciplinary experts identify more strongly with the interdisciplinary purpose of research than with their own disciplines. The focus here is not on disciplinary boundaries and differences, but on topics of common interest to experts in different fields. Topical interest and problem-orientation are, irrespective of disciplinary training, simply the focal point of discussion. The disciplines have become avenues of convergence at the focal point.

These concepts make up a scale to measure the increasing determination and skill needed to integrate knowledge across disciplines and sciences, and they apply to different modes of integration (see Figures 3.1 and 3.2). This build-up of increasing competence is a direct consequence of pushing back frontiers and redrawing the boundaries of the old monodisciplinary structure. The degree of specialization may—at least in theory—reach a point where the integrative potential is transformed into an *interdisciplinary community*, indicative of a situation in which the disciplinarians traverse with relative intellectual ease between the range of disciplines and make disciplinary inputs to most or all of them. This is at best a rare ability attributed to Renaissance Man, of whom there are extremely few, if any at all. In most cases, such integration is beyond the reach of the average researcher, irrespective of his or her specialist brilliance. It is a mode of competence/skill that will relate to *all-encompassing cross-disciplinarity* (see Figure 3.4). However, this is not to say that new steps will not be added to the scale of increasing integration in future. Further, there are procedures for bringing the various steps of the scale to fruition. How then to develop the methodological skill stemming from the various modes of interdisciplinary sensitivity?

Psychologist Daniel Kahneman, a Nobel laureate in economics, may have come up with an alternative to the traditional conflict-ridden format of critique-reply-rejoinder, which has repeated itself many times over in multiple scientific discussions over the decades (see Chapter 1). At the outset, Kahneman takes personal pride in never having taken part in scientific controversies or launching attacks on someone else's work. In his experience, scholars who spend time and effort in the legitimate enterprise of criticism rarely adopt the most charitable interpretation of the works with which they disagree. As a substitute for the traditional format and to avoid the risk of widening the gulf of disagreement between the 'contestants', he has long championed a different procedure which he has labelled *adversarial collaboration*. This alternative form involves a good faith effort on the part of contestants to engage in debates by carrying out joint research. This procedure, which is based on *interdisciplinary curiosity*, implies forcing oneself to listen to what the other party has to say under the auspices of formalized project cooperation. Since the contestants are not expected to reach complete agreement at the end of the exercise, "adversarial collaboration will usually lead to an unusual type of joint publication, in which disagreements are laid out as part of a jointly authored paper." In Kahneman's experience, the adversarial collaborations in which he has taken part have all ended with some new facts being accepted by all, narrower differences of opinion, and considerable mutual respect.[35] This mode of discussion and the mutual respect it produces constitute the very essence of the many offshoots of interdisciplinary sensitivity discussed above. It should be remembered, however, that adversarial collaboration is a measure to be applied both in disciplinary and interdisciplinary research, but that practising it across rather than within disciplinary boundaries is probably the most challenging application of all, not to say across the boundary between academe and society.

Another way in which to develop these skills further is to establish permanent networks of interdisciplinarians. For such networks to function as effective integrators of knowledge, no discipline or type of knowledge should be regarded as superior to any others and none of the participants should claim command of the expertise of the other domains. Through level playing field cooperation of this kind, participants will engage in 'negotiations' on how to converge, then sort and put the bits and pieces of knowledge together to form a comprehensive whole (see synergies below).[36] At the same time, their interaction—directly and indirectly—will refine, strengthen and develop the competence stemming from interdisciplinary sensitivity.

Juxtapositional Depth and Breadth

The juxtaposition of disciplines makes two methodological contributions. First, it guarantees that the depth of research is preserved by the representation of individual disciplines. Thus, the benefits of specialization are preserved and continued. Second, the juxtaposition adds breadth of research in that the number of disciplines involved widens the horizontal scope of the object of study. It is additive. Together, these two aspects form a methodological measure, *juxtapositional depth and breadth,* which is both similar and different from monodisciplinarity. It adds an integrative potential. Although the additive element is much more salient than the integrative, both aspects make simple multidisciplinarity different from monodisciplinarity and extended monodisciplinarity.

Interdisciplinary Exchange and Adjustment

Interdisciplinary sensitivity or more integrative versions of the same concept presupposes an awareness of the idiosyncrasies, concepts, tools and techniques of other disciplines. Such awareness may result in one discipline borrowing concepts, methods, procedures and theories from others, also called *shared methodologies.*[37] If these borrowings/sharings are adjusted to comply with the object of study of the borrowing discipline, an *interdisciplinary exchange and adjustment* has taken place, enhancing the integrative potential of the borrowing discipline. Boundaries have been crossed and additional interdisciplinary insight has been gained. In this scheme of lending and borrowing, neither discipline is being modified, and their epistemic natures and cognitive structures have not been challenged.[38] Boundaries have been crossed but not revised or broken down. It is, for example, considered quite natural for political scientists to borrow concepts from sociologists and to emulate economic concepts in trying to understand political developments, or for sociological concepts to be extended and applied to political analysis.[39] Concepts such as *role, reference group, mobility, status, modernization* and *self*, as well as *decision-making, action, information* and *communication* are examples of cross-disciplinary concepts. As the methodologies cross the vertical pillars of the disciplines, they promote theoretical convergence.[40] It is this process of convergence that makes interdisciplinary exchange and adjustment a methodological tool of interdisciplinarity. Concepts form an integral part of scientific methods, since they represent the final results of the development of human thought in the past; they may even be inherited and are in any case indispensable tools for doing scientific work in this day and age.[41]

Borrowing concepts, methods, and theories does not enrich the lending discipline. The use of mathematics and statistics by social scientists has not contributed significantly to mathematics and statistical analysis as disciplines,[42] but it has replenished the methodological inventory of the social sciences: Today, mathematics and statistics play a significant role in both the natural and social sciences, underscoring that interdisciplinary communication often assumes a mathematical form.[43] Even humanistic disciplines borrow methodological devices from the natural sciences. By the application of *pollen analysis* (biology), the *C-14 carbon dating* (nuclear physics)[44] and *blood sample analysis* (medicine),[45] archaeological studies have made great strides towards a more precise and theoretically vivid understanding of pre-historic times. Shared or borrowed methodologies have, for example, facilitated the establishment of many touching points between human and physical geography.[46] The natural sciences can hardly avoid using the qualitative methods of the humanities and adapting the interpretation of the social sciences. In this way, qualitative assessments have also become part of the methodological business of the hard sciences (see Chapter 4). According to Gary King et al., the differences between quantitative and qualitative traditions are only stylistic and as such, methodologically unimportant. All good research can be understood to derive from the same underlying logic of inference.[47] *Boolean algebra,* a tool that moves "beyond qualitative and quantitative strategies"[48] may serve as an example of this underlying logic.

Boolean algebra is a truly quantitative measure which is applied through comparative studies to preserve the holism of individual cases and identify their similarities and differences. It views cases as combinations of values and compares different combinations holistically. This feature makes the approach ideal for identifying patterns of multiple conjunctural causation.[49] It starts by assuming maximum causal complexity and "then mounts an assault on that complexity."[50] The approach simplifies complex data structures in a logical manner and treats cases as whole entities rather than as collections of parts. Boolean techniques, according to Charles C. Ragin, have a strong inductive element because they proceed from the bottom up, simplifying complexity in a methodical, stepwise manner. These techniques start with a bias toward complexity, examining every logically possible combination of values, and simplify this complexity through experiment-like contrasts—procedures which approximate the logic of the ideal social science comparison.[51] The approach provides a tool that combines the interpretive understanding offered by case-oriented approaches, the in-depth analysis of a selected number of cases and a holistic understanding of each case. In this way, it examines cases, uses categorical variables, looks at different combinations of conditions, is applied to categorical dependent variables and involves data reduction. It provides an avenue for addressing large numbers of cases without sacrificing complexity. Thus, social scientists are allowed to be broad and theoretical without resorting to vague, imprecise generalizations. These assumptions seem to stand the test of research. As stated in an empirical study of the interplay between technological change and public policy, the Boolean approach allows the researcher to preserve the detailed knowledge of the subtleties of single cases while also allowing for the identification of more general patterns as the number of cases analyzed grows.[52] This is not to say that the approach is fully developed. Researchers applying the method admit that it has not yet been developed to its fullest potential and that more time and work are needed to make it a mature methodological tool of convergence.[53] The transdisciplinary approach combining the specificities of cases with the universalistic orientation of the sciences may contribute to the further refinement of the Boolean method. The two approaches share the same method, but serve different purposes.[54]

In terms of theoretical approaches, complex system science can neither do without nor avoid the integrative potential of discipline-based, synthetic, inter-field and seminal theories. As will be argued below, these theoretical categories are to a certain extent *discipline-neutral,* also bridging the gorge between the hard and soft sciences.

Of the three branches of learning, the social sciences borrow the most. This gives the social sciences an extra integrative boost in interdisciplinary projects, and may in the long term offer a bridge between the humanities and the natural sciences (see Chapter 4).

In terms of methodological inventory, simple multidisciplinarity harbours the following elements: *interdisciplinary curiosity, holistic problem-orientation, aggregative approach* (attitudes) and *interdisciplinary practice* (skill), which, in terms of ability of integration, holds a slightly stronger potential than that of *interdisciplinary sensitivity, juxtapositional depth and breadth, interdisciplinary exchange and adjustment and, as an example, the Boolean approach* (tools). These elements extend the methodological grasp of individual disciplines and make simple multidisciplinarity an integrative effort, although with modest synthetic ambitions.

Synergetic multidisciplinarity is distinctly more than an offshoot of monodisciplinarity. It is a juxtaposition of disciplines working in concert to achieve the integration of knowledge from two or more disciplines to address a particular research problem. Apart from being additive, it holds a higher integrative potential than its simpler cousin. In terms of methodological elements, it is composed of the same elements that make up simple multidisciplinarity but adds two more accretions: *interdisciplinary competence*, which is a more synthesizing skill than interdisciplinary sensitivity, and *synergetic discourse*, which is a fairly powerful tool of convergence.

Synergetic Discourse and Synergetic Expressions

As pointed out by Ted Anton, most science is usually haphazard, creative and accidental.[55] To this list, one might even add an element of serendipity. Werner Heisenberg believes that it is probably true, quite generally, that in the history of human thinking the most fruitful developments have frequently taken place at those points where two or more different lines of thought meet, be it accidentally, haphazardly or deliberately.[56] At these venues, *synergetic effects*, i.e. fresh, multidisciplinary, adjusted new insights—empirical as well as theoretical—generated by the discourse between disciplines, are likely to emerge. To put it differently: The discourse provides *extra-disciplinary insights* and adds structure to the knowledge-evasiveness at the tangential points of disciplines. In this way, the whole becomes different from the sum of its parts, and no matter how comprehensive disciplinary understanding may be, "there will always be unknowns beyond the sum of current knowledge."[57] Emmanuel Kant called this approach a *discursive method,* which means an intellectual movement in two directions between the *analytical method*, which relates to experience, and the *synthetic method,* which requires intellectual construction. In elaborating on this approach, Johan Hjort concludes that the discursive method is not only a movement made once in each separate direction, but should be regarded "as a constantly alternating movement in both directions, finally arriving at a state of equilibrium (a unity) which is yet not absolutely permanent, but depends upon the analyses and syntheses which have taken place in each particular case."[58] In this way, the method is dialectic, forging unified understanding by way of two processes. Thus, the 'unknowns' beyond the sum of current disciplinary knowledge are what synergetic discourse provides in terms of additional scientific insight.

Synergetic effects are reported to emerge by means of multidisciplinary discussions, including adversarial collaboration, between members of interdisciplinary teams working together to illuminate a particular problem. Some call this interaction *invisible college,*[59] while others call it *internal education.*[60] Both refer to a continuous cooperative process of interaction between team members where a variety of integrative techniques[61] are applied over an extended period of time to improve the collective and individual ability of the members to work and integrate across boundaries. The process is a kind of intellectual playfulness generating a cooperative environment, reducing disciplinary competition and facilitating the sharing of data, concepts, theories, methodologies and insights.[62] In this context, theory and discourse gradually become boundary concepts across a range of disciplines.[63] This process is not invisible, as assumed in the concept of invisible college, or internal, as indicated in internal education, but a methodological measure deliberately used to produce synergies. To underscore its scientific methodological nature, we will designate this process, or even better, this tool: *synergetic discourse.* It takes place internally between team members, has an educational effect, and aims at generating synergies across spheres of knowledge. But the process of succeeding is in no way automatic and will not always pay off.

As has been pointed out, when we gather for multidisciplinary exchanges, we tend to speak in tongues. "It sounds good and it feels good to put up a show of togetherness. We pretend to understand each other, but the substance often escapes us through lack of comprehension."[64] Intellectual comprehension is a product of individual will and an interest in looking beyond one's own professional confinements—a kind of curiosity that straddles boundaries and works through three core principles to succeed: *maturing and deepening, cooperation and interplay,* and *creativity.* 'Maturing and deepening' entails working towards excellence in stages (interdisciplinary sensitivity, practice, etc.) whereas 'cooperation and interplay' indicate that the parties learn from each other as they work together (invisible college, adversarial collaboration), while creativity refers to the act of crafting multiple elements into an organic whole (*dual* and *triple competence* (see below).[65] All the elements are part of the synergetic discourse, which is a methodological measure to be applied by the 'polygamous marriage' of different disciplinarians and stakeholders.[66]

The experience of synergetic discourse is shared by multiple teams, including the *Climate Impacts Group* of the University of Washington in Seattle, which is composed of researchers from the physical, biological and social sciences who work together to understand the impacts of climate variability and change in the north-western part of the United States.[67] The Group has created its own dynamic process, asking questions across the disciplines, interchanging ideas and input to realize the overall programme objectives. The discourse started out as a simple multidisciplinary endeavour and gradually became interdisciplinary in integration. The interaction process of the Climate Impact Group translated specialized knowledge into a synthetic product, which came as a 'pleasant surprise', i.e. an eye-opener, to all the members of the group.[68] Their experience was that the cooperation of the team moderated "the tribal antagonisms that differentiate disciplines."[69] The experiences of the Climate Impact Group are also shared by others.

The Stanford University project on the *Services of Nature*, aimed at developing a scientific basis for the policies needed to sustain Earth's life-support systems, offers a similar description: "At first we didn't agree on anything, but gradually we learned the language and the assumptions of each others' disciplines. It was fun being a part of. The social dynamics played a prominent role in creating ideas."[70] In a somewhat similar study, it was reported "that the slow, sometimes difficult, process of interdisciplinary adaptation led to broader perspectives on their research projects and to novel insights which could not have occurred otherwise."[71]

In an Australian multidisciplinary project *on the feasibility of conducting a trial involving a heroin prescription for heroin addicts,* a team of more than 100 researchers from 14 different disciplines "provided a sounding board to refine ideas."[72] The group concluded that the power of multidisciplinary research was "in its ability to bring together different viewpoints and arguments and to subject conflicts to close scrutiny. If all sides do not feel they are being dealt with fairly and respectfully, the approach is devalued."[73] All the researchers involved had to break down important barriers by linking their science with their inner sense of wonder (interdisciplinary sensitivity).[74]

In the course of conducting some forty workshops over eight years, a programme on *Writing-across-the-curriculum* at Michigan Technological University gave rise to a most interesting result: "Many faculty members who did not know each other or did not understand each other's discipline (after having worked together) now feel a common bond. . . . The workshops actually remind(ed) some people why they became college teachers in the first place—before they retreated to separate buildings, isolated offices and competitive research. . . . The net professional effects of this cooperative effort were numerous: joint teaching experiments, collaborative grant proposals, and co-authored articles, (which) helped to unite a dozen teachers in a collaborative effort providing both intellectual and social cohesion among the participating writers."[75] "The remarkable fact was how well most of the mixed disciplines people (got) on most of the time . . . and sometimes better than we guessed."[76]

Another team, relating regime theory to the fields of international relations and law reported that: "they (the participants) may not yet be speaking with one voice, nor should they. But each side is finding something to say, in a deepening and mutually profitable conversation."[77] A similar experience was reported by Norwegian researchers: "there have been magical moments when our results melted together to form exciting contributions illuminating the topic, which more than justified all the efforts."[78]

A Swedish project on communication concluded that the group "can probably become a leading centre in communication theory, widely construed," not least because there are few other institutions where scholars are thinking so intensely about so many facets of communication as the group under review.[79]

An experienced political scientist at the University of Chicago reporting about his joint works with an international lawyer indicated that he, as a consequence of the cooperation, gradually had changed his professional focus of interest from that of power relations to that of 'international legislation.' Together they had elaborated and developed the concept of 'Soft Laws', i.e. basically declarations of principles of state behaviour as a means of regulating international relations. Although, these 'laws' are not, in a strict legal sense, part of international law or being enforced by anyone, they appear to have a regulatory effect on state behaviour. This conceptual observation led the two parties to focus their joint research on the role of values in international politics. Since this collaboration had been going on for some time, the political scientist admitted he had become more value and law oriented in outlook, whereas the lawyer most likely had become more power-oriented and even somewhat more cynical.[80]

A Norwegian team of scientists conducting a multidisciplinary project on forest management and societal developments in the highlands of Madagascar experienced that their understanding of the approaches and contributions of other disciplines was best when conducting field studies that lasted for many weeks and months. According to the team of researchers, this was because research is a social activity and because crossing boundaries requires not only intellectual but also social effort.[81] Here, social interaction lowered the hurdles between the disciplines.

In a project on communication across disciplinary boundaries at the Norwegian University of Science and Technology, a multidisciplinary group of researchers, concluded that in order to understand what was going on in their joint project, they had to search behind the disciplinary terminology of the others. This could be done only by investing time in exploring each others' points of view and thus to develop mutual understanding and respect. In so doing, they gradually formed a *creative action community* that emphasised loosing up, being informal to the point of engaging in humour and opening the door to a certain kind of reckless lack of respect for established truths. In this setting, so called *creative weirdos* became instrumental in building sufficient collective generosity for the individual members of the group to accept and dare to make mistakes. In this laid back atmosphere, the group members were far more open to different outcomes and learned to view cases from a variety of angles, rather than just the one dictated by their respective disciplinary training. Prior to reaching this stage of "communicative harmony," the group went through a gruelling process of collective frustration due to lack of progress and tangible results, which was exacerbated by their lack of a common frame of reference. Finally, the project simply stalemated. To get around the impasse, the group decided to restrict the agenda of their meetings. One meeting was to deal exclusively with strategy, another with idea generation, a third with reviews etc. Most importantly: If they failed to agree on the content and objective of a planned meeting, it was never conducted. This restructuring of the agenda-setting eased communication in a way that surprised participants as it was "close to intimidating." By overcoming the bureaucratic meeting culture practiced at the outset of the project, the group gradually developed a linguistic tool belonging to all. As a consequence, they ultimately ". . . got the good feeling one gets when the fog burns off."[82]

In 2000, a group of physicists and technology experts from the Norwegian University of Science and Technology, the University of Oslo and the Institute for Energy Technology formed a research group called *Complex*. In their activity report for 2006, they state that in their study of complex systems they have experienced large synergies as a result of combining perspectives from different topics and systems. For instance, patterns and structures emerging on scales larger than that of their basic constituents were being studied through multiple systems, such as simple and complex fluids, i.e. clays, fluidized powders, dry granular flow, two-phase flows in porous media, magneto-active fluid-particle systems and surface deposition structures. Another example where the "combination of different studies (yielded) more than the sum of the components" was that of breakdown processes: Fracture in brittle materials, avalanches formed by flux quanta in superconductors, and instabilities in the human economy are processes that are extremely diverse, yet shed light on each other through their similarities. The group concluded the report by saying that they were able to "not only combine scientific topics but also experimental equipment, computational techniques and manpower across the institutions of its members. The ability to do this in a stable way allowed (the group) to create a scientific environment which could not have been established within a single Norwegian university."[83]

The experiences of the Wissenschaftskolleg zu Berlin comply. Every year the Kolleg identifies 40 of the most accomplished and creative scholars from around the world and puts them together in very agreeable living and working conditions. In selecting the scholars, the Kolleg attempts to give fair representation in each year's class of 40 fellows to as many different disciplines of scholarly activities as possible. The useful but unintended consequence of this is that it also ensures that each scholar has few or no peers "to trim away shoots of thoughts sprouting outside the narrow radius of acceptability." This provides an opportunity to present one's own work and ideas to scholars of a different breed, training and turf without the presence of the old guard. The conclusion drawn by an ethnologist taking part in the activities of the Kolleg is succinct and unambiguous: "I know of no better method of fostering unhindered creativity," he said, admitting that his stay at the Kolleg started a personal intellectual evolution that made him "a very different biologist"[84] from what he had been previously.

All the above examples relate to studies on academic interdisciplinarity. This is because transdisciplinary projects until recently have been in short supply. In 2008, however, reports on the synergetic potential of transdisciplinary studies were published in the *Handbook of Transdisciplinary Research*. In summarising the experiences of 19 individual transdisciplinary projects, the editors conclude that despite multiple stumbling blocks in transdisciplinary practice, there are some promising lessons to be learned as well. One is that transdisciplinary research has a potential to stimulate innovation in the participating disciplines, and has proved particularly successful in combining different means of integration, i.e. developing joint theoretical frameworks, applied models and concrete common outputs. There is also strong evidence suggesting that attention to values and stakes (a core difficulty in this kind of research) should be dominated by mutual learning attitudes and not by positions. Furthermore, training and education are best developed in close connection with the disciplines of origin. The similarities to academic interdisciplinarity are striking, which may be why the editors express their conviction that this kind of research "forms a major avenue for enhancing science based contributions to solve complex problems in the life-world" . . . "and has demonstrated a potential to stimulate innovation in a broad range of disciplines."[85]

The challenge of all these and other programs has been focused on finding those moments of interdisciplinary insight that occur when one sees the world through someone else's eyes, if only for a moment.[86] When so doing, "there seem to be negligible problems related to myths and prejudices across disciplines. Those aspects are being joked with, but when one gets to know each other and each others' disciplines this becomes of no importance."[87] On the basis of this clarification, teams may begin to develop a vocabulary and rules of measures[88], and as the group(s) became more integrated and the multidisciplinary models developed became more complex, each refinement in one field forced a refinement in the others.[89] What seems to happen is that combining tools create "a synergy of increased observing power."[90]

Although conducted on different continents at different times and to some extent in different countries, the phrasings and experiences cited above have a striking resemblance: they all refer to a process that (often) started out in disciplinary competition and conceptual confusion, and ended up in a joint product spawned by cooperation, discourse and adversarial collaboration. The joint products—the synergies—came in two forms, either as *measures of de-education* from disciplinarity or as *offshoots to scientific progress*. The measures of de-education tended to 'break down barriers between disciplines', 'resolve disciplinary competition', 'provide internal education', 'forge professional cohesion', 'surmount negligible problems related to myths and prejudices across disciplines' and 'form new social connections and dynamics.' Whereas the scientific offshoots emerged as 'refinements of ideas', 'broadening of perspectives', 'novel insights', 'profitable conversations', 'increased observing power', 'closing of language gaps', 'new vocabulary and concepts', 'interchange of disciplinary viewpoints', 'new rules of measures and modes', 'merging of results', 'combining tools', 'sharing of data and methodologies', ' focus alignments', and 'refinement in one discipline led to refinement in others.' Some of these phrasings are just different wordings for one and the same phenomena, but most of them point to a variety of different synergies in each category. If we had added more experiences, the likelihood is that we could have listed several more synergies (see Chapter 5). The point here is not to make a complete and final list, but to illustrate the synergetic potential offered in interdisciplinary research.

The Preconditions of Synergies

Usually, it does not take long before members of interdisciplinary groups experience synergies in their work. The Climate Impact Group had worked together for about three years at the time of my interview with the head of the programme. Members were amazed by how much they had gained scientifically from their discussions during a relatively short period of time.[91] However, bridging the gap from a disciplinary to an interdisciplinary platform is usually more time-consuming. In a Swedish study spanning a period of more than 11 years that was conducted at the University of Linköping, an International Review Group concluded that four to five years of disciplinary interaction was too short a period to create a new interdisciplinary vocabulary and/or measures and modes of analysis.[92] To overcome the pressure of time and to develop a common conceptual model, the group members made a concerted effort to participate in each others' teaching as much as possible during the first five years, and tried to continue to do so also in the ensuing years.[93] In so doing, they actually broadened the scope and endeavours of their synergetic discourse and increased their interpretational power across disciplines. We can call this *extended synergetic discourse* (see Figure 3.2). They themselves called it an invisible college. Despite the fact that the members came from widely different disciplines, they jointly perceived that there was sufficient commonality of values and interests at the outset to facilitate collaboration. One reason for this was that it was "OK to be interested in fields apart from one's own and try out new perspectives." Most importantly, there was no heavy burden of tradition or 'old-guard' faculty that one had to watch out for. They experienced a kind of frontier spirit.[94] When the old guard faculty is not in a position to exercise their disciplinary authority and control/power through what is referred to as the "sociology of science"[95], the young faculty yields more easily to interdisciplinary curiosity and finds the courage necessary to indulge in the forbidden fruits of unconventional science.

Within the research teams in this programme, the disadvantages of interdisciplinary research were mainly voiced by the most recent additions to the staff, i.e. by those least socialized to the setting and the work.[96] It took some synergetic discourse, and consequently time, before they experienced that many of their qualms were exaggerated and not entirely founded in reality. Thus, the benefits of interdisciplinary research are *process and time dependent*. The participants need a period of grace to socialize and internalize what interdisciplinary research has to offer, and they need the educational part of the synergetic discourse to make headway in de-educating them from the most common confinements and perceptions of their own disciplines.

The evaluation also shows that many of the qualms aired against the programme by Swedish universities and research institutions before the programme was started actually failed to materialize. The criticisms or doubts expressed fell into three categories: 'not borne out', 'to early to tell', and 'may be a problem'. The Review Group concluded that the great bulk of these doubts were not borne out, while several that 'may be a problem' were correctly identified. The latter included the failure to build up the necessary number of professorships according to plans. Foremost in the 'too early to tell'-

category was the concern expressed about the marketability or placement of graduate students after completion of the programme.[97] More importantly, none of the concerns related to the approach and methodology of the programme was reported to borne out. In general, the more time invested in synergetic discourse, the more returns can be expected in terms of synergetic offshoots of scientific progress. The time dependency of synergetic effects resembles the time dependency associated with rungs on the ladder of interdisciplinary sensitivity. In the course of practice and time, interdisciplinary sensitivity develops into, in sequence, interdisciplinary practice, competence, identity and ultimately interdisciplinary community. The concept of synergies should be rated along a similar continuum. For the purpose of this book, we distinguish between three forms in terms of time dependence: *short-term synergies* (1–2 years) will, in terms of beneficial offshoots, breed profitable conversations, and new ideas and perspectives that transcend the reach of individual disciplines. In terms of measures of de-education, short-term synergies will start to resolve interdisciplinary competitions and break down barriers. The beneficial offshoots of *medium-term synergies* (3 to 5 years) will provide new ideas and perspectives and start building common concepts and an emerging theoretical foundation of interdisciplinary usefulness (see Chapter 5). In this time span, new professional and social connections have been formed. The last version, *long-term synergies* (6 years+) will, in both categories, provide an array of different effects, and the longer the cooperation the more comprehensive and deeper the effects will be. This is the version related to extended synergetic discourse that is needed to cope with the trickiest of complex system challenges, for instance, the functioning of the ecosystems of planet Earth. New ideas, concepts, theories and methods born in the interfaces of disciplines increase man's ability to address complex problems, be they knowledge-instrumental or practical-instrumental. At the same time, it relates to the need to form networks among interdisciplinarians as referred above. The more one interacts over an extended period of time, the more competence will be gained in interdisciplinary ability. Interdisciplinarity is thus a speciality built on top of the disciplinary training acquired in universities, but it is a competence forged by practice, not by lectures.

Synergetic discourse may result in a shared conceptual model straddling disciplines and sciences that will serve as a cornerstone of project development and integration. What most project experiences express is that the disciplines grow towards each other in the "sense of forming an interface of theories and subject matters. Through dialectic conversation among disciplines, a new picture or mutual understanding of subject matter may develop."[98] Heinz Pagels concludes that when disciplines interact, a process of horizontal integration is theoretically under way. This integration alters the architectonics of

knowledge by strengthening connections outside of what is commonly regarded as the discipline proper.[99] To achieve this, the disciplines must demonstrate *theoretical relevance*, i.e. how disciplinary thinking increases the understanding of the problem, *conceptual relevance*, i.e. how disciplinary terms
can be used to communicate meaning across disciplines, and *methodological
relevance*, i.e. how disciplinary methods can be applied to acquire data for a
more comprehensive understanding of the problem under study. Applying a
common language understandable to all participants, irrespective of disciplinary training, i.e. a discipline-neutral language, is regarded as a fourth prerequisite.[100]

Along with the synergetic discourse go the tools of modern science: Today's technical tools, e.g. computers, remote sensors, artificial intelligence,
automated DNA sequencers, telescopes, transgenic mice, magnetic and radi-
ation imaging, can be combined to create deeper, broader and more synergetic
effects. These tools enable researchers to move faster and delve deeper into
his or her questions than ever before.[101] Believing in the symbiosis between
computing and complexity, Peter Coveney and Roger Highfield are convinced that the essential role played by the computer explains why the study
of complexity has been overlooked for so long by so many people. They conclude that the science of complexity is intricately entwined with and crucially
dependent on computer technology.[102] The drive at present is to reproduce living complexity on computers to come to grips with more of nature's complexity than ever before. According to Coveney and Highfield, today scientists are looking at previously little understood facets of nature to extend the
computer's range and power, drawing, for example, on the growth of crystals,
magnetic properties and exotic alloys, and the annealing of metals for inspiration.[103] They contend that fast and powerful computers allow biologists,
physicists and computer scientists pondering complexity now to explore complexity in its full glory, throwing light on questions that once lay exclusively
in the provinces of philosophy and mysticism.[104]

It has also been asserted that architecture is important to breed cooperation
across disciplines. For example, the University of Chicago is highly esteemed
and renowned for the quality of its interdisciplinarity. One of several stimu-
lants to that effect is the relatively small size and architecture of the campus.
The university is built around a square where faculty cannot avoid running
into one another when leaving or arriving at the buildings. The individual
buildings host many disciplines whose representatives share tearooms and
other interactive facilities. Most faculty also live in designated areas close to
campus. Thus the housing and physical features of the university force interaction, and eventually cooperation among departments.[105] To be physically
together—forced or voluntary—and to be lodged in the same building com-

plex are also factors considered important for fulfilling the integrative objectives of the Research Council of Norway (RCN).

In 1993, the RCN was established through the amalgamation of five separate research councils[106] into one entity to be accommodated under the same roof. A main reason for this move was to provide a basis for producing useful research policy advice to the government "based on an holistic national perspective," to integrate high-quality basic and applied research and to meet the social and industrial needs of society.[107] The achievements of the Council were scrutinized by an International Evaluation Committee which concluded that the experiment should continue. In the evaluators' collective mind, this is so because in many areas the distinction between the fundamental and the applicable is narrowing and the ability to work both within and between disciplines is becoming more important. "This mean we need nimble researchers and agile research funders, able to bring together the powers of the most distant and dissimilar disciplines."[108] One means of achieving this is physical or architectural, i.e. for different departments to be located in close proximity to each other within the same building complex. Thus, the ideal of synergetic discourse is in itself a complex matter involving multiple elements such as multidisciplinary teams, interdisciplinary curiosity, unfamiliar methodology, the willingness to invest in time and education, computer technology, architecture etc.

SIMPLE, SYNERGETIC AND ALL-ENCOMPASSING CROSS-DISCIPLINARITY

All versions of crossdisciplinarity are different integrative expressions of academic interdisciplinarity. In general, cross-disciplinarity is more comprehensive in scope and vision than synergetic multidisciplinarity. It holds out the promise of 'overarching synthesis', in which the disciplines become subordinated to that larger framework.[109] To underline the subordination of disciplines, concepts such as *supra-disciplinary* and *meta-disciplinary* have also been used. In terms of integrative potential, cross-disciplinarity comes in three versions: Simple, Synergetic and All-encompassing.

All-encompassing cross-disciplinarity is completely *integrative*.[110] This version implies that the disciplines merge into a higher unity/entity in terms of sharing a joint set of concepts, theories and methodology. It is the absolute synthesis of disciplines,[111] stitched together by a full-fledged theory and a joint set of interdisciplinary concepts. This version equals the naturalist position, asserting the *unity of sciences thesis,* which reduces the sciences to one basic science, either mathematics or physics, and/or the *consilience theory,* which, if proven correct, may close the gaps or blank spaces between disciplines and "encircle the whole of reality" (see Chapter 4)[112] It depicts a *theory of everything*.[113] These theories are claimed to straddle all disciplines and all sciences, to embrace *total reality* (see Chapter 4) and to harvest the most long-term synergies in terms of de-education from disciplinary confinements and offshoots of scientific progress from an extended synergetic discourse. They represent a thinking "that would unite all the known laws of the universe into one all-embracing theory that would literally explain everything in existence."[114] These theories are interdisciplinarity in the extreme, also including stakeholder knowledge. At present, these theories are nothing but working hypotheses to be verified or disproved in future. As discussed in Chapter 4, Ken Wilber and others believe such a theory is logically undoable, whereas Edward O. Wilson refuses to concede as regards the vision of total unity of knowledge. After all, if theories of this kind is being applied in empirical research, they most likely will be composed of *interdisciplinary curiosity*, *holistic problem-orientation*, *the aggregative approach*, *interdisciplinary exchange and adjustment*, *juxtapositional depth and breadth*, *interdisciplinary community*, *extended synergetic discourse* and what is below defined as *quadruple competence*, and *a bridge community of interdisciplinary peers* (see Figures 3.1 and 3.2).

Simple and synergetic cross-disciplinarity differ from their futuristic sibling in that they are already being applied, in whole or in part, in the practice of disciplines such as medicine, veterinary science, agriculture, geography, oceanography, political science, etc. These topical fields are defined as ordinary disciplines, but should rather be considered multidisciplinary disciplines because they are composed of elements from several domains and subject areas (see Chapter 4). As has been noted about these disciplines, "Behind the subject façade, interdisciplinarity is flourishing."[115] Their composition has created a new methodological measure for further integration of research: *dual, triple and quadruple competence*.

Dual, Triple and Quadruple Competence

For the sake of illustration, let us use medicine and engineering as examples. Health problems are no more purely biological than their solutions are purely medical, social, psychological, pharmaceutical, or therapeutic.[116] They are inter- and transdisciplinary. According to Gunnar Skirbekk, this fact has methodological implications. His assumption is that the social and natural science disciplines involved in medicine combined are, at one and the same time, more than a juxtaposition of disciplines, and less than a disciplinary synthesis. His reasoning goes as follows: What we have in medicine, is an *in-*

tegrated multidisciplinary activity, which can be called interdisciplinary in a qualified sense. We have a profession, with joint institutions, from education to research. By being socialized into this interdisciplinary community, one is not learning to speak in 'one tongue', as if the medical profession was merged into a single realm of semantics. But by learning discipline by discipline, and by walking between the disciplines of medicine first as a student and later on as an executive physician, also eventually as a medical researcher, one gets the needed 'dual competence' which is a prerequisite to be able to walk between and see across disciplines.[117] There is a need to develop the ability to learn from and eventually contribute to wider cultural conversations. Baruch Blumberg claims that in his medical experience the complexities of a phenomenon can be understood, at least in part, by repeated observations of a whole organism or a population of organisms under a wide range of circumstances.[118] In this line of argument, cross-disciplinarity is not defined by the criteria of shared concepts and theories, but by a most developed sense of *interdisciplinary identity* and a *dual competence* which refers to long-term learning about many disciplines by walking between them and seeing across them. Keith Clayton's conceptual distinction between the "concealed reality of interdisciplinarity" and "overt interdisciplinarity" is of relevance here.[119] The former is just another way of expressing the combined abilities of 'interdisciplinary identity' and 'dual competence. In dual competence, mainstream disciplinarians get "to know the enemy" (read: other disciplines) by moving towards more in-depth knowledge of other disciplines, their general structures, principles and way of thinking.[120] Dual competence refers to *integrative learning and the practice of many disciplines*. Cicero's concept of the *doctus orator*—the man who combines extensive knowledge of all sciences with a wide experience of the problems of everyday life—comes to mind as a means to describe a person who harbours dual competence and an interdisciplinary identity. So does the concept of a 'Renaissance Man.'[121] Dual competence requires the ability to identify which disciplines are relevant and what it is that they might have to offer, as well as enough knowledge about each discipline to be able to have a meaningful dialogue with the experts and to be able to identify the experts with whom to have a dialogue; a good understanding of the 'cultures' of different fields and empathy with their concerns.[122] In the process of seeing and walking across disciplines, dual competence and interdisciplinary identity will develop as a consequence.

In reference to area specialists, Julie Thompson Klein takes this reasoning one step further: "Needing disciplinary, linguistic, cultural and interdisciplinary skills, area specialists face the demands of dual, triple even quadruple competence."[123] The number of skills, including extra-disciplinary skills such as culture, will determine whether competence is dual, triple or quadruple.

Dual competence can be broaden, deepened and further improved in a never-ending sequence. By definition, it is open-ended—a continuous cultivation of skills. Developments within family medicine may serve to illustrate the point. This branch of medicine is holistic in orientation by the continuous walking of its practitioners across disciplines. For a long time, traditional family medicine was the very opposite of medical specialization—its antitheses. In recent years, family medicine has been defined by many universities and hospitals all over the world as a branch of medical specialization in its own right. Here, wholeness has become an area of specialization as a multidisciplinary discipline or speciality.[124] The same goes for the poly-technical ideal of engineers whose specialization in one field is combined with a broad ability to resolve a whole range of technical problems outside their own field. Thus, this ideal and its application show clearly that it is possible to combine specialization with generalization.[125] To many, this may seem as a contradiction in terms, whereas in reality it is just a transformation of dual competence into triple and/or quadruple competence. There is no reason why other problem areas should not reap the same benefits as those in the field of medicine and engineering. The approach is not problem-bound but problem-applicable.

Dual, triple and quadruple competence can be represented by *bridge scientists*[126] who have the training to see and walk across disciplines, that is, specialists in the transition zones and generalities of disciplines. A bridge scientist will tend to emphasize problems of language, translating the formulation of a research problem from one monodisciplinary perspective into another, in an attempt to sort out paradigmatic conflicts and explain differences in evaluation standards.[127] Knut Holtan Sørensen introduces the term: *polyvalent specialist,* to depict the competence of researchers to enter into cooperation with specialists from other disciplines and to produce synthesized knowledge on the basis of their ability to explain their own unit of topical specialization, including limitations as well as merits, to others. The concept of polyvalence is taken from chemistry, reflecting the ability of certain chemical substances to bond with other substances.[128] What first and foremost characterizes a bridge scientist and/or polyvalent specialist, is his or her ability to bring together delimited entities. If a project can avail itself of many polyvalent specialists, they will form a *bridge community of interdisciplinary peers.* Such a community will facilitate and accelerate the integration of research.

Simple and Synergetic Cross-Disciplinarity

Both enjoy dual competence, the latter more so than the former. To make a conceptual distinction, it is reasonable that the dual competence of synergetic cross-disciplinarity is more triple than dual. Cross-disciplinarity as defined by the example of the multidisciplinary discipline of medicine is thus a cooperative exercise. The participants identify and confront differences in perspectives and approaches not in order for one discipline to be 'better', according to some criterion, but for each to learn from and contribute to the others, and hence also to become more aware of the merits and limitations of their own discipline. Due to its dual or triple competence, this form of research integrates several aspects of one discrete discipline with that of several others.[129] It represent a conscious attempt to integrate material from all involved disciplines into "a new, single, intellectual coherent entity."[130] This requires an understanding of the epistemologies and methodologies of other disciplines and a team effort to build a common vocabulary. In either of the two versions of cross-disciplinarity, joint authorship of reports across disciplines is common, and the analysis of one discipline draws on that of the others to integrate and accumulate a knowledge base which becomes the property of all. They differ, however, in the elements of methodology they entertain. Apart from the disciplinary tools, simple cross-disciplinarity may be composed of the following measures: *interdisciplinary curiosity, holistic problem-orientation, aggregative approach* (attitudes), *dual competence, interdisciplinary competence* (skills), *interdisciplinary exchange and adjustment, juxtapositional depth and breadth, synergetic discourse*, and *seminal theories* (tools). Synergetic cross-disciplinarity embraces the same elements, but substitutes interdisciplinary competence with *interdisciplinary identity* and dual competence with *triple competence,* and adds *synthetic and inter-field theories* as new bridge-building tools (see Figures 3.1 and 3.2).

Synthetic, Interfield and Seminal Theories

Synthetic Theories

Defined as a consistent set of ideas and/or variables about how phenomena work, scientific theories are generally diffused through and across disciplinary boundaries. They posses integrative power, i.e. the ability to piece perspectives, knowledge, ideas and concepts together into a whole. Even theories developed within the confines of a particular discipline—discipline bound theories—have been adopted by other disciplines, helping the 'borrowing' discipline to conceive of phenomena in new and broader ways (see below).[131]

   In the early 1970s, Jean Piaget made the case that a theory of interdisciplinarity should be based on common structures to be developed from the holistic perspective of systems theory concerned with patterns and interrelations of wholes.[132] Piaget acknowledged that when it comes to building blocks, a system in the social sciences is no different from a system in the natural sciences, or for that matter between natural science disciplines. As a forerunner to Piaget, the Norwegian biologist Johan Hjort observed in the early 1920s that the word *organization* is a biological expression for the same thing as is known in physics and chemistry as a *system*. He continued: "the two concepts, 'organisation' and 'system' will reveal to us the unity in our mind from which they both arise."[133] Thus, the very concept of systems theory refers to one theory and many terms equalling the concept of system. The theory is system-applicable, implying that the theory is composed of a set of concepts jointly used by a number of disciplines and sciences.

Although somewhat different in content between the disciplines, the concepts have a common denominator to be applied for interdisciplinary research. From this perspective, systems theory holds synthetic power. So does structural theory which is concerned with the interrelatedness of the cognitive and biological components underlying human thought. These theories have, according to Julia Thompson Klein: "been used to strengthen theory in one discipline, to unify a single discipline, to provide an integrative methodology or theory of cluster of disciplines, and even to function as a unified science by integrating all disciplines around a single transcendent paradigm."[134] Theory is important as combined storage and bridging device, storing the summarized results of work in one area in the form of ideas which can be transferred across theoretical bridges to other fields.[135] System theory is not alone, however, in hosting the converging power of theories.

Richard Peet focuses on the integrative power of *social theory*. According to him, it occupies the middle ground between abstract philosophy and empirically based theory. In Peet's mind, the middle ground of social theory provides a storage and bridging ability when it comes to summarizing the results of work in one area in the form of ideas which can be transferred across theoretical bridges to other scientific fields.[136] This is so because social theory is closer to the social, political and cultural realities of real life societies than is the case with philosophy.[137] Because theory is inherently inductive and constructed from empirical sources, it has direct contact with the occurrences, events and practices of experienced reality.[138] From its in-between position in the hierarchy of generalizations, social theory is a theory of relevance to a variety of social sciences, uniting across boundaries—also those separating academia and society. Against this backdrop, Peet asserts that most social science disciplines are making social theory their primary task, seeking to add, refine and develop it rather than focusing on their own disciplinary ends.[139] But the social sciences are not the only ones with bridge-building theories.

Stoichiometry is one of several examples of natural science theories that study the balance of multiple interacting elements in complex ecological and biological systems. It provides a theoretical framework that connects and touches on a wide variety of topics such as molecular and cell biology, life history evolution, nutrition, secondary production, nutrient cycling, food-web-structure and stability, global change and others.[140] The span of its disciplinary coverage is reflected in the subtitle of a book published by Robert W. Sterner and James J. Elser entitled: *Ecological Stoichiometry: The biology of elements from molecules to the biosphere.*[141] As suggested in the subtitle, ecological stoichiometry involves research on basic mechanisms having effects on higher levels of ecosystems. In this way, stoichiometric theory is a device for building bridges between reductionism and holism, i.e. between lower levels and higher levels of ecosystems.[142] After years of intense study, the authors came to appreciate the many ways in which ecological stoichiometry helped them understand patterns and processes in the natural world, incorporating data from complex systems such as species and habitats. This exercise challenged them to stretch intellectually beyond their normal literature comfort zone, but they feel the outcome was worth the effort[143] One of the major goals of their book is to make the core concepts of ecological stoichiometry accessible and interesting to researchers in other fields and disciplines. Recent findings indicate, for instance, that concepts in ecological stoichiometry can be broadened for use also in evolutionary studies, for integrating ecological dynamics with cellular and genetic mechanisms, and for developing a unified means for studying diverse organisms in diverse habitats. This broader approach would then be considered 'biological stoichiometry'.[144] Stoichiometric theory also offers an ecological perspective on tumour dynamics.[145] Thus stoichiometric theory seems to possess the same interchangeability between natural science disciplines as social theory does for social science disciplines.

Interfield Theories

A distant relative of synthetic theories is what Darden and Maull call *Interfield theory,* which seeks to identify relationships between phenomena studied by different fields of inquiry. A field of inquiry is defined by *a central problem, facts related to the problem, techniques and methods applicable to the problem,* and *goals and factors that provide clues about how the problem is to be solved.* Sometimes, if not always, *concepts, laws* and *theories* related to the problem may be applied to realize the exploratory goals of research. Thus synthetic theory may or may not be part of an interfield theory. The motivation to develop an interfield theory arises when researchers recognize that the phenomena in which they are interested are connected to phenomena in other disciplines (the hybridization of specialities). The pooling of disciplines contained in cognitive science is an example of interfield theory.[146] Here, the broadly based disciplines of psychology, artificial intelligence, linguistics, anthropology and philosophy are united by a common focus, unit of topical specialization or a field of inquiry known as cognition. Given that the goal of an interfield theory is to identify these relationships, there is no need to derive a theory of one field from that of another. An interfield theory is designed to reveal relationships between phenomena in different fields, for example, to identify in one field the physical location of an entity or process discussed in another, "frequently revealing a part-whole relation between the entities studied in the two fields. It may also identify an entity characterized physically in one field with the same entity characterized functionally in another, or it may locate in one field the cause of an effect recognized in the other field."[147] If the field of inquiry is a machine, we take it apart, see how the parts work, and then attempt to rebuild the machine out of different parts. The parts of the machine are similar to the units of topical specialization of cognitive science. To succeed in building interfield theories, no attempts should be made to make the theory of one discipline work for another, but rather to draw useful connections between the investigations of each discipline. Such connections may generate true interfield theories in which researchers in one discipline can enrich their own understanding of the object of study by discovering how the phenomena are related to other phenomena. Phil Hubbard et al. show how concepts derived from other disciplines have been invoked and developed by geographers, using appropriate case studies to clarify the distinctive contribution of particular ideas.[148]

Interfield theories differ from synthetic theories in that there are no pre-developed overlapping concepts bridging disciplines to address their field of inquiry. Contrary to synthetic theories, they are developed and forged as the research progresses. They are the immediate generalized products of research—theories developed through spontaneous combustion. William Bechtel stresses that interfield theories are developed and accepted because they facilitate inquiry that was previously not possible. Thus the account of interfield theories is designed to describe theories actually produced by scientists. It may therefore offer a more useful tool for understanding cross-disciplinary inquiry. Another distinguishing product of interfield theories is that the end product is typically a theory that span fields rather than two theories related by a derivative relation.[149] The focus on 'fields of inquiry' also makes interfield theory different from the positivist unity of science thesis in that it is not simply a matter of justifying theories and relating them to one another logically.[150]

Darden and Maull did not relate their reasoning on theory development to the phenomenon of scientific overspecialization. The assumption of this book is that the relationships to be identified in interfield theories are likely to be discovered as the disciplines branch into units of topical specialization. As noted above, the trading of disciplines takes place when a unit of topical specialization in one discipline has significant features in common with a related unit in another discipline. Thus, the Gribbin syndrome of *knowing all about nothing* also seems to have a bearing on the formation of interfield theories. In this vein, overspecialization as well as synergetic discourse are helpful measures in the building of interfield theories.

Although theory-building is not a main concern of transdisciplinary research, it is interesting to note that the procedure applied to forging interfield theories is similar to that assumed to be working across the academe/ society boundary in societal problem solving. This procedure is couched in these terms: Meeting the knowledge demands of transdisciplinary research requires "grasping the relevant complexity of the problems, linking abstract scientific reflection with relevant case-specific knowledge, and constituting knowledge with a focus on problem solving for what is perceived to be the common good."[151] Thus transdisciplinarity is not alien to theory. Actually, it combines ideographic concerns about problem solutions with nomothetic expectations of generalised knowledge.[152] Wolfgang Krohn, a noted transdisciplinarian, seemingly has no hesitation about claiming that knowledge production without contributing to theoretical knowledge can hardly be acknowledged as a scientific endeavour.[153] On these grounds, Funtowicz and Ravetz believe that transdisciplinarians will be able to produce a new synthesis of theory and practice[154]—of generalities and specificities. To be accepted as a 'new science', transdisciplinarity may have something to gain from studying the process and content of interfield theory building.

Seminal Theories

There are also those theories that may prove neither synthetic nor interfield theories, i.e. theories that carry a yet to be unleashed potential for pooling clusters of disciplines together, i.e. seminal theories. To illustrate, let us take the examples of game and regime theories. Game theory has been applied in economics, finance, psychology, biology, political science, law, and military planning and strategy, while regime theory has been used in sociology, economy, geography, political science, international law and biology. Since these theories are applicable to all these disciplines on an individual basis, the likelihood is that they may also be used to pool the same disciplines together in a unified theoretical perspective. Some of the bridging, polyvalent potential of regime theory has already been tested. In 1989, Kenneth W. Abbott published an article in which he argued that regime theory holds a rich potential for closing the gap between international relations theory and international law. He even went so far as to call for a 'joint discipline' between law and political science.[155] Since that time, several interdisciplinary projects between the two disciplines have been conducted with remarkable results. In a most thorough stocktaking article on the growing body of interdisciplinary literature, Anne-Marie Slaughter et al. conclude that international relation theory and international law have rediscovered each other and that "a new generation of interdisciplinary scholars has emerged, acknowledging that the disciplines represent different faces of and perspectives on the same empirical and/or intersubject phenomena."[156] On the basis of their assessment of the inventory around which international relations and international law scholars converge, they, in line with Kenneth Abbott, suggest the establishment of 'joint disciplines' between the two domains when it comes to international governance, social construction and liberal agency. They assert: "This focus on substantive themes cross-cuts established paradigms and self-defines disciplinary boundaries and leads to six clusters of research questions on which collaborative research agenda might be built. These six clusters fall under the headings of regime design, process design, discourse on the basis of shared norms, transformation of the constitutive structures of international affairs, government networks and embedded institutionalism."[157] The authors project that the application of regime theory will generate practical as well as theoretical insights, implying that the seminal power of regime theory will add to the inventory of synergies. These extra-theoretically bound synergies are important to distinguish synergetic cross-disciplinarity from simple cross-disciplinarity.

These three categories of theories all possess synthetic power. Synthetic theories are based on a common denominator, *system*, which is a feature of the social as well as the natural world. As such, synthetic theories may be applied to pick up the bits and pieces of complex systems spanning the human-natural divide. Interfield theories also bridge this divide, but offer a supplement to the approach of synthetic theories in that the former assumes no pre-defined theoretical outlook or common denominator at the outset. Theories built within a variety of fields of inquiry develop gradually as the empirical research process evolves and they differ between the fields. Thus interfield theories carry thematic flexibility and complementarity to the integrative potential of the other categories of theories. Seminal theories are thematic and problem-oriented. Since problems rarely come in discipline-shaped forms, these theories are by definition established at the cross-section of disciplines as a common property of fields.[158]

Although some synthetic and seminal theories have been applied to the natural as well as the social sciences, the assumption is that they are restricted in their scope and will only cover limited segments of reality. Their grasp is limited to subject matters, for example to systems, but they do not apply to those parts extending beyond systems, e.g. cognitive psychology. In this way, synthetic, interfield and seminal theories cover only clusters of phenomena, and not all phenomenon. They are not theories of everything. Of relevance in this respect is the fact that there exists a complex tension between the unifying forces of theory and the unitary forces of disciplinary identity.[159]

To gain more in-depth knowledge about the extras of wholes, some of which may never be fully comprehended, there is a need to increase the integrative power and scope of all available means of integration. The rich stock of *discipline-straddling theories* may increase their collective integrative ability by combining applications that will serve as complimentary and supplementary means of convergence, whereas *discipline-based theories* may unleash some unrealized interdisciplinary potential through wider applications.[160] The potential of the latter has to a certain, but limited extent been demonstrated. One example: The *theory of turbulence* is a theory of physics used to study phenomena such as turbulence in plasma and neutral fluids.[161] By way of experiment, it was briefly applied in linguistics to study language change and in the neurosciences to study information processing. The linguist concluded that "the nature of language is antithetical to the notion of turbulence,"[162] whereas the psychiatrist suggested ". . . new directions for research, particular in drug mechanisms and concepts of disorder on the basis of the possibility that deregulation of information turbulence underlies some brain disorders."[163] Thus, applications of discipline-based theories to unfamiliar fields may prove useful to understand integrative limitations as well as straddling potentials. Discipline-based theories may be discipline-straddling without anyone being aware of it.

The Concept of Interdisciplinarity

As indicated in Chapter 1, the notion of interdisciplinarity is an umbrella concept, subsuming all other modes of knowledge integration—transdisciplinarity included. As such, it spans all possible combinations of disciplines working together across disciplinary boundaries and the academia/society divide, ranging from simple multidisciplinarity to all-encompassing cross-disciplinarity. It embraces everything from the simple communication of ideas to the mutual integration of organizing concepts, methodology, procedures, epistemology, terminology, data and organization of research and education in a fairly large field.[164]

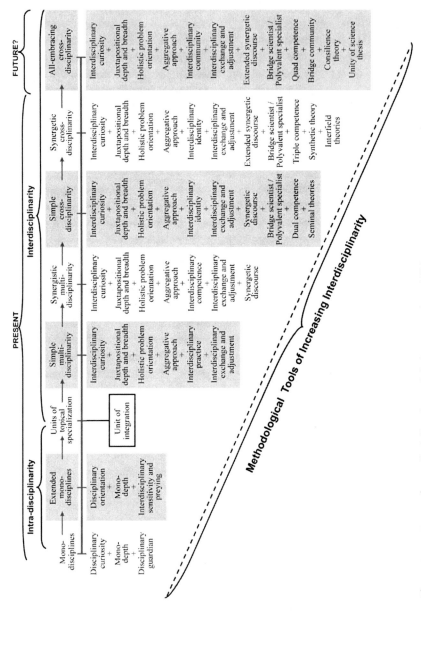

Figure 3.1.   Types of Interdisciplinarity and their Integrative Abilities.

TYPES OF INTERDISCIPLINARITY METHODOLOGICAL ELEMENTS	Monodisciplinarity	Extended monodisciplinarity	Simple multidisciplinarity	Synergetic multidisciplinarity	Simple crossdisciplinarity	Synergetic crossdisciplinarity	All-encompassing crossdisciplinarity
Scientific attitudes	Disciplinary curiosity	Disciplinary Orientation + Interdisciplinary preying	Interdisciplinary curiosity + Holistic problem-orientation + Aggregative approach	Interdisciplinary curiosity + Holistic problem-orientation + Aggregative approach	Interdisciplinary curiosity + Holistic problem-orientation + Aggregative approach	Interdisciplinary curiosity + Holistic problem-orientation + Aggregative approach	Interdisciplinary curiosity + Holistic problem-orientation + Aggregative approach
Scientific skills	Disciplinary guardian	Interdisciplinary sensitivity	Interdisciplinary practice	Interdisciplinary competence	Interdisciplinary identity + Dual competence + Bridge scientist / polyvalent specialist	Interdisciplinary identity + Triple competence + Bridge scientist / polyvalent specialist	Interdisciplinary community + Quadruple competence + Bridge community
Scientific tools	Mono depth	Mono depth	Juxtappositional depth and breadth + Interdisciplinary exchange and adjustment	Juxtappositional depth and breadth + Interdisciplinary exchange and adjustment + Synergetic discourse	Juxtappositional depth and breadth + Interdisciplinary exchange and adjustment + Synergetic discourse + Seminal theory	Juxtappositional depth and breadth + Interdisciplinary exchange and adjustment + Extended synergetic discourse + Synthetic theory + Interfield theories	Juxtappositional depth and breadth + Interdisciplinary exchange and adjustment + Extended synergetic discourse + Consilience theory + Unity of science thesis

Figure 3.2. Types of Interdisciplinarity and their Methodological Expressions.

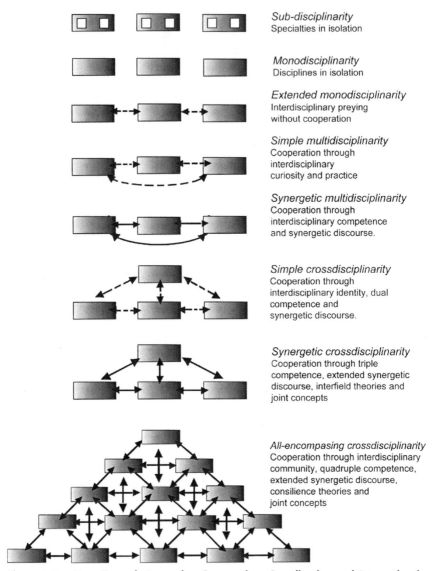

Figure 3.3.   Steps Towards Increasing Cooperation, Coordination and Integration in Research.

FIGURES 3.1, 3.2. AND 3.3: WHAT DO THEY REALLY SAY?

Certain comments must be attached to the interpretation of the scientific tools listed in Figures 3.1, 3.2 and 3.3, and discussed and defined in the text. As visualized in Figure 3.1, there are multiple tools listed under each category of integrative types or modes, from simple multidisciplinarity to all-encompassing cross-disciplinarity. A rigid and even logical interpretation of the three figures would be that all tools listed under each category are necessary to achieve the integrative level of the respective mode. However, this is not necessarily the case. If we compare the tools listed under synergetic multidisciplinarity and simple cross-disciplinarity, they are *similar* in that they both harbour juxtapositional depth and breadth, interdisciplinary exchange and adjustment, and synergetic discourse. At the same time, they *differ* when it comes to seminal theory. Thus, seminal theory is what is needed to increase integration from the level of synergetic multidisciplinarity to the level of simple cross-disciplinarity. In comparing simple cross-disciplinarity with synergetic cross-disciplinarity, three new tools are listed under the latter, i.e. extended synergetic discourse, synthetic theory and interfield theories. Applying one of these tools will suffice to increase integration from the former to the latter level of synthesis. The two others are listed because they are assumed to hold integrative power equal to the one applied, but are only applicable alternatives to the former. The application of more tools will not increase the level of integration beyond the level achieved by just one of them. Together, they present a pick and choose situation, and either one will do.

It may also be possible to apply, for instance, extended synergetic discourse under a different mode than synergetic cross-disciplinarity to achieve simple cross-disciplinarity. If this happens, there is probably no need to apply seminal theory, but the 'risk' is that the level of integration is closer to synergetic cross-disciplinarity than to simple cross-disciplinarity. If so, this is not a problem because the more integration one gets, the more wholeness one gets, and more wholeness is preferable to less wholeness in interdisciplinarity. The tools are to a certain extent interchangeable between the modes. Alternative combinations deviating from those listed in the figures are possible, and even likely. Consequently, the three figures should not be interpreted rigidly. In practice, deviations are likely to occur. What is more: When new and improved methodological tools are developed in the future, the constellations of measures will also be changed. What will persist, however, is that tools of varying integrative power will be used interchangeably, creating new and different modes of synthesis than those depicted in Figures 3.1, 3.2 and 3.3.

When it comes to the other two elements of scientific methods, *skills* and *attitudes*, there are fewer variations and combinations. As seen in Figure 3.2, the same attitudes applies to all the modes—from simple multidisciplinarity to all-encompassing cross-disciplinarity. This is so because the basic difference between the attitudes of monodisciplinarity and interdisciplinarity revolves around exclusiveness and inclusiveness. The former is exclusively preoccupied with disciplinary depth, whereas the latter includes all relevant disciplines to achieve wholeness in depth and breadth. In other words, disciplinary curiosity is up against interdisciplinary curiosity.

When it comes to scientific skills, variability is somewhat more pronounced. This is so because the skills refer to the training and experience of the researcher. The experience may be long or short, and the training, extensive, limited or comprehensive. Thus, competence ranges from disciplinary guardianship to bridge community, and from interdisciplinary sensitivity to interdisciplinary community.

SUMMING UP

Figures 3.1, 3.2 and 3.3 succinctly sum up the main conclusions of this chapter and convey 12 important messages:

- *Interdisciplinarity arises between units of topical specialization and not between entire disciplines;*
- *The hybridization of specialities is the process that facilitates academic interdisciplinarity and synthesis in science and scholarship;*
- *Interdisciplinary methods are composed of three interrelated elements: attitudes, skills and tools;*
- *The methodological measures illustrate the interrelatedness between disciplinary and interdisciplinary research.* Irrespective of which approach is applied, juxtapositional breadth and depth, and holistic problem-orientation are parts of the methodological inventory of interdisciplinary research which also makes use of all the disciplinary tools involved in any one project. Specialization has also produced interdisciplinary sensitivity as a useful multidisciplinary measure to be employed for interdisciplinary purposes. In these respects, the disciplines cannot be replaced as the sole providers of discipline-based interdisciplinary methods. In conceptual terms, the exclusion of disciplines from the concept of interdisciplinarity would leave us with nothing but the prefix 'inter'. Either way, interdisciplinarity without disciplines is theoretically and conceptually a contradiction of terms and a practical absurdity. The one depends on the other. Intradisciplinarity is part of interdisciplinarity, and the monodisciplines are the mother of interdisciplinarity.

- *The methodological measures illustrate the uniqueness of the methodological inventory of interdisciplinary research.* Interdisciplinary research, although dependent on the disciplines, has its own methodological inventory to be applied at the border zones of disciplines. It accounts for holistic problem orientation, interdisciplinary exchange and adjustment, hybridization, juxtapositional depth and breadth, interdisciplinary practice, competence, identity and community, synergetic discourse, extended synergetic discourse, dual, triple and quadruple competence, synthetic, interfield and seminal theories, aggregative approach, bridge communities and adversarial collaboration. These elements of research have been developed in the practice of interdisciplinary research.
- *The methodological measures illustrate the time and process-dependency of interdisciplinary research.* Several of the methodological elements have been developed and improved over time. Interdisciplinary sensitivity developed through many stages to become interdisciplinary identity, and is in due course expected to go through many intermediate stages to become an interdisciplinary community. In a similar manner, synergetic discourse can be further developed to become extended, the double competence will get deeper and broader as time passes and dual competence may develop into a bridge community of interdisciplinary peers. The methodological improvement of interdisciplinary research is thus a product of time and practice. It is dynamic and process related, and realistic expectations are that these, and not yet identified measures will be further developed and multiplied.
- *The methodological measures illustrate the passing of sliding stages between the various categories of interdisciplinarity.* The differences between them are one of degree, not of kind. Dual competence gradually becomes triple competence, interdisciplinary sensitivity becomes interdisciplinary identity and synergetic discourse becomes extended. The differences are gradual as one moves from one stage to the next. New stages can only be identified on the basis of increasing integrative competence.
- *The methodological measures illustrate that the various modes of interdisciplinarity are 'ideal types,' not necessarily real life types.* In practice, all the methodological elements of, for instance, simple cross-disciplinarity may be in place with the exception of interdisciplinary identity, which has been replaced by interdisciplinary competence. Do we then have a sixth category of interdisciplinarity or synergetic multidisciplinarity, or is it still cross-disciplinarity? It seems overly academic to spend time deciding what category of interdisciplinarity we then encounter. In practice, we have what we have of methodological tools, and that will set the stage for the degree of integration for which we may realistically hope. It is also possible to

have dual competence represented in projects that opt to achieve synergetic multidisciplinarity, for example. To a certain extent, the methodological measures identified are interchangeable between the various modes of interdisciplinarity and can be mixed differently in real-life projects. Figures 3.1, 3.2 and 3.3 should therefore be regarded as an 'ideal type' in the tradition of Max Weber. What we label the integrative modes is less important than what we achieve as regards degree of synthesis.

- *The methodological measures show that interdisciplinary activity has developed tools of research to be applied by all sciences, disciplines and specialities.* It has provided something that was not there in the Golden Era of monodisciplinarity. Without having focused on the trading zones of specialities, the sciences as a whole would not have gained what they now may achieve collectively. The science community as a whole has been enriched in its ability to unmask the interactions of multiple compounded realities.
- *The methodological measures show that the interpretational power of interdisciplinarity increases with an increasing number of methodological measures and with the application of those measures that hold the highest integrative potential.*
- *The methodological measures show that there are bridge-building concepts and theories located in the trading zones between specialities of many disciplinary origins* (see Chapter 4).

NOTES

1. Jacoby (65).
2. Fler-og tverrfaglighet i miljø- og utviklingsforskning (02), p. 5.
3. Hansen and Simonsen (04), p. 15.
4. Schumm (98), p. 10.
5. Gribbin(99), p. 1.
6. Autonomization "concerns the way in which a discipline, a profession, a clique, or an 'invisible college' becomes independent and forms its own criteria of evaluation and relevance." See Latour (99), p. 102.
7. Latour (99), p. 102.
8. Weinberg (93), p. 19.
9. Hjort (20), p. 4.
10. Amundsen (02), pp. 6–7.
11. Capra (00), p. 131.
12. Barnes, Bloor and Henry (96), p. 155.
13. Thompson Klein (96), pp. 55–56.
14. Pilet (81), p. 634.
15. Østreng (05), pp.6–9.

16. Thompson Klein (96), p. 43.
17. Thompson Klein (90), p. 43.
18. Dogan (01), pp. 14851–14855.
19. Piaget (70), p. 524.
20. Dogan (01), p. 14854.
21. Dogan (01), p. 14854.
22. Thompson Klein (96), p.53.
23. Thompson Klein (96), p. 237.
24. Thompson Klein (96), pp. 42–46.
25. Skirbekk (94), p. 2.
26. Thompson Klein (90), p. 56.
27. Thompson Klein (90), p. 58.
28. Thompson Klein (96), pp. 113–114.
29. Kockelman (79), p. 131.
30. Greenblatt and Gunn (92), pp. 190–192.
31. Skirbekk (94), p. 2.
32. Wisted and Mathisen (95), p. 34.
33. Brock, Comitas, Sigurd, et al. (86), pp. 34–35.
34. Brock, Comitas, Sigurd, et al. (86), p. 76.
35. Kahneman (03), p. 729.
36. Sørensen (99), p. 125.
37. Herbert and Matthews (04), pp. 9–10.
39. Thompson Klein (96), p. 62.
39. Thompson Klein (90), p. 86.
40. Thompson Klein (90), p. 26.
41. Heisenberg (99), p. 99.
42. Thompson Klein (96), p. 62.
43. Thompson Klein (96), p. 64.
44. Lillehammer (96), pp. 25 and 37.
45. Krag (96), pp. 39–40.
46. Herbert and Matthews (04), p. 9.
47. King, Keohane and Verba (94), pp. 4–5.
48. Ragin (89), p. vii.
49. Ragin (89), p. 101.
50. Ragin (89), p. x.
51. Ragin (89), p. 101.
52. Christiansen (01), p. 67.
53. Interview with Atle Christer Christiansen at the Fridtjof Nansen Institute on 3 April 2002.
54. Krohn (08), pp. 369–383.
55. Anton (00), p. 177.
56. Heisenberg (99), p. 187.
57. Blumberg (95), pp. xi-xii.
58. Hjort (20), p. 129.
59. Walmsley (92), p. 196. See also: De Mey, (92), pp.135–138.

60. Presentation by Professor Ed Miles at the University of Washington on 28 March, 2000.

61. For a discussion of these techniques see Thompson Klein (90), pp.188–196.

62. Berge and Powel (97), p. 19.

63. Thompson Klein (96), p. 167.

64. Fladmark (98/1), p. 228.

65. Thompson Klein (96), pp. 221–222.

66. The Concept of polygamous is cited from Thompson Klein (90), p. 190.

67. Impacts of Climate Variability and Change (99), p. 109.

68. Presentation by Professor Ed Miles at the University of Washington on 28 March, 2000.

69. Expression used by Ziman (00), p. 211.

70. Anton (00), p. 131.

71. Brock, Comitas, Sigurd, et al (86), p. 61.

72. Bammer (97), p. 34.

73. Bammer (97), p. 35.

74. Anton (00), p. 173.

75. Fulwiler (86), pp. 243 and 246 respectively.

76. Fulwiler (86), pp. 238 and 246, respectively.

77. Slaughter, Tulemello and Wood (98), p. 393.

78. Nyhus (97), p. 9.

79. Brock, Comitas, Sigurd et al. (86), p. 68.

80. Interview with Professor Duncan Snidal at the University of Chicago on 25 October 2000.

81. Edwards, Armbruster, Blaikie, et al. (04), p. 98.

82. Evensen (04), pp. 134, 136, 139, and 143.

83. Activity Report 2006, Complex (06), p. 5.

84. Godagkar (07), p. 167.

85. Wiesmann, Biber-Klemm, Grossenbacher-Mansuy, et al. (08), pp. 433–441. The citation is from page 441.

86. Hess (97) p. 178, p. 5.

87. Wisted & Mathisen (95), p. 67.

88. Anton (00), p. 132.

89. Anton (00), p. 135.

90. Anton (00), p. 177.

91. Interview with Professor Ed Miles at the University of Washington on 28 March 2000.

92. Brock, Comitas, Sigurd, et al. (86), p. 60.

93. Brock, Comitas, Sigurd, et al. (86), p. 33.

94. Brock, Comitas, Sigurd, et al. (86), p. 61.

95. The concept of "sociology of science" refers to the influence that older, established scientists have over the careers of younger scientist." See Smolin (07), p 267.

96. Brock, Comitas, Sigurd, et al. (86), p. 76.

97 Brock, Comitas, Sigurd, et al. (86), p. 13.

98. Thompson Klein (96), p. 65.

99. Pagels (88), pp. 41–42.

100. Berge and Powel (97), p. 27.

101. Anton (00), p. 177.

102. Coveney and Highfield (95), p. 88.

103. Coveney and Highfield (95), p. 89.

104. Coveney and Highfield (95), p. 277.

105. Interview with Professor Duncan Snidal at University of Chicago on 10.25.2000.

106. The following councils were merged: *the Royal Norwegian Council for Scientific and Industrial Research, the Norwegian Council for Science and the Humanities, the Agricultural Research Council of Norway, the Norwegian Council for Fisheries Research,* and *the Norwegian Council for Applied Social Science.*

107. Arnold, Kuhlman and Meulen (01), p. iii.

108. Arnold, Kuhlman and Meulen (01), p. 122.

109. For a discussion of this concept see Thompson Klein (90), pp. 65–73.

110. Thompson Klein (90), p. 66.

111. Skirbekk (94), p. 3.

112. Wilson (99), p. 293.

113. Wilber (00), p. 189.

114. Wilber (00), p. x.

115. Clayton (84).

116. Thompson Klein (90), p. 140.

117. Skirbekk (94), p. 3.

118. Blumberg (95), p. ix.

119. Cited from Bruun, Hukkinen, Huutoniemi et al. (05), p. 25.

120. The phrasing "getting to know the enemy" is from Thompson Klein (90), p. 72.

121. Thompson Klein (90), p. 23.

122. Bammer (97), p. 33.

123. Thompson Klein (96), p. 110.

124. Collini (98) p. lvii.

125. Sørensen (99), p. 9.

126. Nelson (00), pp. 25–29.

127. Thompson Klein (90), pp. 131–133.

128. Sørensen (99), pp. 126–127.

129. Berge and Power (97), p. 5.

130. Thompson Klein (90), p. 57.

131. Hubbard, Kitchin, Bartley, et al. (02), p. 4.

132. Piaget (72), pp. 127–139.

133. Hjort (20), p. 165.

134. Thompson Klein (90), p. 29.

135. Peet (98), p. 6.

136. Peet (98), pp. 6–7.

137. Hubbard, Kitchin, Bartley et al. (02), p. 4.

138. Peet (98), p. 5.

139. Peet (98), pp. 6–7.
140. Elser (04).
141. Sterner and Elser (02).
142. Hessen (04).
143. Sterner and Elser (02), p. xvii.
144. Elser, Sterner, Gorokhova, et al. (00), pp. 540–550.
145. Elser, Nagy and Kuang (03), pp. 1112–1120.
146. For a discussion of this aspect see Bechtel (98), pp. 110–117.
147. Bechtel (98), p. 98.
148. Hubbard, Kitchin, Bartley, et al. (02), p. 93.
149. Bechtel (98), p. 99.
150. Bechtel (98), p. 97.
151. Pohl, Kerkhoff, Hirsch Hadorn et al. (08), p. 414
152. Krohn (08), p. 374.
153. Krohn (08), p. 370.
154. Funtowicz and Ravetz (08), p. 367
155. Abbott (89), p. 335.
156. Slaughter, Tulemello and Wood (98), p. 393.
157. Slaughter, Tulemello and Wood (98), pp. 367–397.
158. Østreng (05), pp. 14–15.
159. Hubbard, Kitchin, Bartley et al. (02), p. 62.
160. Østreng (05), p. 15.
161. Truelsen (05), pp. 121–125.
162. Andersen (05), p. 177.
163. Williams (05), p. 169.
164. Berger (72), p. 25.

Chapter Four

The Science War, Positivism, Post-Positivism and Interdisciplinarity

"There has never been a better time for collaboration between scientists and philosophers, especially where they meet in the borderlands between biology, the social sciences and the humanities. We are approaching a new age of synthesis, when the testing of consilience is the greatest of all intellectual challenges.

Edward O. Wilson, 1999[1]

Philosophy is about the rigorous interrogation of the definitions and assumptions of thought. It aims at the logical clarification of thoughts for the purpose of making them clear and giving them sharp boundaries. As such, philosophy is more a method of analysis than a study of any particular issue or empirical subject matter.[2] The philosophy of science has been characterized as the 'theory of theories' and as an important tool of science to structure and provide direction to thought.

If we disregard for a moment the hotly debated *unity of science thesis* of positivism and the vague and ambiguous concept of holism, there is at present no consensual coherent philosophy of interdisciplinary science. This implies that interdisciplinary research is to a large extent deprived of a meta-theoretical perspective from which one can examine and potentially evaluate one's own enterprise.[3] In practical terms, this means that researchers have no coherent logical and methodological analysis of aims, methods, criteria, concepts, laws and theories to lean on, and to help them understand the meaning and logical structure of interdisciplinary science.[4] This is all the more surprising since following a line of thought in philosophy often leads into areas claimed by other disciplines.[5] As a result, academic theorists have long considered the philosophy of science to be interdisciplinary in orientation as well as in substance.[6] The first to advocate philosophy as a unified science was Plato, who contended that philosophers were capable of synthesizing knowledge.[7] This notwithstanding, the fact of the matter is that the community of philosophers has not developed the inherent potential of their own craft to cope logically with all the complexities of synthesis.

Edward O. Wilson reacts to this omission on the part of philosophers by accusing some of them of having actually thrown up their hands and declaring that the borderland between the natural and social sciences is too complex to be mastered by contemporary imagination and may forever lie beyond their reach.[8] In the non-positivist philosophy of science, interdisciplinarity is the misbegotten child, making the synthesis of knowledge potentially more uncertain, volatile, evasive and less structured than that of the other practices of science.

This is not to say that interdisciplinary researchers are completely deprived of meta-theoretical guidance. In the actual practice of research, scientists stick to the rules and means provided by their respective disciplinary philosophies of science derived from their own training and schooling as researchers. The paradox is nonetheless apparent: *interdisciplinary research* is basically conducted on the basis of *disciplinary approaches and philosophies*. As long as research on the tangential points of disciplines has little practical guidance from general theory, interdisciplinary research may not be sufficiently guided by the philosophy of science. This undoubtedly explains why interdisciplinarity has been variously defined, applied and perceived in the course of past century as a methodology, a concept, a process, a way of thinking, a theory, a philosophy, an amorphous belief system, a panacea to resolve pressing societal problems and a reflexive ideology. The philosophical confusion of what interdisciplinarity is all about is prevalent, not least among practical researchers.

The purpose of this chapter is not to muddle too deeply into the unfinished business of philosophers. The objective is rather to extract some of the insights stemming from the conflicting philosophies of the 'science war' and the emerging post-positivist era, and relate them to the working conditions that apply in the trading zones of disciplines. The question posed is simple, but crucial: What prerequisites can be identified from these two incidences that can highlight the philosophical context and ramifications of interdisciplinarity? Or—to what extent does the philosophy of science matter to interdisciplinary research? Which scientific philosophical paradigms seem most appropriate to promote and/or impede interdisciplinary research: positivism or the still unfinished business of post-positivism?

THE SCIENCE WAR

The 'science war' was fought between positivist philosophers who had their heyday in the early half of the 20th century, and their critics, the looming post-positivists, on the question of what criteria constitute a science. As such, this issue concerned the very soul of and raison d'tre for all scientific activity. The two camps used conflicting scientific paradigms as weapons, professional journals as battlefields, auditoriums to win adherents and conferences to form alliances. The war grew utterly bitter and took many casualties in terms of lost careers and the marginalization of ideas.[9] Even courtrooms were considered potential arenas for the resolution of scientific disagreements.[10] If the science war was about anything, it was largely about the power and authority of science: "the fury of the scientific community stems from its recognition that the traditional legitimacy of science is eroding; and the authority of science has haemorrhaged beyond repair."[11] For many decades, the conflict simmered just below the surface. It surfaced in the 1950s when the 'first salvo' was fired, and reached a preliminary climax in the 1970s when positivism was forced to pay off some of its old scores.[12] A new heightened climax saw the light of day in the mid-1990s when a wave of attacks on prominent figures in science studies became particularly intense.[13]

The Science War: Origin and Vein of Thought

The conception of *classical science*, also called dogmatic realism,[14] is based on two premises. The first is related to the *Newtonian model* which was founded on the presupposition that science can achieve absolute knowledge about the universe and that there is no need to make a distinction between the past and the future because the 'divine mechanics' of the universe has coexisted in perpetuity. The situation is stable and not subject to change. The second premise stems from the French philosopher Rene Descartes who articulated the concept of *Cartesian dualism,* which made a fundamental distinction between nature and humans, between tissue and soul, between the physical world and the social/spiritual world. In this way, two distinct premises were defined in respect of the acquisition of knowledge: The *scientific approach*, inspired by Galileo and Newton, aspired to unravel the timeless universal laws of nature, and the *philosophical approach* of Descartes aspired to explain human behaviour and existence through the thought processes of philosophers.

Inspired by Galileo, Newton laid down the principles of the modern scientific method of investigation by comparing theories and models with experiments and observations of the real world. The defining criteria of science during the Newtonian era were narrowed down to three: the ability to *predict,* to unmask *cause-effect relationships* and to discover *universal laws.* The reference was to the earth sciences, which by definition were considered 'real sciences.' All disciplines that claimed to belong to the echelon of real sciences had to prove their claim. On the basis of experimentation and observation, science was supposed to reveal general laws based on quantitative, 'objective' measurements, predicting future outcomes and developments in a precise way. The laws, once unmasked, would provide Man with a useful means of prediction. Their object of study—the universe—was postulated as being tidy and governed by repetitive laws and processes that existed and could be understood without subjective human interpretations—the universe was perceived as a linear system. Any method that was not totally person-independent, interest-neutral and value-free had to be discarded. True knowledge corresponded to the objective world, and the methods applied had to be unbiased and non-interpretive to produce the truth. The ability to bypass the brain by the application of objective methods would distinguish a true scientist from a philosopher who just thought and wrote about his own thinking.[15] "Science does not think,"[16] Heidegger once touted. Thinking introduces subjectivity, which has no role in science.[17] Absolute certainty and objectivity on the part of the 'outside world' could only be achieved if the brain was surgically removed from the rest of the body and put 'in a vat,' dismembered from the subjective filters of the human brain. Only such a disembodied observer looking at the world from *the inside out* and linked to the outside by "nothing but the tenuous connection of the *gaze* will throb in the constant fear of losing reality."[18] The subjective elements of composite humans stood in the way, and had to be overcome to reflect reality in its intrinsic objectivity. The Newtonians left the impression that there were no assumptions in their physics which were not necessitated by the experimental data. This occurred when Newton himself suggested that he had made no hypotheses and that he had deduced his basic concepts and laws from experimental findings.[19] By each generation of true scientists standing on the shoulders of previous generations, science would progress harmoniously and without conflict.

If nature occasionally acted in an unpredictable way, it was because science had not yet uncovered the totality of interacting variables causing the unexpected outcome. To the scientist, the individual observation had no value in itself. Its only significance was as an example of its universal validity. Those disciplines that did not comply with these criteria were either labelled unscientific, pseudo-scientific, junk science or inferior to the natural sciences.[20] In the main, the disciplines of the social sciences and the humanities did not comply with these criteria and became the speculative outcasts of the science community.

The Unity of Science Thesis and Interdisciplinarity

Out of Newtonian imagery grew the *unity of the science thesis*, also called the *theory reduction model*,[21] claiming that all true sciences defined by the three criteria would ultimately be reduced to one 'mother discipline.' Because the universe, of which humans are an integral part, is orderly and governed by exact laws, it could, in the Newtonian perception, be broken down into entities that could be measured and arranged in relation to each other. Here, entities at any given level are composed of entities from lower levels. For example, atoms make up molecules, molecules in turn make up cells, and cells make up organs.[22] The interrelationship between the social and natural sciences was depicted as follows: Societies are made up of people whose brains consists of nerves, which are composed of atoms that can be reassembled into nerves, which make up brains, which guide people, who organize society. From this perspective, any science could be reduced to another, and all the sciences were ultimately reducible to a single discipline, physics and/or mathematics.[23] If the social sciences proved to be real sciences, they could be reduced to psychology; psychology reduced to biology; biology reduced to chemistry; chemistry reduced to physics or a combination of fundamental sciences such as physics and chemistry. Thus, the social sciences and the humanities were reducible to the laws of physics and/or mathematics through a long and complicated chain of reductions.[24] Everything was perceived as interconnected and had its origin in the physical world: "just as *Applied Mathematics* links *Abstract Science* to the *Physical Sciences*, so *Bio-physics* attempts to link the *Physical* and *Biological Sciences* together. Applied Mathematics and Bio-Physics are thus the two links between the three great divisions of science, and only when their work has been fully accomplished, shall we be able to....conceive all scientific formulae, all natural laws, as laws of motion (see Figure 4.1)."[25] The unity of science thesis required a unity of science method and theory because everything had the same origin and was part of the same integrated whole.

In 1892, Karl Pearson offered the following account of the interrelationship between knowledge and method: "The field of science is unlimited; its material is endless, every group of natural phenomena, every phase of social life, every stage of past and present development is material for science. The unity of all sciences consists alone in its method, not in its material. . . . The facts may belong to the past history of mankind, to the social statistics of our great cities, to the atmosphere of the most distant stars, to the digestive organs of a worm, or to the life of a scarcely visible bacillus. It is not the facts themselves which form science, but the method in which they are dealt with."[26] He concluded, "The scientific method is one and the same in all branches, and the method is the method of all logically trained minds."[27] The

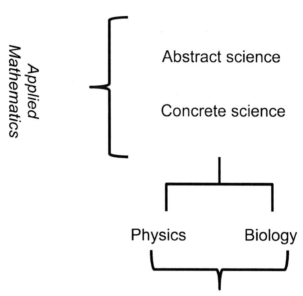

Figure 4.1. The Unity of Sciences by Way of the Formal Sciences and New Science Constructs (Disciplines). *Source: Karl Pearson: The Grammar of Science, Everyman's Library – 939 – J.M. Dent & Sons LTD, London, 1892, p. 333.*

unified method referred to by Pearson was clearly articulated by the French revolutionary Marquis de Cordorcet, who called for the science of man to be mathematical in approach, and to carry the 'torch of quantitative analysis' to illuminate the moral and political sciences.[28] For reasons that remain elusive to scientists and philosophers alike, the correspondence of mathematical theory and experimental data in physics in particular is uncannily close. It is so close as to compel the belief that mathematics is in some deep sense the natural language of science.[29] Sir James Jeans once remarked that in his opinion God—the creator of the Universe—had to be a mathematician.[30] Darwin partly complied, stating that those having an under-standing of "the great leading principles of mathematics . . . seem to have an extra sense."[31] If mathematics is the language of physics, the assumed mother of all disciplines, it had to be the language of all the sciences.

The unity-perspective also implied that theories straddling disciplines could not be contradictory, even if the different sciences were not yet reducible to one discipline. Theories at all levels of science are basically used for the same purpose: *to explain and predict events in nature*, and should therefore be completely commensurate with one another.[32] Theories that seek to characterize phenomena at higher levels are in fact describing the behaviour of entities made up of lower level parts. The task of the unity of science thesis is to show how higher level theories might be related logically to lower level theories. Thus, reductionism has a downward orientation because the goal is to relate higher level theories to lower level ones.[33] In line with this argument, the concepts, methods and theories of physics and/or mathematics were the common property of a unified science. Thus, the modern dream of a final all-encompassing theory—a theory of everything—begins with Isaac Newton.

Unification of knowledge is the heart of interdisciplinarity. No integration theory can be more all-embracing than that of the unity of science thesis. This being the case, one would expect most 'interdisciplinarians' to invest their hearts, souls and minds in support of the realization of positivist ideas. To put it differently, positivism seems the ideal scientific paradigm for interdisciplinarity. Upon closer examination, however, such an expectation is built on loose, sandy ground that is subject to easy erosion.

The Unity of the Science Thesis and the Hierarchical Ordering of the Sciences

It soon proved difficult to realize the unity of science thesis. Steven Weinberg comments: "After all, the world is a complicated place. As scientists learned more about chemistry and light and electricity and heat in the eighteenth and nineteenth centuries, the possibility of an explanation along Newtonian lines must have seemed more and more remote."[34] The disciplines were simply too different, and what was perceived as 'the imperialism of physics' caused the other branches of learning, in particular the humanities, to counteract and close their ranks in opposition to the unity of science thesis. Instead of unity, the schism between the sciences grew deeper over time.

One view purported by the opposition was that reducing the theories of one discipline to those of another would eliminate the first discipline. Although this is not necessarily the case, there was a sense in which a successful theory reduction would undercut the status of the discipline whose theory was reduced, and "thereby place the practitioners of the reduced discipline in the role of applied scientists who are simply working out the applications of more basic principles for scientific domains."[35] In other words, the unity of science thesis implicated a hierarchical ordering of the sciences. The top stratum was shared between the *formal sciences* (mathematics and logic) and the *experimental natural sciences,* and the various disciplines of the natural sciences were ranked in relation to each other on the basis of their deterministic and predictive abilities. Physics, the "kind of gold standard against which weaker and debased forms of science could be measured"[36] ranked above chemistry, chemistry above biology, etc. The humanities were ranked in the lowest strata, while the social sciences were sandwiched in-between the humanities and the natural sciences. The in-between position of the social sciences was due to the fact that the latter was somewhat closer to the natural sciences than to the humanities in approach and methodological orientation, which made them figure as a test ground for quantitative methodology.[37] Equality in scientific value was not perceived to be part of the unity of science thesis. Stratification and subordination were the gauging criteria.

This thesis caused a split within the social sciences. Gradually, influential social scientists became convinced that the future of their science was to adopt the logic of physics and mathematics. In their minds, the behavioural sciences needed to move on and to benefit from the Newtonian paradigm. Advocates of this view were called 'social science positivists' and their new-found perspective was denominated 'naturalism.' Social science naturalists went to great lengths to convince their peers that all the social sciences would become quantitative in approach and method in a few years. Their premise was that social realities can be measured and that it should be possible to achieve universal agreement on their validity. The assumption was that the behavioural sciences would basically be based on figures, graphs, tabulations and quantifications. The time of written accounts and interpretations had long passed. At best, the usefulness of those sources/measures would be drastically diminished, and only be used as supplementary tools to explain the 'mathematics of social life.' Social science *interpretation* would ultimately be replaced by quantitative *explanations.*[38] This methodological conviction also carried overtones of convenience: The quantifications provided rhetorical devices of persuasion by which the scientific authority of their assertions could be reinforced.[39] Graphs and equations became the ultimate means of combating the underdog position of the social sciences, i.e. of being promoted to the status of real science. The naturalists sought the neutrality of the positivist shield.[40]

To the social science naturalists, Newton had won the day, and they were convinced they had won the future. They were convinced that a paradigm shift had occurred for the social sciences. The unity of science thesis was regarded as the dawning of a new era for the soft sciences. From then on, *reliable knowledge*, obtained by objective Newtonian methods, was ranked above *perceptional knowledge,* obtained by the subjective thinking and interpretation of philosophers, social scientists and humanists. Quantitative methods were set off against qualitative methods.

Those rejecting the naturalist view that social realities can be adequately studied by extending the methods and doctrine of physics to human beings and society have come to be called 'anti-naturalists.'[41] The most hard-to-convince of the anti-naturalists counter-attacked and formulated their *disunity of the science thesis*, underlining the differences between the social and natural worlds and the scientific value of multiple approaches in methods, concepts and theories. They underscored the differences rather than the similarities, the split rather than the unity.[42] Their assumption was that the ability of the social sciences to predict future developments on the basis of causalities formulated as scientific laws was limited. Fronts were established and, for a long time, little diplomacy and communication were practiced between them. The emotional intensity of the positions made the parties go their own ways and the anti-naturalists to be in total defiance of any unity of science model.[43]

As observed many times over, war has never been a situation in which people can think subtle thoughts, but it has always offered a license for taking shortcuts, seizing any expedient at hand, and riding roughshod over all the values of discussion and argumentation. The science war was no exception.[44] In retrospect, it can be concluded that by using the natural sciences as a model, the naturalists created expectations that have proved difficult to effectuate, i.e. to draw conclusions of universal relevance, application and validity. One of the expectations was that emulating the natural sciences would improve the ability of the social sciences to predict developments; another was that it would provide better, more reliable governance. Both were based on the anticipation of quantifiable preciseness, but none of them materialized to meet the expectations. What happened was the very opposite.

In the 1970s and 1980s, the social sciences were accused of being exclusive and parochial in approach—of being Eurocentric, gender- and bourgeois-biased. In the eyes of their critics, the social sciences had neglected a whole host of different groups, regions and cultures in their research. Women, the world outside of the western hemisphere and minority groups had not been on the research agenda of western social science.[45] It turned out that parochialism rather than universalism had become the hallmark of western social science. Against this backdrop, there was a need for redefinition and adjustment to strike a reasonable balance between methods of interpretation and quantification.

Today, esteemed physicists admit that the application of the concepts of classical physics to chemistry, for instance, was a mistake. The physicist Fridtjof Capra has no hesitation about declaring that physics has lost its role as a model for other sciences.[46] Being more cautious, Werner Heisenberg claims that nowadays one is less inclined to assume that the concepts of physics, even those of quantum theory, can be applied everywhere in biology or other sciences. "We will, on the contrary, try to keep the doors open for the entrance of new concepts even in those parts of science where the older concepts have been very useful for the understanding of the phenomena."[47] John Searle is convinced that science needs to abandon once and for all the idea that the social sciences are like physics and we are waiting for a

set of Newtonian laws for the mind and society.[48] Others take the criticism in
a different direction, claiming that physics, due to a most restricted area of
contact with other disciplines, has less of an interdisciplinary nature than biology and sociology, for instance.[49] Here, contrary to the positivist position,
physics is conceived to be among the more solitary and secluded of scientific
disciplines, possessing only a most limited capacity to integrate inputs from
other fields. The intellectual future of interdisciplinary studies depends, according to Giles Gunn, on avoiding the temptations to treat the methods of
one field as sufficient to interpret the materials of many, and the materials of
one field as a mere epiphenomena of the subjects of another.[50] Even Karl Popper, a staunch defender of the unity of science thesis, admits that there are significant differences between and within the natural and the social sciences,
and that it may not be obvious what the unified methods are.[51] Thus, Newto-
nianism has lost some, if not all, of its traditional grip on the other sciences.

Cartesian dualism, which has had a tremendous impact on scientific thinking over the centuries, has received a devastating blow from unexpected quarters. The assumption that mind and matter were completely different in their
essence and not could act upon each other made Werner Heisenberg declare
Cartesian dualism to be invalid in modern physics.[52] In the modern parlance,
the mind is *holistic* and *interwoven* with the mechanics and physics of the
body. Paul Davies argues that the brain consists of billions of neutrons,
buzzing away, oblivious to the overall plan. This is the physical, mechanical
world of electrochemical hardware: "On the other hand we have thoughts,
feelings, emotions, volitions and so on. This higher level, holistic mental
world is equally oblivious of the brain cells; we can happily think while being totally unaware of any help from the neutrons. But the fact that the lower
level is ruled by logic need not contradict the fact that the upper mental level
can be illogical and emotional." This is to say that the essential ingredient of
the mind is not the *hardware,* the stuff your brain is made of or the physical
processes that it employs, but the *software,* the organization of information or
the 'programme.' And Davies concludes: "The two-level (or multi-level) description of mind and body is a great improvement on the old idea of dualism
(mind and body as two distinct substances) or materialism (mind does not exist). It is a philosophy that is rapidly gaining ground with the emergence of
what are known as the cognitive sciences: artificial intelligence, computing
science, linguistics, cybernetics and psychology. All these fields of inquiry
are concerned with systems that process information in one way or another,
whether Man or machine."[53] The natural sciences do not simply describe and
explain nature, Werner Heisenberg argues, they are part of the interplay between nature and ourselves. They describe nature as exposed to our method
of questioning: "This was the possibility of which Descartes could not have

thought, but it makes the sharp separation between the world and I impossible."[54] The dualism is artificial and shows the grave defects of the Cartesian partition.[55] Heisenberg admits, however, that the Cartesian partition has penetrated deeply into the human mind during the three centuries since Descartes and that it will take a long time for it to be replaced by a really different attitude towards the problem of reality.

On these grounds, Edward O. Wilson urges the parties of the science war to end a century of misunderstanding. From his vantage point, the science war has run its exhausting course and is no longer anything but an old game turned stale. It is time to call a truce and forge an alliance.[56] The demand is for a new scientific paradigm.

THE GIST OF POST-POSITIVISM AND THE DIFFERENCES IN ITS INTEGRATIVE POTENTIAL

When Thomas Kuhn published his book the *Structure of Scientific Revolutions* in 1960, he launched a significant attack on the core premises of positivism. Kuhn challenged the positivist assumption that science offers a steadily accumulating body of knowledge, claiming that scientific disciplines go through distinct stages and that the nature of research in the discipline varies between stages.[57] Science is history-dependent, as is the progress of science. While the positivists had been preoccupied with what science *logically should be*, the post-positivists focused on the *actual practice of scientists* as documented throughout history. Having criticized the positivists for failing to describe real science, post-positivist philosophers proposed to develop their analyses not from general logical considerations but on the basis of careful examinations of the actual processes of science, particularly as revealed through its history.[58] Here, the message of pure logic is replaced by the message of practice, and the message of wishful thinking, by the message of harsh reality. This is not to say that the logic of positivism cannot be challenged by the logic of post-positivism. If the positivist claim that only testable phenomena can make meaningful and solid knowledge is correct, Mark Blyth reasons, there is an immediate problem: *the statement cannot be tested.* "As such, it is unfalsifiable by its own logic and positivism becomes self-refuting."[59]

Post-positivists acknowledge that social and historical factors do influence the process of science, maintaining that judgments must be pragmatic based on what has been successful science. Some post-positivists even hold the position that their endeavours may give rise to normative judgments about the best ways to pursue scientific inquiry. This redistribution of emphasis is demonstrated not least in post-normal science and transdisciplinarity, which

seeks the integration of knowledge not only within the sciences but also beyond academia—in society. Breaking out of the scientific family in search of
relevant knowledge for solving societal problems makes post-normal science
a radical breach with the traditional conception of what science is and is not.
Here, two aspects of reality are on display. First, the reality of how science is
being conducted, and second, how the reality of non-scientific knowledge impacts on generalised scientific knowledge. Leading post-normal scientists
claim their own practice to be 'a new philosophy of science'[60], but they has-
ten to add that it is in need of further development and refinement.[61] Bruno
Latour argues that scientists have started the process of extracting themselves
from the most blatant defects of positivism, but without yet having found a
better alternative. William Bechtel claims that although the popularity of positivism has declined in recent decades, it continues both to set the agenda for
many ongoing philosophical discussions and to provide a criterion that many
scientists still use to judge what is good science.[62]

Post-positivism is in an initial stage of evolution. This is reflected not least
in the differences of viewpoints between post-positivists when it comes to the
integrative potential of the new emerging paradigm.

Degrees of Interdisciplinarity: Differences
of Post-Positivist Opinions

Robert Hollinger holds that the classical unity of science thesis has few, if any,
adherents anymore.[63] Edward O. Wilson takes issue with the position that the
unity of science thesis is idle and that it has suffered a decisive defeat.[64] He
still holds the deep conviction that the world is orderly and can be explained
by a small number of natural laws, and that the social sciences are intrinsically
compatible with the natural sciences. He even takes one extra step, claiming
that the arts and religion should be unified with all the sciences under the banner of science. In this respect, Wilson is utterly provocative and much more
ambitious in his integration objective than positivists, who disregarded the religions and the arts as not being compatible with the strict methodological requirements of true science. The measure on which to forge unity is, according
to Wilson, *consilience theory*. This theory assumes that all tangible phenomena, from the birth of stars to the workings of social institutions, are based on
material processes that are ultimately reducible to the law of physics. In support of this idea, he argues that humanity is kin to all other life forms by common descent. For centuries, consilience has been the unifying measure of the
natural sciences, and he invokes evolutionary biology and psychology as being the best poised to serve as bridges from the natural sciences to the social
sciences and humanities.[65] What is needed is to transcend the social and natu-

ral demarcations to include biology and psychology, i.e. the hereditary attributes of human nature. This is not to reduce the social sciences or the humanities to the natural sciences, as proposed by the unity of science thesis, but to bridge the natural and social disciplines so that they can work together to explain the indiscriminate whole of elusive man, including the social as well as the genetic origins of behaviour. The goal is to merge the major branches of knowledge under the unified banner of science. Upon presenting the consilience theory, Wilson anticipated that it would cause reactions, not least among the ranks of social scientists and humanists. He was right. After decades of in-fighting in the science community, reviving the core postulate of the unity of science thesis is like waving a red cloth in front of agitated bulls.

John Ziman claims that the tenet of science progressing towards producing a complete and comprehensive scientific world picture, which will constitute the ultimate reality, is beyond human reach. In his mind, scientific knowledge excludes many human aspects, such as moral and aesthetic values, which are just as real as physical data and biological traits. It is a metaphysical goal which is not attainable even in principle. Nor is it attainable in practical terms. Scientific knowledge is always expanding too rapidly to be grasped as a whole.[66] Wholeness will inevitably lag behind disciplinary progress, and will continuously have to be updated without ever being able to present the final unified picture.

Ken Wilbur, himself a leading theorist on integral scientific thinking, agrees with Ziman's basic reasoning. An all-embracing theory is not only "beyond any one human mind; it's that the task itself is inherently undoable: knowledge expands faster than ways to categorize it. The holistic quest is an ever-receding dream, a horizon that constantly retreats as we approach it, a pot of gold at the end of a rainbow that we will never reach."[67] The quest for an all-embracing wholeness is nothing but playing God.

Steven Weinberg expresses conditioned doubts, assuming that if history is any guide at all, it seems reasonable to assume that physicists will find the final laws, for instance, within high-energy physics. In his judgment, scientists are already beginning to catch glimpses of the outlines of a final theory.[68] He admits that it may be centuries away and that it may turn out to be totally different from anything we can imagine at present. Although he believes in ultimately finding those laws, he does not completely rule out the possibility that scientists will find that there is no law at all, or that humans are simply not intelligent enough to discover or to understand the final theory.[69] Whatever the outcome, Weinberg's point is that within sections of reality, if not total reality, science may discover the final laws. Future generations of scientists will decide what that something will be. At present, convergence of disciplines is a proven exercise to make reality a little more whole and a possible avenue to unravel the working of complex systems. Physicists hold out the likelihood that there are fundamental laws about complex systems, and that these laws are new kinds of laws. They are laws of structure and order and scale, and they simply vanish when you focus on the individual constituents of a complex system, just as the psychology of a lynch mob vanishes when you analyse the individual participants.[70] Complex science is thus assumed to have the ability not only to produce empirical insight that is different from the sum of its individual parts, but also new laws that can only be revealed through holistic research.

The philosophical disagreement poses the question of whether it is possible to agree with the assumption that all-embracing integration is impossible and still believe in holism as a viable practice of science. The answer is simple, but crucial: Complexity arises at many levels in nature and society, fashioning pattern within pattern and endless tiers of design, which implies that a little bit of wholeness is better than none at all: "An integral vision offers considerably more wholeness than the slice-and-dice alternatives. We can be more whole, or less whole; more fragmented, or less fragmented; more alienated, or less alienated—an integral vision invites us to do a little more whole, a little less fragmented, in our work, our lives, our destiny."[71] It stems from this, that although the vision of a unified all-embracing theory of complexity may seem unrealistic, the search for common denominators of complex systems is in itself of great scientific significance. A search of this kind can either be called *extended reductionism* or *limited holism,* in that they are both *inclusive*—integrating across disciplinary boundaries—and *exclusive,* restricting the empirical focus to a definite set of defined variables. This perspective is "a leitmotif of the science of complexity as we see it today."[72]

The Newtonian idea of a 'unified science' is fundamentally different from that of unifying a few disciplines.[73] As was discussed in the Chapters 1 and 3, wholeness can be created over a wide range of issue areas. As correctly observed, there may be limits to the capacity of humans for integrative thought, but there is no evidence that they have yet been reached.[74]

Positivism and Post-Positivism on Four Dimensions

Historically, the positivist ordering of the sciences and disciplines in hierarchies turned out to be an obstacle rather than a promoter of synthesis. In John Ziman's pen, the core content of the new era is that there are many different 'sciences—physical, biological, behavioural, social, human, medical, engineering and so on'—but they are all varieties of the same cultural species. The cultural species referred to is *'academic science,'* i.e. basic science in many fields designated as equal in value, ranking and quality.[75] It is only a question of emphasis whether one regards these changes as indicating that, rather than two cultures, there are in fact two hundred and two cultures or that there is fundamentally only one culture.[76] Here, the new thinking has made a full turn: there are no longer superior or inferior sciences and/or disciplines—they are of the same academic breed, but different in content, complexity and orientation.[77] The social sciences and humanities aspire to understand the *language of man*, whereas the natural sciences aspire to read the *language of nature*. The following shifts in the direction of philosophy were taken on four dimensions stressed during the positivist era:

- Nature versus man
- Qualitative versus qualitative methodology
- Laws versus fact-finding sciences—theory-building versus particularistic sciences
- Facts versus values—objectivity versus subjectivity

Nature versus Man

Man is the object of study in the social sciences, the humanities, and also in some of the natural sciences, e.g. biology and medicine. Two perspectives on man's behaviour prevail: The one does not recognize the possibility of human variation (naturalists); the other does not see any limits to mankind's ability to transform itself (anti-naturalists). Both make a valid science of man impossible.[78] Because humans are self-defining animals, the science of man is largely *ex post* understanding, making hard predictions about future behaviour impossible. Changes in his or her self-definition go hand in hand with what man is, such that he has to be understood in different terms.[79] Thus, the focus of the social sciences is on man, who is guided by variations in intentions, self-definition, will, cultural stimulus, values, religion, rationality, emotional feelings and spontaneity. These characteristics are elusive and fluctuate with individual moods, inclinations and characteristics, cultural and political contexts, and time.

Man is exempt from the ironclad laws of ecology that bind other species, and there are few limits on human expansion that our special status and ingenuity cannot overcome. We have been set free to modify the Earth's status.[80] Man has free will, and even though society tries to restrict and control his behaviour by enacting and enforcing laws, man is in principle nature's only unpredictable creature. At best, society has succeeded in restraining man's unpredictability and conditioning him to be predictable within the framework of changing circumstances, yet it has not eliminated his elusive nature and behaviour. As has been noted, if mankind is ruled by natural forces, we cannot develop a theory of human action based on free will.[81]

Man can be studied, but his elusiveness makes some of the social sciences more *fact finding* and *particularistic* in approach (e.g. history) than others (e.g. sociology). The elusiveness of man is what define the boundaries, framework and methods of the social sciences, causing innovation, transformation and change, not only in social life, but also increasingly in the basic laws of nature (global warming, depletion of the ozone layer, etc). Man is, according to the anti-naturalists, different from nature as an object of scientific study. The distinction often referred to in this respect is between *human action* and *bodily movements*. Actions are intentional and performed to achieve a particular purpose, while physical movements are mechanic and therefore belong to the sphere of nature. In other words, man has an intention beyond the physical occurrence of a move-ment.[82] The emphasis on predictive explanation in physics seems impossible for many of the phenomena studied by the social sciences. The elusiveness of man implies that qualitative methods based on interpretation are among the more im-portant scientific tools to grasp the full spectre of behaviour.[83]

Qualitative versus Quantitative Methodology

Since social and natural realities differ in substance, the methods applied to understand them also must differ. No carpenter would use a hammer to saw lumber, or for that matter, a saw to pound a nail in the wall as long as both tools are readily available for both functions. Success in workmanship as well as in science depends on choosing the right tools for the job. Man has to be studied using methods able to catch the elusiveness of behaviour, and nature has to be studied by methods able to grasp the content of conditioned regularities. Un-derstanding behaviour is more like reading a text than performing experi-ments, and vice versa. The social sciences and for that matter the humanities cannot be measured by the standards of a science of verification. An important difference between these sciences is to be found in the nature of explanation.[84] In pure contempt for and opposition to the unity of science thesis, the question of method is still rarely raised within the field of literary studies. The view-point held is that literary work belongs to the realm of meaning, not to the realm of facts, and that the reading of a text requires not just the capacity to make observations in the text but also the capacity to respond to it as a human being. However, as has been argued by post-positivist scholars of literature, one can share the view that the unity of science ideology is seriously flawed and still think that the question of method ought to be at the centre of the meta-discussions and self-reflection of literary scholars.[85] In other words, forget about the negative connotations of positivism and take on board its preoccu-pation with the use of methods in analysis.

Any attempt to transfer 'superior' methods from a 'model science' to an 'in-ferior science' for the sake of making the latter more scientific is nothing but derailed and misconceived science. Gary King et al. flatly denounce any such positivist claim, demonstrating that neither of the two traditions (quantitative and qualitative) is superior to the other.[86] In their reasoning, the difference between them is one of style and technique rather than of kind. All good research can be understood, indeed is best understood, as deriving from the same underlying logic of inference.[87] To be scientific is to fulfil the purpose of science, i.e. to obtain knowledge made public by the consensus of the collective judgment of the bulk of one's disciplinary peers. Thus, the application of quantitative methods, inspired by, and borrowed from, the natural and formal sciences, makes the social sciences better equipped, sophisticated, suitable and able to fulfil their overall purpose. Instead of rejecting each other's tools and approaches, as the disunity of science proponents did, the call is now for combining methods to optimize science as a whole.[88] As argued by Steven Weinberg, the formal science of mathematics is never the explanation of anything—it is only the means by which scientists use one set of facts to explain another, and the language in which scientists express their explanations.[89] What comes to mind is J.E. Cohen's famous statement about the relationship between mathematics and the Life sciences: "Mathematics is Biology's next microscope, only better; Biology is Mathematics' next physics, only better."[90]

The increasing focus on the rich, unexplored potentials of the individual sciences at the expense of the unity of science thesis may result in a revived appreciation of the value of the diversity of sciences, disciplines and methods. As has rightly been pointed out, there is no reason to suppose that convergence of scientific explanations must lead to convergence of scientific methods.[91] In an interdisciplinary perspective, all disciplines involved are on an equal footing and attract the same kind of respect across the whole range of sciences. The working principle between them is that of *division of labour* through *cooperation* rather than subordination of the one to the other. Thus researchers involved in interdisciplinary projects, although specialists in their respective fields, are discipline-straddlers in attitude. As a consequence, interdisciplinary research has to be defined through the common lenses of *all* the involved disciplines or, as suggested by Stanley A. Schuum, by an all-embracing *scientific approach* rather than and all-embracing method and theory.

Stanley A. Schuum contends that because of the diverse and multiple problems faced by different disciplines, we must instead consider a scientific approach which is less involved with procedure and method, and more with *the state of the mind* of the investigator.[92] Schuum postulates that the approach has five characteristics that are applicable to all sciences:

1. *The approach should be systematic.* Here, the requirement is to proceed in an orderly fashion to collect relevant data.

2. *The approach should be characterized by a lack of biases.* The work must be carried out in an objective manner. According to Schuum, this is not to say that a researcher should be criticized for the biases of his profession, but that he or she should be aware of them and make all possible efforts to correct and minimize their subjective impact on the result.[93]
3. *The approach should be characterized by absolute intellectual honesty.* Here, no attempt to fabricate and manipulate the data is allowed.
4. *The approach is based on laws and therefore hypotheses can and should be tested.* Metaphysical explanations are not acceptable.
5. *The goal of the approach is to develop broad generalizations in order to classify and express relationships with definiteness and precision.* This involves the identification of cause and effect, if appropriate.

Schuum concludes that because of the great number of types of science it is only the scientific approach that will be applicable to any science, whereas a strictly defined scientific method will not.[94] As discussed in Chapter 3, applying a simple scientific method to study complex systems invites failure. Interdisciplinarity equals a variety of methods. In this perspective, Schuum's approach stands out as the only unifying scientific device applicable to all disciplines, whereas a method is to a certain extent science-specific and only as powerful as the objectivity of the individual using it. The approach provides unification through self-correction in both quantitative and qualitative research, and helps develop a rational scientific approach to complex systems.[95] Stanley Schuum goes on to point out that errors and fraud will eventually be revealed with considerable satisfaction by a colleague or former student. Consequently, it is in the self-interest of the scientist to apply the scientific approach. Self-correction ensures the advance of science, and it should enforce objectivity and honesty in research. Thus, the unification of the sciences is not provided by the *single method conception* of positivism, but rather by the *single approach conception* of post-positivist thinking. It should be noted, however, that Schuum's approach is tailor-made for academic interdisciplinarity, and not fully for transdisciplinarity in that points 4 and 5 above are coined in the language of science and are not based on the specificities of stakeholder concerns.

Law versus Fact-Finding Sciences—Theory-building
versus Particularistic Sciences

In wars, truth is always the first casualty. The science war was no exception. At the height of the battle, the truth about the state of affairs within and between the sciences was distorted and exaggerated beyond reasonable recognition. One of the exaggerations was that the natural sciences were exclusively concerned with law-finding and theory-building, whereas the social sciences were exclusively fact finding and particularistic in approach. According to the post-positivists, the truth is to be found somewhere in between these two extreme positions. The division is not only *between* the sciences but also *within* them. In the natural sciences, the division occurs between biology and natural history, for example, and in the social sciences between sociology and history. No one would seriously dispute the fact that sciences such as chemistry and physics are *nomothetic* attempting to discover regularities and universal laws which, in principle, should be applicable everywhere at all times, while sciences such as natural history and human history are *ideographic* sciences concerned with local, particularistic events of a fact-finding nature. This being said, there are examples on both sides of the social/natural science divide that do not comply with this strict dichotomy. First, there are some predominantly fact-finding natural sciences such as geology and palaeontology. Then there are some law-finding social sciences such as political science, sociology, economics and linguistics. What is more, within certain disciplines, e.g. geography, there are those who argue convincingly that geography is a typical ideographic science, whereas others, equally convincing, claim that it is nomothetic.[96] The crux of the matter is that there are no clear or absolute patterns subdividing the sciences on the basis of the two concepts.

Yet what the dichotomy between nomothetic and ideographic sciences overlooks is that behind every fact there is a causal structure waiting to be explored and interpreted.[97] In this vein of thought, the production of facts is essential to constructing general abstractions, whereas theories are built on the collection of facts directing attention to some particulars rather than others. Social scientists study particulars, but good social science attempts to go beyond the particulars to build more general knowledge. Generalizations do not eliminate the importance of the particulars on which generalizations must rest. In fact, the very purpose of moving from the particular to the general is to improve our understanding of both.[98] The two elements interact and ideographic research is the backbone of nomothetic research in both the social and the natural worlds. However, it should be added that discovering patterns in the social world is extremely complicated because the number of variables is so huge. The social science's method of seeing social patterns repeating themselves is by constantly comparing research material from different cases.[99] As pointed out by John Hughes and Wes Sharrock, had the social sciences been measured against one or another of the natural sciences apart from physics, e.g. astronomy, botany or geology, rather than against the abstract picture of 'science' painted by philosophers of science, then the status of the social sciences *as sciences* might have seemed a good deal less problematic.[100] This raises the question: What are the criteria that constitute a law/theory? Are laws different in the social and natural spheres? If so, what are the differences? If not, what are the differences between the social and natural sciences?

The concept of a law refers to *statements that express regularities*.[101] The only reason we have for believing in the law of causation is that we observe certain regularities or sequences. We observe that, under certain conditions, A is always followed by B. We call A the cause and B the effect, and the sequence A-B becomes a causal law.[102] As a comment to this sequence, William T. Stace argues that neither scientist nor philosopher knows why anything happens or can explain anything: Scientific laws do nothing except state the brute fact that 'when A happens, B always follows.[103] On this basis, he concludes that the concept of causation, however interpreted, is invalid.[104] Bruno Latour agrees, asserting that one should abstain from using causality to explain anything: "Nowhere in this universe does one find a cause, a compulsory movement that permits one to sum up an event in order to explain its emergence."[105] According to Stace, this reasoning is based on the premise that even though scientists talk about electrons, protons, neutrons and atoms, they have not really observed them in reality. These phenomena are basically conceptions, from which one cannot validly infer that they exist. The fact that the object under study cannot be seen means that its exact nature can only be guessed,[106] and that for instance gravitation is not a thing, but a mathematical formulae existing only in the heads of mathematicians. Rudolf Carnap calls these regularities, *theoretical laws*, which are laws that are not based on observables and which cannot be measured in simple direct ways. They are hypothetical and abstract. Consequently, neither Newton's nor Einstein's laws are strictly speaking laws of gravitation, or *empirical laws*, confirmed directly by empirical observations.[107] They are only laws of moving celestial bodies, which tells us *how* these bodies will move, not *why* they move as they do. The concepts and formulae of scientists are nothing but mental constructions and fictions, ingeniously worked out by the human mind to help humankind organize its experiences.[108] What we observe is nothing but courses of events allowing us, on the basis of our experience, to conclude that in all likelihood, we are witnessing cause-effect relationships.[109] Albert Einstein made the same deduction. There is no logical path to the discovery of universal laws. They can only be grasped by intuition, based on what may be translated to be 'emotive' experience, or 'informed imagination' and sensation. Natural scientists of the present refute the most simplistic view of causality. In quantum physics:

"The discovery that individual events are reducibly random is probably one of the most significant findings of the twentieth century. . . . But for individual events in quantum physics, not only do we not know the cause, there is no cause.

The instant when a radioactive atom decays, or the path taken by a photon behind a half-silvered beam-splitter are objectively random. There is nothing in the Universe that determines the way an individual event will happen. Since individual events may very well have macroscopic consequences, including a specific mutation in our genetic code, *the Universe is fundamentally unpredictable and open, not causally closed.*"[110]

As has been observed, there is a large variation in the level, or even possibility, of causal understanding. Whereas causality in the natural sciences is connected to a mechanistic understanding of the underlying principles, causal understanding in the social sciences is a far more difficult issue and often very controversial. For example, while a malfunctioning heart can be studied in a laboratory with large numbers of specific tests, the same cannot be done with a dysfunctional marriage. "In fact, scientific knowledge does not represent a consistent and non-contradictory view of the world in any complete sense. The scientific world view may be a success, but its power of explanation is still fundamentally limited."[111]

This is why old theories, although held in high regard at their respective times, are often repudiated as false, and consequently replaced or supplemented. The theory of relativity has, for instance, been superseded by the so called general theory of relativity, and few physicists doubt that, in turn, the general theory in due time will be improved upon. This development finds expression in Thomas Kuhn's historic scheme of scientific revolutions in which new theories are incommensurate with their predecessors insofar as they use incompatible concepts to describe the world and make inconsistent claims about it.

Nancy Cartwright takes this reasoning one step further arguing that *true laws* are scarce and that many phenomena which have perfectly good scientific explanations are not covered by any laws, but by generalizations that hold only under special conditions. To qualify as real laws, scientific theories based on laws must tell us both what is true in nature and how we should explain it. Cartwright holds the position that mankind does not know whether it exists in a tidy or untidy universe, i.e. if we are governed by laws and to what extent and degree. In physics, she says, there are no laws without exceptions. Every theory we have proposed in physics is known to be deficient in specific and detailed ways.[112] This, in her mind, is also true for every precise quantitative law within physics theory. Even the best natural laws are known to fail, and most laws in the physical world hold only under special circumstances and should rather be labelled generalizations or *ceteris paribus laws* that, by implication, are not true laws. As strange as it may sound, it can be demonstrated that, for instance, three planets interacting by gravitational forces can exhibit regular motion for extended periods of time, until one of them suddenly tears loose and can leave the system: "Any prediction made on past observations suggesting regular motion for all future times will be in error. For systems like these, chaos is manifested by lacking predictability: we can observe and describe the system at any instant, but we can only do so with a finite accuracy. Even the slightest error or inaccuracy of our description of a state of the system at a given reference time can result in a prediction of future state which is in grave error, even though we purport to know the basic laws of motion exactly.....the basic models are still so as to ensure that identical conditions lead to identical results."[113] As suggested by Stephen Hawking, our understanding does not advance just by slow and steady building on previous work. "Sometimes as with Copernicus and Einstein, we have to make the intellectual leap to a new world picture. Maybe Newton should have said 'I used the shoulders of Giants as a springboard."[114] Thus, within the natural sciences, there is a need to make a sharp distinction between *ceteris paribus laws*, which are generalizations and frequent in number, and *true laws* which are real, but few in numbers.

Nancy Cartwright takes this reasoning a step further, claiming that natural objects are much like the people in societies. Their behaviour is constrained by some specific laws and by a handful of general principles, but is not determined in detail, even statistically. What happens in the natural and social worlds on most occasions "is dictated by no laws at all."[115] The implications being that the differences between the sciences are those of degree, rather than of kind, and they apply both *within* and *between* fields. The natural sciences are not alone in seeking to produce general laws, replicable results, and cumulative knowledge.[116]

Facts versus Values—Objectivity versus Subjectivity

The *objectivity-subjectivity*, *fact-value* dichotomy has in the course of the past 30 years come under increasing scrutiny. In positivist thinking, the concept of objectivity refers to the discrepancy between what is *observable* and what is *appraisable*, what is *description* and what is *emotion*. Positivists believe the words *is* and *ought* belong to different worlds, so that sentences which are constructed with *is* usually have a verifiable meaning, and those constructed with *ought* have not.[117] In the positivist view, impersonal description of observables make science objective, whereas human interpretations of texts filtered through layers of emotions and values make it subjective. Under real life conditions, is it possible to make value-free science, and, if so, is this dichotomy valid as a demarcation line to distinguish between the natural and social sciences?

Part of this question was answered long before post-positivism was launched as a competing scientific paradigm to positivism. At the height of logical positivism in the first half of last century, a few secluded voices were heard claiming that even "empiricism naturally operates with human thoughts," and that thoughts make their subjective marks on science when, for instance, scientific problems are being formulated and when its results are being discussed. By the same token, it was openly acknowledged that the mistrust of positivists to the constructive powers of the mind had often proved an obstacle to scientific progress.[118] The method of *judgment* had, according to Johan Hjort, been used in the applied natural science in the early part of the 1900s in spite of its subjective inclination. Although subjectivity causes difficulty and in many cases uncertainty, making special demands upon the accuracy of the observations and upon the planning of the work, this should not, in Hjort's mind, be "any ground for rejecting in principle the task of making observations in such a way. . . . Least of all should dogmatic (materialistic) science arrogate to itself the right to condemning such methods as being opposed to the great general conception of scientific method."[119] The few rebel 'post-positivist voices' of the early era of logical positivism came up with a unified position: *all sciences are value-laden and subjective, although too different degrees and in different manners.* In light of the flaring emotions of the science war, it may come as a surprise that this early position no longer seems overly controversial among philosophers of science.

This conclusion should not be confused with the radical post-modernist position which denies the existence of a universal human nature and contends that the truth is absolutely relative and personal. Post-modernism, which tends to provoke enthusiasm and condemnation in equal measure, is a theory of absolute subjectivity, excluding any sort of objectivity. There is simply no truth outside of interpretation. Modern science in all branches is based on the presumption that certain regularities and patterns of human and natural behaviour exist, and that these 'laws' can be unmasked by scientific procedures and methods. If all is in flux, as pre-supposed by the most radical post-modernists, nothing can be discovered by scientific method short of the flux itself. Post-modernists consider science a 'social construct,' no more objective than stories, novels and folk tales. The German social theorist Wolf Lepenie's claims that one must no longer give the impression that science represents a faithful reflection of reality. In his mind, science is nothing but a cultural system exhibiting an alienated interest-determined image of reality specific to a definite time and place.[120] Other post-modernists call for a blurring of the genres, implying that physics and history are just different types of texts or forms of writing.[121] In essence, post-modernism is based on the notion that there is no single form of knowledge that is necessarily superior or dominant to another, reasoning that no voice should therefore be excluded from the dialogue. According to Phil Hubbard et al., in post-modern approaches, organized science is replaced by "post-science which acknowledges the position of scientists as agents and participants in the creation of knowledge rather than a neutral observer of given 'truth.' Postmodernism thus offers 'readings' not 'observations,' 'interpretations' not 'findings,' seeking intertextual relations rather than causality."[122] The implication of this is that scientific knowledge is no different from practical and philosophical knowledge and knowledge based in mystique, myths and religion. The four modes of knowledge-seeking are equally good or equally bad (see Chapter 1). In any event, absolute subjectivity rules.

In this way, radical post-modernism is destructive for many of the assumptions that give modern science a pride of place in our world today.[123] This is why people like Ken Wilber state that postmodernism, previously the 'megahip' item, is now met with a slow yawn and casual scorn: "postmodernism is *so* yesterday (and isn't that ironic?)"[124]

The radical version of post-modernism is not subscribed to in this book, which is based on the post-positivist notion of public knowledge, accepting both the existence of subjectivity and objectivity in the body of scientific knowledge. From this post-positivist position, it seems logical, however, that all aspects of post-modernism should not automatically be discarded. There can be little doubt that many of the key ideas associated with post-modernism have become influential, often even for those who are sceptical of the perspective as a whole.[125] The post-modernist tenet that culture is a source of subjectivity is agreeable to most post-positivists. The schism occurs between the two schools of thought when post-modernists go to the extreme, claiming that science is nothing but subjectivity lacking the ability ever to reflect objective reality. Most will agree that the Earth is round (as an objective scientific fact) without refuting the possibility that there may be elements in the shape that are still misconstrued by scientists (in subjective terms). At the same time, there may be cultures that insist that the Earth is flat (in cultural terms) without rocking the boat of scientific insight. In this connection, interdisciplinarity stands out as a vehicle to sort out, reconcile and merge parts of conflicting philosophies and theories—a supplementary device to avoid and reduce theoretical conflicts and philosophical extremism. The core idea of transdisciplinarity is to traverse the boundary between science and society for the purpose of making practical use of their comparative advantages in knowledge production and problem resolution.

Post-normal science embraces the premise that science does not deliver certainty, nor is it free of values, personal judgements or institutional agendas. The faith in the truth and objectivity of science is overthrown, according to Funtowicz and Ravetz.[126] Despite this, they do not agree with post-modernists. Post-normal science takes a position between traditional 'objective' science and the all-embracing 'subjectivity' of post modernism, claiming that: ". . . the demise of the long-standing dogmatism of science need not lead to postmodern anarchy of nihilism or of relativism about facts, values and reality. We can grant that scientific knowledge, like any other image of reality, simultaneously reveals, distorts and conceals. The great philosophical challenge of our time is to comprehend apparently contradictory, but actually complementary, aspects of knowledge. With a holistic, systems conception of knowledge itself, it is possible to undertake the reconstruction of our philosophy of scientific knowledge, with its varied dimensions of knowledge, power and experience." In this way post-normal science is a special branch or version of post positivist thinking, designating holism as a means to reconstruct a new philosophy of scientific knowledge.

Post-positivists do not go to the extreme, as do the post-modernists. To the former, it is objectivity in the positivist version of the concept that is impossible. Søren Kjørup points out that a description will always be selective and will always contain elements of an appraising and emotive character, i.e. it will necessarily always contain traces of the 'subject' who is writing, and will always have 'subjective' traits. However, this is not a problem for science, in his mind: *Research is a human activity, and this is why we find traces of humans in its results.* Or, in the words of Lee Smolin, "Science . . . is the way it is because of the way nature is—and because of the way we are."[127] Physics and chemistry, "mathematics and logic, bears the fingerprints of their distinctive cultural creators no less than do anthropology and history."[128] In this version of the concept, objectivity is not in direct opposition to subjectivity, but to deliberate distortion of facts and results.[129]

According to Heisenberg, we can never speak about nature without at the same time speaking about ourselves. The implication is that a shift in thinking has taken place from *objective science* to *epistemic science* depending on an understanding of the process of knowledge.[130] Thus, what constitutes public knowledge in one epoch is not necessarily that of another. Scientific knowledge is by its very nature "wrong, always wrong, because the world is too remote from ordinary experience to be merely imagined."[131] Thus, public knowledge conceived as *the* truth is often subjective and time-dependent. The concepts of science are neither absolute nor everlasting.[132] Building knowledge and truth is a never-ending process. Science stands out as an endless frontier.[133] Besides, research data are nothing but samples of reality, based on, and filtered through, the mental images engulfed in the theoretical models of their time. The basis of the samples is historically constructed, and will always change with the changes of the world.[134] Scientists, like the rest of humankind, have roots in certain social settings and, as such, carry with them assumptions, values, emotions and prejudices which directly or indirectly colour their interpretation of the world under study, be it social, humanistic or natural. Thus, the possibility to make 'impersonal descriptions of observable phenomena' is nothing but an ideal never to be achieved in its entirety under real life conditions. In light of quantum theory, the suggestion has been made that it is impossible to make a distinction between reality and our knowledge of reality, between (objective) reality and (subjective) information. This is so because we cannot refer to reality without using the information we have about it, suggesting that the two concepts in a deep sense are indistinguishable.[135] It derives from this that there will always be research that will make the established truth more truthful, and this will go on in perpetuity, and relates to all branches of learning. According to Werner Heisenberg, insistence on the postulate of complete logical clarification would make science impossible: "We are reminded here by modern physics of the old wisdom that the one who insists on never uttering an error must remain silent."[136]

In what way do biases make their way into the research process and colour the results to reflect the idiosyncrasies of the scientist? What values that the researcher cannot correct himself are rooted directly or indirectly, intentionally or unintentionally, in the research process?

Values Affecting Research

The term 'value' is elastic indeed, and many political, moral, social and religious beliefs may be accommodated under the concept. To make a definitional shortcut, we will resort to the distinctions made by Ernan McMullin: *emotive values*, *characteristic values, epistemic values* and *pragmatic values*.

- *Emotive values* refer to differences in attitude and behavioural responses between humans. These values are highly subjective in that they build on and correspond to such features in the individual as attractions, emotions and feelings. In many respects, these expressions are alien to the work of the sciences.[137] Emotive values are generally a threat to the integrity of all sciences, but may find their way more easily into the social sciences than into the natural sciences because the former is more reliant on interpretation as a method of science. The quality of interpretation depends on the strength of our emotive filters. As a rule, 'strong filters' will actively seek to filter out more of the most strongly felt internalized values, whereas a 'relaxed filter' will have many emotive values to sift through and colour interpretation.

- *Characteristic values* focus on the properties of the object of study. These properties may either be *objective*, in terms of being real and measurable, or *subjective* in terms of being assessed against other properties of the same object. An example: Speed is a desirable characteristic in wild antelope because it aids their survival. This is an objective characteristic of the properties of the animal. If, however, we were asked to judge whether the speed characteristic is more of less valuable to the antelope than other characteristics, the assessment will be subjective and less measurable. A distinction should therefore be made between *value characteristic* and the *judgment of value characteristics*. Both find their way into the sciences, but it is possible to maintain a distinction between what is factual and what is mixed with judgment.

- *Epistemic values* refer to those values that are implicit in scientific practice, and that have been shaped through experience over many centuries. McMullin labels them epistemic because they are presumed to promote the truthlike nature of science, and, if pursued, will help toward the attainment of certain objective knowledge about the world we seek to understand. These are the kind of values that "permeate the work of science as a whole, from the decision to allow a particular experimental result to count as 'basic' or 'accepted,' to the decision not to seek an alternative to a theory which so far has proved satisfactory."[138] These values have been characterized as internal to the scientific process, referring to the logical and moral status of those research norms that are necessary to make science a truth-pursuing activity.[139] To a large extent, epistemic values define and constitute what is normal science within the ruling scientific paradigm.

-The fourth value is called *pragmatic* rather than epistemic, even though the borderline between them may be hard to draw exactly. Pragmatic values are derived from the finiteness of the time and/or resources available to conduct research. It is essential to the process of science that pragmatic decisions are made upon the temporary suspension of further testing of a theory, for example.[140] Let us see how these categories of values are making their way into the scientific process.

John Ziman states that science stands in the region where the intellectual, the psychological and the sociological coordinates of man interact.[141] In other words, because science is conducted by internal and external forces, the elusiveness of man is an unavoidable object of all scientific activity. No one can escape his quite human self, not even researchers trained to correct for their own subjectivity by applying scientific methods and procedures. The mere search for knowledge expresses a value, and thus the distinctions between reliable and unreliable knowledge claims, between good and bad methods, are partly normative judgments.[142] Any decision to do research is thus based in an emotive value.

It is now public knowledge that even laws and theories shape, constrain and colour observations. Theories and laws direct the vision and restrict the focus of the involved scientists. Scientists are often so blinded by their success in building theories and discovering laws that their brains, psyches and peer relationships are, at best, restricted, at worse, they prohibit them from taking account of anomalies, i.e. those observations that raise doubts about the validity of the theory or law they are exploring. Karl Popper's statements that "no beliefs are immune from error,"[143] and that "all theories are born refuted,"[144] should act as constant correctives for any nomothetic science. A researcher has no other way to know whether one additional observation will verify or falsify his or her law/theory, than by continuing to make observations indefinitely.[145] Scientific theories are constructed specifically to be blown apart if proved wrong and, if so destined, the sooner the better.[146] In practice, the inclination of many researchers is rather to stop making fresh observations based on the belief or hope that new observations will only confirm what has already been established as law, or they are forced to do so due to lack of time and funding. In both cases we are talking of pragmatic values which apply to all branches of science.

Richard Rudner addresses the epistemic value of research, contending that value judgments are essentially involved in the procedures and methodology of science. To illustrate his point, he refers to what has become the everyday experience of researchers: *no scientific hypothesis is ever completely verified.* When considering a hypothesis, a scientist must decide whether the evidence is sufficiently strong or the probability is sufficiently high to warrant accepting the hypothesis.[147] Some elements of falsification will always be present when a hypothesis is tested. The elements of falsification will always present to question the correctness of the 'objective' conclusions drawn. This requires that value decisions be made: How much verification is sufficient to accept the anomalies that could potentially disprove the hypothesis? The scientist has no objective criteria for making an impersonal decision on such a question. He has to rely on his professional judgment to weigh the elements of falsification against the elements of verification, and then make a decision based on value judgments often established in previous practice without any rigid theoretical justification. Here, epistemic values are accompanied by pragmatic as well as characteristic values involved in deciding on the methodology of research. For the scientist to close his eyes to the fact that even scientific method intrinsically requires value judgments, is to leave an essential aspect of scientific method out of the equation: "What is called for . . . . is nothing less than a radical reworking of the ideal of scientific objectivity. The slightly juvenile conception of the cold-blooded, emotionless, impersonal, passive scientist mirroring the world perfectly in the highly polished lenses of his steel-rimmed glasses—this stereotype—is no longer, if it ever was, adequate."[148]

The natural sciences rely on interpretations, as do the text-dependent social sciences and humanities. Meteorology is an illustrative example of the blend of objective observations and interpretations. Despite belonging to the realm of the hard sciences, no one watching a weather forecast on TV would ever dispute the fact that meteorologists' interpretations are often a far cry from how the weather actually turns out to be. Weather forecasting is often considered to be as much an art as a science.[149] This is indicative of two factors: First, certain natural sciences rely heavily on interpretation as a scientific method and, second, the number of interacting variables and the complexity of their interconnectedness make interpretation a difficult endeavour in both the social and natural sciences alike. This fact should be taken advantage of to form an alliance between the natural and social sciences to increase their joint interpretational power, in the opinion of Edward O. Wilson.[150] None of the branches of learning is value-independent or free of subjective judgments and biases. In Peter Killeen's words: "Bias does not discredit truth; bias complements truth."[151]

This does not mean that all sciences are equally vulnerable to subjectivity. The humanities and social sciences seem more susceptible than the natural sciences to the influence of most of the categories of values discussed above.[152] Again, however, the difference is one of degree rather than of kind. The social and natural sciences are more similar when it comes to subjectivity and objectivity than the most ardent naturalists will ever admit or accept. One of the important achievements of the looming post-positivism has been to show that objectivity in research is blurred by subjective judgment and that the truth is relative, time-dependent and constituted in the consensus among disciplinary peers. The ideological and value-laden nature of all sciences has been proven beyond doubt. Bruno Latour puts it bluntly: "Yes, we have lost the world. Yes, we are forever prisoners of language. No, we will never regain certainty. No, we will never get beyond our biases. Yes, we will forever be stuck within our own selfish standpoint. Bravo! Encore!"[153] Even with all their tools of 'objectivity,' natural scientists will never be able to escape completely from the narrow focus of human intentionality, which is the main focus of the social sciences. John Pearle argues that the fact that the social sciences are powered by the mind is the source of their weaknesses vis-à-vis the natural sciences. At the same time, he hastens to add that this is also precisely the source of their strength as social sciences: "What we want from the social sciences and what we get from the social sciences at their best are theories of pure and applied intentionality."[154] The roles of imagination, of metaphor and analogy, of category transforming speculation and off-beat intuition have come to the fore much more. Stefan Collini draws the unavoidable conclusion: "more now tends to be heard about the similarity rather than the differences of mental operations across the science/humanities divide . . ."[155]

This is not to say that there are no methodological means available to enhance the scientific reliability and validity of science in general and the social sciences in particular. A common method is to use *triangulation strategies,* i.e. cross-checking information and conclusions by the use of multiple research procedures, sources and theories. R. Burke Johnston outlines four forms of tri-angulation. The first type is *method triangulation,* the use of multiple research methods to study a phenomenon (Juxtapositional depth and breadth). *Data tri-angulation* is the use of multiple data sources within a single research study (multiple observations, multiple interviews, etc.), as well as collecting data at different times, at different places and with different people (see Boolean al-gebra). The third type, *investigator triangulation,* involves taking into account the ideas of and explanations by other researchers studying the same phenom-enon (see Synergetic discourse). The last type, *theory triangulation,* involves examining how the phenomenon is being studied and explained by different theories (see interfield, systemic and seminal theories).[156] By these strategies, subjectivity can be reduced, but not totally eliminated.

THE SCIENTIFIC COMMUNITY—TWO CULTURES OR ONE?

As the science war dragged on, representatives of the 'warring' parties started to question the infallibility of their own positions. First, the fronts gradually began to dissolve, then indirect mediation stemming from post-positivist thinking began to be felt and, last but not least, there may be a historic shift of scientific paradigm. Let us discuss these steps in more detail.

The Dissolution of Fronts

The dissolution of the fronts started back in the 1950s, when the naturalists in the social science community gradually became convinced that they had some-thing to gain and learn from the other side. The declared enemy was believed to be partly right. They became 'renegades,' and started to adjust the other side's tools and reasoning to fit the needs and purposes of the social sciences. They started the process of *interdisciplinary exchange and adjustment,* which has become one of the measures to promote interdisciplinarity in research.

At the same time, long-lasting discontent in the natural sciences with some of the Newtonian presuppositions, not least the limited ability of the older the-ories to provide adequate solutions to complex phenomena became more ar-ticulated and visible. Developments in quantum physics and chaos theory en-ticed some to declare explicitly that the mechanistic model of the properties and behaviour of matter, which had been dominant since Newton, was now dead.[157] There were also mathematicians who had come to the conclusion that qualitative interpretations would, in certain instances, provide more precision and accuracy than some of the quantitative methods would. More importantly, these scientists emphasized the significance of time and change. According to them, the Newtonian perspective of the world as a linear, deterministic and predictable machine or system was much too static and exaggerated. The world is both unstable and complex, and irregularities play a far more significant role than originally thought. The chaos theory, the ruling paradigm at present, concerns the irregular, unpredictable aspects of nature. It comprises non-linear causality in which small deviations at the outset decide the difference between a preferable outcome and a catastrophe. In a chaotic world, science can never find a precise, accurate analytical mathematical solution that will enable it to predict how a particular planet or moon will move in its orbits forever. For the planets of the solar system, there is always the possibility that the orbits may not stay the same forever and that they may change in literally unpredictable ways.[158] The quantum cognition is that science can never come up with the exact 'answer' that fits perfectly the behaviour of the objects in the real world. In its search to capture and understand complexity, the chaos theory cuts across disciplines like physics, mathematics, meteorology, biology, etc.[159] From this perspective, chaos is the science of processes, of movements, that is, of something that is in the making instead of something that is. It indicates randomness, but what comes out of chaos is that certain processes in nature can produce marvellous constructions of great complexity without randomness.

Nature can no longer be regarded as passive, and some have even concluded that complex systems are self-organizing (note Nancy Cartwright's assumption that we do not know if we live in a tidy or untidy universe. Also note the concept of complex systems and the Gaia hypothesis discussed in Chapter 2). These new perspectives do not necessarily reject the core premises of Newtonian physics as such. What they do is to rock the boat of old dogmatism. To illustrate the point: Newtonian physics describes the movements of the planets, but not the evolution of the planetary system. It describes systems in equilibrium, but not systems far from being in equilibrium, which are far more common than systems in equilibrium (note the reasoning of W. T. Stace above, concerning the invalidity of causal laws). This implies that it is not enough, as presupposed in traditional Newtonian physics, to know the laws of nature and the original prerequisites to predict the future condition of a particular system. In a non-equilibrium system, the future is uncertain and the preconditions irreversible. Science is therefore not in a position to formulate *definite laws*, only *possible laws* (See distinction above between *ceteris paribus laws* and *real laws*). The complexity of the global climate is such that a gradual increase in the levels of greenhouse gases will not always result in a gradual shift in climate: it can trigger a sudden climate reversal within a single lifetime. Even the behaviour of some of the simplest of mechanical systems cannot be described in the complete and deterministic Newtonian manner previously thought possible. There is no simple algorithm to apply. Instead, we must try to understand the world in more global terms through the *interactions* between its components. Instead of attempting to take a deterministic, mechanical view of the world, we need a higher-level perspective if we are to make sense of it.[160] This also relates to the functioning of the social world. For instance, the complexity of a modern industrialized economy is such that it will seldom respond to the elementary manipulations of a country's treasury secretary alone. Real-world complex systems, be they natural or social, do not behave with clockwork regularity, and precise long-term forecasts about them are frequently balderdash.

The Mediation of Post-Positivism

The next step was the mediation of the post-positivists who showed that the distinction between the two scientific cultures defined by C.P. Snow, philosophically speaking, was less pronounced, less watertight and more blurred than originally thought of in the 18th century and onwards. Brian Fay and Donald Moon suggest that the true nature of the social sciences could never have been understood by any of the protagonists of the science war.[161] In their mind, the social sciences qualify as a true science for two reasons: First, they have a specific and unique object of study: *intentional phenomena* that have to be identified and understood in terms of their meaning. Second, the social sciences try to develop systematic theories to explain the underlying causal interconnections among phenomena of widely divergent natures. The reason why the most fanatic naturalists and anti-naturalists failed to provide an adequate account of the social sciences was that each of them focused on only one of these features.[162] If the condescension of the two positions had been allowed to prevail unabated, the likelihood is that the old trenches of the science war would have been dug deeper, broader and with more ramifications. At the close of mediation, the creed is that there are many different sciences—physical, biological, behavioural social, human, medical, engineering and so on—but they are all varieties of the same cultural species"— *academic science*.[163] Post-normal science adds a new dimension to this equalized culture by declaring societal stakeholders (practical knowledge) to be part of an extended peer community.

The Shift of Paradigm?

Today, natural scientists work to expand and include the dynamics of irreversibility and probability in their models. The natural sciences are preoccupied with joining the ideas of natural laws with the ideas of events, news and creativity. On the basis of this change/development in the practice of the earth sciences, the early defection of some scientists from both camps and the rethinking of post-positivists, the earth sciences have moved closer to the 'soft' social sciences. The focus of the natural sciences is gradually shifting away from the search for new fundamental laws and towards a new kind of synthesis for understanding complex systems. Fritjof Capra describes the new inclination as follows: "As we study the various models of subatomic physics we shall see that they express again and again, in different ways, the same insight—that the constituents of matter and the basic phenomena involving them are all interconnected, interrelated and interdependent; that they cannot be understood as isolated entities, but only as integrated parts of the whole."[164] The Gulbenkian Commission calls for the new emphasis in social science research to be placed on the *complex*, the *temporary* and the *unstable*, which corresponds with a growing interdisciplinary movement.[165]

Complex systems have long been a main concern of the social sciences, addressing complex, interactive, interwoven hierarchical organizations, structures and processes in a time-related perspective. Thus, scientific analysis based on a non-equilibrium dynamic, with emphasis on multiple futures, complementary and choice, historical dependence and, for some, inherent insecurity, complies with the most important social science traditions.[166] Natural scientists now talk about the significance of time, which has been a topic of prevalence in the soft part of the social sciences for a long time. Meanwhile, representatives of literature studies which used to be the most ideographic in orientation and hostile to the positivists are now making 'theory' a key concept in their research.[167] Many humanists have, in this respect, accepted the positivistic viewpoint to copy the methods of the natural sciences.[168] Some even compare their own activities with those of chemistry, physics, biology and natural sciences.[169] Given the slow but gradual convergence of their respective perspectives, the three branches of learning have been challenged to create a coherent, consistent and logical system of general ideas of the world from which every part of our experience can be understood and interpreted.[170]

The gradual but still reluctant and slow playing down of the differences has not, as suggested by the unity of sciences thesis, resulted in a mechanistic perspective on man, but has rather affected the perspective of nature as active, creative and complex—as non-linear and concerned with behaviour far from a state of equilibrium.[171] This cognition is more similar to elusive man than to the positivist conception of law-governed nature. In line with this development, positivism has been discarded as impossible; a myth that does not apply to any of the sciences, not even to the natural sciences, because its doctrines have been proven wrong.[172] The concern of the emerging post-positivist era is no longer with the differences between the sciences, but with 'good' and 'bad' science. Against this backdrop, most post-positivists will concur that the walls between the 'two cultures' have been permeated and that they are gradually moving towards one culture—*academic science*—which finds many expressions both between and within the three branches of learning.

It is still too early to say if the philosophical rapprochement between the sciences in terms of blurred demarcation lines and similar approaches to complex system-problems represents a real breakthrough in their relations. In this day and age, positivist and post-positivists work side by side. Philosophy has undoubtedly moved on to a new mindset, although practice may lag behind. In 1995, the New York Academy of Sciences sponsored a Conference in New York under the title: *The Flight from Science and Reason,* which mobilized a broad coalition of natural scientists, social scientists and other scholars for the defence of positivist science. The conference declared that there was a real threat to science posed by sociologists, historians, philosophers and feminists who work in the field of 'science and technology studies.' It attacked the social theories of science, declared feminist epistemology a 'dead horse' and the criticism of science 'common nonsense,' while most critics of science were described as 'charlatans.'[173] These are mindsets rooted in the most fundamentalist version of positivism. Twenty three years earlier, in 1962, Thomas Kuhn placed reactions like these into their proper historical context: "Normal science often suppresses fundamental novelties because they are necessarily subversive of its basic commitments . . . (but) when the profession can no longer evade anomalies that subvert the existing tradition of scientific practice . . . ," then extraordinary investigations begin to produce a new set of assumptions on which the profession can conduct its research.[174] A shift of paradigm alters the fundamental concepts underlying research and inspires new standards of evidence, new research techniques, and new pathways of theory and experiment that are radically 'incommensurate' with the old ones.[175]

In pure opposition to moves like the New York Conference, Fritjof Capra declares that a new paradigm has already materialized and is defined by six distinctive features:

1. In the new paradigm, the belief is that while the properties of the parts certainly contribute to the understanding of the whole, at the same time the properties of the parts can only be fully understood through the dynamics of the whole.
2. In the old paradigm, the thinking and focus were on structures, whereas in the new paradigm, process is primary and every structure observed is a manifestation of an underlying process.

3. In the new paradigm, there has been a shift from *objective science* to *epistemic science* (see above).

4. In the new paradigm, the foundation of knowledge is not thought of as being solid. The acknowledgement is based on historic evidence that the foundations have shifted repeatedly, and have been completely shattered from time to time.

5. The new paradigm recognizes that all scientific concepts and theories are limited and approximate and subject to change. The shift is from certainty in terms of truth to approximate descriptions.

6. Capra's last criterion differs from the others in that it does not express an observation, but rather an advocacy. In the old paradigm, science and technology were based on the belief that an understanding of nature implies man's domination of nature. In the new paradigm, the belief is based on cooperating with nature on the basis of non-violence, i.e. behaving in an ecological manner.[176]

Capra's contention is that the shift of paradigm is from positivism to post-positivism (as defined by his six criteria), and that Kuhn's revolution has already taken place.

Others are arguing that the social sciences and the humanities are moving towards a *post-disciplinary landscape or era*. According to its proponents, this shift is due to two transboundary scientific phenomena labelled the '*spatial-turn*' of the humanities, social and cultural sciences, and the '*cultural-turn*' of geography to engage in various strands of social theory. The former implies that researchers involved in sociological, cultural, and literary studies have come to appreciate the geographical concept of space in explaining social and cultural phenomenon. The latter, on the other hand, refers to the usefulness of social theory in culture analysis by geographers.[177] Thus, the 'two turns' stand out both as multidisciplinary input to disciplinary enrichment and as a fresh measure of scientific convergence. Geographers claim that the cultural turn has meant a strong intervention of interpretative theories, methods and ideas into geography—a field previously influenced heavily by tasks of mapping, describing societies spatially, and economic theories. At the same time, the argument is that geography seems to borrow far more than it is borrowed from, and that the spatial work of other disciplines turns out to be of inherently geographical interest.[178] The premise underpinning this reasoning is that the boundaries between disciplines and sciences are being dissolved, i.e. blending into a post-disciplinary landscape on the backs of two converging vehicles, spatial and social theory. This is a controversial point of view.

To acknowledge the potential cross-fertilization of the two turns is not, however, the same as to accept the assumption that science is heading towards a post-disciplinary landscape. It has, for instance, been argued that it is unlikely in the foreseeable future that the two turns will produce a post-disciplinary landscape. At best, this landscape will be partial, as disciplinary identities will continue to be maintained for both administrative and political reasons. Accordingly, we will continue to have separate and demarcated schools of knowledge.[179] Furthermore, interdisciplinarity is incomprehensible without the existence of disciplines. The crux of interdisciplinarity is about securing scientific depth through disciplinary work and wholeness through convergence of specialities in an interrelated research process. In this perspective, the concept of post-disciplinarity is a contradiction in terms, undermining the very foundation of what spatial and social theory may produce in terms of interdisciplinarity. Rather, the new landscape is one of increasing interdisciplinarity supported by disciplines (and eventually stakeholder knowledge), the two turns and other methodological vehicles of synthesis (see Chapter 3).

Controversies aside, what can be said with relative certainty, is that serious doubts have been articulated about the old paradigm, the alleged differences between the sciences and the distinctiveness of their demarcation lines. A shift of paradigm seems at least to be in the making philosophically, whereas positivism enjoys less cultural authority than it once did.

THE RELEVANCE OF THE SCIENCE WAR AND POST-POSITIVISM TO INTERDISCIPLINARITY

In an interdisciplinary perspective, the core question of the science war was to measure up the social sciences and the humanities with the yardstick of Newton to see if they could qualify as true sciences. Since the criterion was taken lock stock and barrel from the discipline of physics, the natural sciences were automatically exempt from submitting proof of their eligibility. They slipped through the eye of the needle of acceptance as a unified block in spite of the acknowledged differences in their criteria to qualify, for instance in their differences with a view to predictive power. The unfairness of this situation could hardly avoid negative reactions among those whose academic and scientific credentials were questioned.

Patronizing schemes seems to have the inherent quality of rejecting cooperation from down-up because the underdogs feel underestimated and humiliated, and from the top-down because alpha females are reluctant to waste time on activities perceived to be inferior to their own abilities. Thus, for the most ardent of naturalists and anti-naturalists, the science war, had a most detrimental effect on their ability to foster cooperation between disciplines.

The boundaries separating the sciences were transformed to become un-bridgeable moats of demarcation, and their trading zones deprived of inter-disciplinary presence. The effects are still being felt and expressed.

Many practicing researchers dismissed the philosophical debate of the science war with a shrug of shoulders. The feeling that it was irrelevant to prac-tical research was widespread. This was the war that consumed the philo-sophical journals, but hardly concerned anyone else.[180] Theory founded in logic stood against experience founded in actual research, and no one tried to bridge the gap. Most practical scientists went on conducting their research un-affected by the rattling of swords, following the rules and procedures previ-ously developed for their respective disciplines.[181] This was understood by many of the philosophers themselves. John Ziman was one of them: "The av-erage scientist will say that he knows from experience and common sense what he is doing, and so long as he is not striking too deeply into the foun-dations of knowledge he is content to leave the highly technical discussions of the nature of science to those self-appointed authorities, the philosophers of science. A rough and ready conventional wisdom will see him through."[182] George Gale followed suit, claiming that these almost arcane discussions, verging on the scholastic, could have interested only the smallest number of practicing scientists.[183]

Although these are probably valid observations, they pose a serious danger because what is controversial in philosophy may be accepted lock, stock and barrel by scientists without the latter recognizing the controversial nature of their own practice. To be ignorant about what is going on in one discipline is not acceptable in the context of interdisciplinarity. That is actually the oppo-site of what interdisciplinarity is all about. Interdisciplinary competence is a prerequisite for the transgression of scientific fields and disciplines.

The courage to defy authorities has, however, time and gain proved an im-portant component in surmounting obstacles and continuing to test out one's own convictions. Quite a few people dared to cross the demarcation lines be-tween the sciences and to defy and pass by the 'off limits' signs put up by philosophers along the borders. The result of these crossings gave way to a fairly rich literature on interdisciplinarity as a craft and an approach,[184] and to many in-tegrated studies in the transition zones between the sciences. This being said, the bulk of the studies were conducted *within* and not *between* the sciences.

Academic Interdisciplinarity within Sciences—Narrow Interdisciplinarity

By keeping interdisciplinarity within the bonds of one's own scientific field and practicing narrow interdisciplinarity, one did not risk to the wrath and criticism of the other sciences. The only inevitable battle was within one's own family of disciplines—with those who refuted academic interdisciplinarity as not being sufficiently scientific, or those who wanted to keep their own field of speciality pure from the sullying impact of others. Thus the science war indirectly forced a lot of researchers to stick to their own turf and not approach those of others. However, this was not necessarily a bad outcome since the 'renegades' of interdisciplinarity learned to crawl in familiar territory before walking across continents of unfamiliar terrain—between the sciences. The crawling could be done with relative ease, because the boundaries between 'related disciplines' are in principle less demarcated and the threshold is lower to cross over than those between disciplines of different breeds. In particular, the high degree of interchangeability of methods, concepts and related theories facilitates crossings. To illustrate this type of kinship, let us address the interdisciplinary relationship between history, sociology, political science and economy on the one hand, and the points of contact between history and literary texts, in particular autobiographical writing, on the other.

In the 1960s, representatives of the former disciplines initiated a process of cooperation and harmonization. At that time, the history profession had been criticized for having been too preoccupied with isolated events and the role of individuals as driving forces of historical change. It was too ideographic and it was argued that the history profession ought to change based on help from adjacent disciplines such as sociology, economy and political science. The social sciences were in a position to offer scientific tools that could reveal the underlying dimensions of historical institutions, events and ideas (dimensions such as economic change, social injustice and mobility, attitudes and behaviour, social protest and election patterns). These were tools that the historians did not have: quantitative methods, analytical concepts such as class, role expectation and status differences, models and social change.[185] As a result, history today tends to be more analytic and thematic than narrative and chronological. The movement of history towards the social sciences has been labelled "New history."[186] Social scientists extended their areas from collected data and adding historical data to verifying and/or disproving the generalizations they were concerned with, without becoming traditional historians. They kept their disciplinary identities since the social sciences kept up their nomographic profile and the historians their ideographic profile. This disciplinary rapprochement contributed significantly to foster '*interdisciplinary exchange and adjustment*' among the three disciplines.

The overlapping between sociology, economy and political science went even more smoothly than with history. The sociologists took the lead and created sub-disciplines like 'political sociology' and 'economic sociology.' As early as in the 1950s, the three disciplines included quantitative techniques and even mathematical models in their methodological repertoires. The result was unavoidable: the methodological, conceptual and procedural differences between the four disciplines were drastically reduced. Over the following decades, they increasingly overlapped in fields of interest and methodology.[187] Their methodological differences have diminished drastically. They all employ quantitative as well as qualitative methods in their respective fields of research. Their cooperative process of interaction was synergetic in that it created *interdisciplinary sensitivity/practice* and *dual competence* in the interfaces between them. Together, they also created *interdisciplinary exchange and adjustment* as a joint methodological tool to be applied by interdisciplinary research between the three fields. As of now, it is difficult to draw clear-cut delineations between them, because each has become more heterogeneous in that all avail themselves of an extended mono-disciplinary approach to each other's disciplinary territory. Historical trend analysis is an example of a blend of two research traditions in which the ideographic approach is used to forge nomographic generalizations of history.[188] What this process of convergence has done is to create a more *integrated science of society*.[189] To establish cooperation between most social science disciplines and history is no longer a looming, insurmountable obstacle. Some of these disciplines have an inherent ability to further and strengthen this convergence by way of their own 'interdisciplinary' definition and composition.

The relationship between history and literary studies, particularly autobiographies, has been a subject of discussion for decades, and to a certain extent still is. The traditional position has been to maintain a clear-cut distinction between fact and fiction, pointing out that historians and literary critics occupy different territories of research with few or no points of contact. This position has recently come under attack from several quarters.[190] Louis Montrose, a vocal representative of *New Historicism,* insists on "the historicity of texts and the textuality of history."[191] The new historicists define their field of study in such a way that both are necessary for the study of either of them. In particular, this relates to autobiographical writing based on documentary experiences that contribute to the understanding of contemporary history, for instance, the study of the Holocaust through the narratives of time witnesses.[192] The new historicists compose their work with an eye to supplementing academic historiography written by scholars who belong to another academic tradition and follow different criteria in their presentations. In the same vein, the two disciplines supplement and overlap rather than conflict. Against this backdrop, Beatrice Sandberg concludes that autobiographical studies have become interdisciplinary in nature, underscoring that the uniqueness of a work of art should be understood not by isolating the text from the context, but by placing it more deeply within it.[193] This new trend is to bridge related disciplines without doing away with them.

Multidisciplinary Disciplines

Certain disciplines are interdisciplinary in the way they are composed, accommodating two or more disciplines previously separated by boundaries. They are *multidisciplinary disciplines* or *hybrid disciplines* without being a contradiction in terms. August Comte called these disciplines "specialties of generalities," arguing that it is necessary to develop a meta-specialty for the study of the specificities of each discipline and for the integration of their results.[194] According to Comte, sociology is one such discipline whose main objective is to bring coherence and coordination into the anarchy of excessive specialization.[195] This assumption on the part of Comte has been experienced time and again by many practicing researchers, reporting that multiple disciplines (not only sociology) hold an integrative power beyond that of multidisciplinarity. Indirectly, this implies that the strict monodisciplinary structure of science is a surprisingly rich source of convergence in its own right. In the light of this, multidisciplinary disciplines have long been used as 'training camps' for researchers seeking horizontal integration between specialties. The purpose of the ensuing subparagraphs is to identify some of the hybrid disciplines and other measures of convergence in terms of theories and concepts bringing 'coherence and coordination into the anarchy of excessive specialisation.'

Political science is an example of a multidisciplinary discipline that has been established in the interface between history, sociology, international law, geography, ethnography, social psychology and economy. This being the case, political analysis applies the methods and concepts of many disciplines interchangeably. At the same time, it has become a distinct discipline in its own right through its defined and delimited focus on *politics* as a social phenomena. The difference between political science, political sociology, political economy, political geography, etc. is one of emphasis rather than kind, and the delimitation between them is indicative of the arbitrariness of disciplinary boundaries. Such kinships also break down many of the terminological problems between disciplines. The differences are comparable with those involved in speaking the Scandinavian languages: They are easily intelligible across the national borders, so few are willing to put much effort into learning the languages of the others to perfection. Many other social science disciplines are also multidisciplinary disciplines, e.g. social psychology was touted as an 'interdiscipline' well before recognized definitions emerged in the 1970,[196] whereas history has been portrayed as one of the busiest areas of cross-disciplinary combinations.[197]

In the positivist tradition, the natural sciences are generally thought of as having a joint inventory of objective scientific tools which are interchangeable. Therefore, the assumption is that the disciplinary boundaries within the natural sciences are easy to cross and are even disappearing, and being replaced by shifting hybrid domains which reach across many levels of complexity, from chemical physics and physical chemistry to molecular genetics, chemical ecology, and ecological genetics. Astronomy, geology and evolutionary biology are regarded as examples of primarily historical disciplines linked by consilience to the rest of the natural sciences, whereas physics and chemistry are linked through the 'bridge law' of thermodynamics.[198] Geology is generally considered a derivative science that utilizes information and concepts from astronomy, chemistry, physics and biology.[199] Eight social sciences have roots in the natural sciences: demography, geography, psychology, anthropology, linguistics, archaeology, criminology and cognitive science.[200] Physics is like a federation of disciplines, incorporating such areas as nuclear physics and solid-state physics, areas that have more in common with chemistry and engineering than with traditional physics.[201] Physics also borders on anthropology and biology. Although biology and physics have traditionally been far apart, academically as well as culturally, there are increasing numbers of physicists taking the step across to biology, and of biologists becoming major consumers of computer power and mathematical models.[202] Climatology has been called a federation of many 'little' sciences because its special subdivisions study subjects as diverse as tree rings, pollen grains, riverbeds, lake bottoms, soils, ice, algae and fungi and historical climatology.[203] By pooling the enormous collective knowledge of biology, chemistry and physics, molecular biology has been developed into an interdisciplinary field in its own right, providing an astonishingly detailed picture of the complexity of life.[204] As a result, sciences are in an unrivalled position today to treat disease, prevent illness and genetically engineer crops.[205]

Interdisciplinarity Between the Sciences—Broad Interdisciplinarity

The trouble spots of interdisciplinary research are in the border zones of disciplines belonging to different sciences—broad interdisciplinarity. This is where the challenges of complex system science lie. The challenges are of two kinds. First, it is there that the methods, theories and paradigms of research are the most different and the languages of Babel the most confusing. Second, it is across this divide that the feelings of antagonism are strongest. Within the humanities, the very concept of positivism never had a positive connotation, and in the heightened atmosphere of the 1970s, it grew to become a term of abuse.[206] The concepts of 'true' and 'untrue,' 'real' and 'unreal' sciences still linger in the air. Teamwork has been compromised by the disdain natural scientists have had for engineers, mathematicians for physicists, pure scientists for applied scientists, physical scientists for social scientists and humanists, and vice versa.[207] At present, there are also reports stressing difficulties in linking, for instance, the social sciences to programmes on global environmental change because the problem was initially defined by physical science alone and not jointly between all involved disciplines.[208]

It is a fact that disciplines that do not lend themselves to quantitative modelling are more likely to be underrepresented in interdisciplinary projects than those that do.[209] They are also more likely to be defined as *feeder disciplines*, submitting scientific input to a *source discipline* entrusted to make the overall and final analyses. The subordination of the one to the other is a remnant of positivist thinking. Breaking the automatic reflexes of this scheme and establishing a better balance between the natural and social sciences will enhance the possibility of succeeding with establishing interdisciplinary research projects.[210] A change along these lines may be in the making: although the social sciences are not, in Newtonian terms, usually classified as scientific, despite of the name, social phenomena are increasingly being brought within the scope of scientific analysis as a result of the methods developed for the study of complexity. Technically speaking, bridges have been identified as crossing over between the sciences. They are said to be of two sorts: First, we have the *bridge theories* not yet tested empirically; second, there are *bridge disciplines* that to a certain extent has been tested in practice and *bridge concepts*.

Bridging Theories

As stated in Chapter 3, scientific theories are in principle diffused through and across disciplinary boundaries. This feature is evident with a view to synthetic, interfield, seminal and social theories, but it also relates to theories developed on the basis of joint research foci between disciplines, e.g. space and culture. To avoid being repetitive, those theories will not be discussed here and this presentation will be restricted to the bridging devices introduced by Edward O. Wilson.

Wilson's reasoning is based on the premise that the differences between the sciences lie in the magnitude of the problem, and not in the principles required for its solution.[211] On this premise, he identifies four bridges to straddle the divides between the sciences. The first bridge is the theory of *cognitive neuroscience*, which contains elements of cognitive psychology in that practitioners aim to solve the mystery of conscious thought through analysis of the physical basis of mental activity. The second is *human behavioural genetics*, which aims at unravelling the hereditary basis of genes' influence on mental development. The third relates to *evolutionary biology*, including sociobiology, whose researchers have set out to explain the hereditary origins of social behaviour. The fourth concerns the *environmental sciences*, which address the natural environment from which the human species have evolved and to which their physiology and behaviour have been adopted. These theories are, in Wilson's imagination, overpasses that make research between disciplines across the natural/social science divide a likely possibility.

All these bridges have not yet been accepted by the epistemic community as being undisputable integrative measures of consensual research. For instance, Wilson's theory on sociobiology[212] resulted in a heated debate between representatives of the social and natural sciences, deepening at least temporarily the gorge between the two cultures. Here an alleged theory of convergence aggravated a long-lasting split that the theory intended to bridge and in the long term even heal. Wilson also provoked some of his fellow biologists who accused him of having committed 'a naturalistic fallacy' by applying Darwinist analysis to an area of little or no relevance. Among other things, he was unjustly criticized for putting genetic determinism up against the free will of man.[213] This debate within and between disciplines is a reminder that there is a complex tension between the unifying forces of theory and the unitary forces of disciplinary identity.[214] This being said, Wilson's alleged synthetic theories may be daring suggestions on supplementary paths to complex system research in the future. This makes them innovative contributions both to meet societal needs through practical-instrumental research and to further science by adding interdisciplinarity to the practice of knowledge-instrumental research.

Bridging Sciences, Formal Sciences and Individual Disciplines

The Gulbenkian Commission indicates that the social sciences as such host a bridge-building quality of their own based on three characteristics. First, throughout the positivist era, the social sciences were sandwiched in-between the two other domains. In this position, the social sciences have communicated in both directions and have become knowledgeable of the needs, positions and frustrations in both camps. Second, they have been the test ground of the methodology of the natural sciences and have become more knowledgeable about the potentials and limitations of the natural sciences for engaging in interdisciplinary exchanges. Third, the social sciences carry much of the methodological tradition of the humanities. In this respect, the social sciences carry three traditions in a unique mix that may prove useful in processes of convergence. They may serve either as a venue where the two others meet, or as a bridge for the two to cross over, or as a facilitator to produce convergence.[215] Thus, the social sciences stand out as a scientific mediator between non-communicative domains able to ease tensions and promote middle ground. The overarching challenge is to 'translate the tongue of the one into that of the other.'

The Gulbenkian Commission was not the first to recognize this transversal ability of the social sciences. Four years after C.P. Snow delivered his Rede lecture, he wrote a paper in which he nominated the social sciences as an emerging, although at that time not yet fully constituted, *third culture*, taking a mediating position in between the two conflicting cultures with the long-term aim of providing a convergence of knowledge. Although he clearly acknowledged that the social sciences were a varied and mixed bag, he recognized the inner consistency between the various disciplines. *All were concerned with how human beings are living or have lived—and concerned, not in terms of legend, but of fact.*[216] As such, the social sciences would, in his mind, ultimately unravel that the mutual hostility between the 'two cultures' was artificial and nothing but man-made, based in mutual jealousy and ignorance.

A different approach would be to use economics at the cutting edge of the social sciences, as the stepping stone to the natural sciences and humanities. The preciseness of economics, secured by mathematical modelling, is constantly battered by 'exogenous shocks'—all the unaccountable events of history, society and environmental changes that push parameter values up and down. That alone limits the accuracy of economic predictions. In economics, the main analytic problem is the translation from individual to aggregate behaviour. For the same reason as natural scientists, "many economists have sought to shoehorn all economics into theories whose merits are their mathematical simplicity and elegance rather than their ability to say anything about the way real world economies work."[217] As has been observed, economists cannot tell precisely whether tax cuts or national deficit reduction is the more effective measure for raising per capita income, or how economic growth will affect income distribution.[218] They cannot predict how beliefs and rumours generated by stockholders can induce fluctuations in prices, stock and currency markets. A most precise observation is that economics is a systematic formalized science, but it is not independent of context or free of history. It is grounded in human practices, but those practices are not themselves timeless, eternal or inevitable. John Searle illustrates the point: "If for some reason money had to be made of ice, then it would be a strict law of economics that money melts at temperatures below 0 degrees centigrade. But that law would work only as long as money had to be made of ice, and besides, it does not tell us what is interesting to us about money."[219] In the context of modelling, the problems of economics are multifarious and complex. Alexander Rosenberg claims that the main problem of economics stems from its attempt to be an empirical science. He argues that economy is best viewed as a branch of mathematics, i.e. as a formal science located somewhere on the cusp between pure and applied axiomatic systems.[220] In general, the role of the formal sciences is to serve as vehicles for the empirical sciences to make observations and report events as accurately as possible. In the positivist era, mathematics was assumed to be the parent science, synthesizing all sciences through the process of reduction. This has proved to be overly optimistic and unrealistic, but mathematics still has the ability to be used by various natural science disciplines, and to bring the quantifiable segment of the social sciences somewhat closer to the natural sciences. The formal sciences build bridges, although not as extensively as believed by the naturalists. If economy was to assume the role of a formal rather than empirical science, each of the three branches of learning would have its own formal science: mathematics would continue to be the prime vehicle of the natural sciences, logic (as an offspring of philosophy) would be part of the humanities, and economics would be part of the social sciences. In such a perspective, each of the branches would have a formal vehicle carrying discipline-specific properties to approach the interfaces and trading zones of their respective disciplines. At the same time, the combination of the three may contribute to bridge the gulf between the three branches of learning. Logic has already become an all-encompassing measure of all sciences and disciplines, mathematics has strung the natural sciences together and reduced the gulf to the social sciences, and economics at the intersection of mathematics and social science, could strengthen the overpass between some of the social and natural sciences, which was never finished by the introduction of mathematics. In combination, the three formal sciences would—theoretically—have the potential to strengthen further the measure of *interdisciplinary exchange and adjustment* between all fields. Further, the three vehicles are united in their ability to provide the most accurate measurements available based on discipline-specific properties. It has been suggested that chairs should be established in the formal science to further synthesis of research. Such chairs could become pivotal institutions and of paramount importance in providing intellectual links between the disciplines.[221] This might be a way to enhance the 'quadruple competence' in interdisciplinary research.

Geography has a long history of refusing to be categorized as affiliated with either of the sciences. At the turn of the 19th century, geographers made strenuous efforts to reduce the distance to the natural sciences by emphasizing physical geography, as well as reducing the distance to the humanities by introducing social geography. In the United States, geography developed towards geology and in some cases even history and anthropology. In Germany, it originated from the earth sciences, in France from history, and in Britain from the needs of managing an empire.[222] In other words, geography is to be found in the faculties of the social, natural, humanistic and economic sciences all over the globe.[223] It is an orphan of the sciences, i.e. a synthetic discipline that tries to capture the complexity of the world differently from that of the more specialized disciplines. In principle, geography promotes both narrow and broad interdisciplinarity, traversing also the gap between academia and society. In principle, geography is like family medicine—a boundary-breaking discipline with a passion for complex concepts like *nature, environment, landscape, region, space* and *place*.[224] Geographers often use the word 'interrelations' to describe their focus and the issues they address. It has become a 'crossroads science,' encompassing subfields such as human, cultural, economic, political, urban and regional geography, as well as biogeography, geomorphology, climatology, environmental science and cartography. Some have actually suggested the plural 'geographies' as a more appropriate label than geography.[225] Or, as noted by Richard Peet: Geography is in a permanent identity crisis because what geographers do is complex.[226] Within the geographical community, there is no clear consensus as to what geographers are, what they do or how they study the world.[227] The complexity of the system of investigation is reflected in this definition: "Geography is the study of relations between society and the natural environment. Geography looks at how society shapes, alters, and increasingly transforms the natural environment, creating humanized forms from stretches of pristine nature, and then segregating layers of socialization one within the other, one on top of the other, until a complex natural-social landscape results. Geography also looks at how nature conditions society, in some original sense of creating the people and raw materials which social forces 'work up' into culture, and in an on going sense, placing limits and offering material potentials for social processes such as economic development. The 'relationship' between society and nature is thus an entire system, a complex of *inter*relations."[228] Understanding this system of relations requires that geographers be sophisticated natural and social scientists able to find ways of combining the two, know the methods, and be excited by the insights of both aspects of knowledge. As such, the study of objective and subjective phenomena and its breadth give room to work in the borderlands between sub-disciplines of geography and between geography and other disciplines.[229] Geography is best characterized as a "*singular diversity* in which different theories and ideas coexist in a creative tension, intermingling and cross-fertilizing to produce new theoretical outlooks. This explains why *the first law of geography* is that "everything is related to everything else, but near things are more related than are distant things."[230] Up until 1945, geography was the only discipline that tried to cover huge expanses of the whole world in its approach.

This synthetic inclination notwithstanding, prominent geographers claim most of their colleagues to spend more time talking about interdisciplinarity than actually doing it.[231] Although social, physical and resource geographers are not always placed in the same playpen,[232] political geography is an example of a sub discipline where politics is linked to the features of the physical world in the same way as the psycho-somatic relationship of medicine links biology to the societal aspects of psychology. Thus, geographers specializing in fields of this kind seem ideally fit to serve as bridge scientists in projects addressing the interrelationship between society and nature, and as partners in synergetic discourse to develop interfield, seminal and social theories.

Psychology belonged to the medical profession for a long time. It was physical and even chemical in orientation, and gradually became part of biology, which has been portrayed as being bounded on the one side by mathematics and the physical sciences and on the other by the human sciences. At the same time, there were psychological theories focusing on the social context of humans as a cause of mental disorders. In social psychology, the preoccupation was with social realities, and it often became included as a sub-discipline of sociology. In this way, psychology straddles many disciplines across the divide of sciences.[233]

In a similar vein, sociology as indicated by August Comte is considered one of literature's 'first cousins,' although some relatives are favoured over others. "The major forms of literary purposes are sociology of literature and literary use of sociological knowledge."[234] The 'Jante Law' (equivalent to a 'who-do-you-think-you-are-law') as used in Axel Sandemose's famous book: *A Refugee Crosses His Foot Path* is a significant piece of sociology and a literary drama of high quality to be studied by sociologists, humanists and social scientists alike. Linguists in the Noam Chomsky tradition have gone one step further, adopting models from biology to explain linguistic variation and change. To Noam Chomsky, language and the ability to speak a language is not something exclusively produced by emulation during early childhood, but an innate ability or instinct related to a biological organ located somewhere in the brain ('the innateness hypothesis').[235] Biology is thus the parent discipline of linguistics, which, in this perspective, turns out to be a special branch of cognitive brain science involving psychology.[236] The basic premise is that the ability to learn a language is innate, as is the ability to get teeth. Humans are not born with teeth but get them later in life because of genetic pre-programming. There is no 'learning' involved in getting teeth. Much the same applies to acquiring a language. We are born with the ability to speak through what Chomsky calls an innate *universal grammar* applicable to all languages. But as has been stressed, learning a language is not only a question of biology, psychology and genetics, it also depends on external stimulus. The process of learning a language is thus based on two preconditions: An innate language ability, which is genetically and biologically determined, and external stimulus, which is socially determined.[237] Prominent linguists claim that physics and chemistry have contributed models of utility to the comprehension of language.[238] In this way of thinking, linguistics is a multidisciplinary discipline that straddles the boundaries between the social, human and natural sciences. This has prompted the conclusion that "Whereas in past time scholars had to rely on evidence from material culture and fossil findings to interpret linguistic evolution and variation, genetic data now provide a rich source of additional information on the history of our species."[239] This is not to say that all linguists are equally enthusiastic above the multidisciplinary inclination of their fellow researchers. The Chomsky school may actually send cold shudders down the backs of more humanistically oriented linguists.[240] The point here is simply to say that there are leading modern linguists who try out and prey on the territory of other disciplines to develop further their own comprehension of language.

In general, it seems reasonable to conclude that disciplines matching in approach and sharing basic ideas about what is important and not in research are those that find cooperation easiest and most rewarding. The same goes for concepts used interchangeably between disciplines and sciences.

BRIDGING CONCEPTS

The ultimate vehicle of science is conceptualization.[241] To have command of definition is to have control of discourse. Definitions are inherently boundary-marking and boundary-making. Accordingly, the ownership of terminology is of enormous consequence in dialogue, for by it both ideas and people can be positioned on particular sides of debates.[242] Ownership of concepts provides intellectual power and defines the course to be set by disciplinary science. In principle, there are two ways of gaining command of discourse through the definitional process of concepts.

The first has been labelled *authoritarian,* where the overall purpose is to define the concepts in the only way 'correct,' i.e. to fit the individual goals and needs of the individual disciplines to the unit of topical study. Disciplinary definitions then "become a sort of criterion for discriminating heroes and villains, for arbitrating modern disputes about methods, and for selecting the *dramatis persona* of the drama."[243] This perspective also produces efforts to reach a high level of precision, directly and/or indirectly, in a claim to definitional ownership. Although it is utterly important to provide science with the most precise conceptual vehicles, the claim of ownership is "to try to force a concealed system of values on others."[244]

The second course of definition is labelled *democratic*, or more precisely, the *pluralistic approach*, acknowledging that one and the same concept may have different meanings in different disciplines and contexts. Here, the purpose is to reach a consensual understanding on why the concepts are used differently and the aim is to achieve better communication and dialogue.[245] Concepts such as *species, social class, environment, landscape, system, space, culture*, etc. are used interchangeably by many disciplines, defined in various fashions and discussed endlessly. Such discipline-straddling concepts are also labelled *chaotic concepts*. Some would argue that concepts of this kind should be avoided since science requires precise, unambiguous definitions. However, as pointed out by Michael Jones, chaotic concepts are also *cherished concepts*, i.e. concepts which are appreciated because they capture the complexity and essence of reality as we perceive it on an everyday basis. "Because these concepts seem to enjoy longevity, they prove that humankind has a need for such complex concepts."[246] To reach consensus on the content of chaotic concepts would straddle disciplinary boundaries and alleviate the difficulties involved in synergetic discourse. These concepts also serve political and applied purposes. This point is clearly illustrated in the international deliberations on environmental sustainability. In 1972, the *United Nations Conference on the Human Environment* in Stockholm successfully defined the terms of the ongoing global environmental debate, but failed to resolve the difficult relationship between the environment and development. It lasted until the establishment of the *World Commission on Environment and Development* before the first justification for treating the environment and development as two intimately interlinked problems developed. Having received the commission's report, in 1989 the UN agreed to convene a global conference to implement the concept of sustainable development in Rio de Janeiro in 1992. Sustainable development was also the key concept at the World Summit that convened in Johannesburg in 2002.[247] Chaotic concepts thus serve both analytical and political purposes, reflecting the needs of science and politics alike. They are post-normal in orientation.

SUMMING UP

Although academic interdisciplinarity no longer has an explicit philosophy of its own, post-positivism is a most congenial foundation on which to conduct interdisciplinary and even transdisciplinary research. It tells us that values and uncertainties are present in all types of research, that elusive man is just a little more elusive than complex nature, that all the methodologies of the sciences are to a certain but restrictive degree interchangeable, that all sciences are both law and fact-finding, particularistic and theory-building, that there is a scientific approach that applies to all sciences, that the vein of thought of positivism is logically impossible and that different kinds of knowledge can integrate. The sciences have moved from a Newtonian situation of 'exclusivity,' 'hierarchical subordination,' and 'differences of kinds,' to a looming post-positivist situation of 'inclusivity,' 'pluralism,' 'equality,' 'permeability of cultures' and 'differences of degree'; The movement is from 'all sciences reducible to physics,' to 'academic science reducible to none, but integrative to multiple moulds of realities.'

Post-positivism has opened up a whole new avenue for straddling the boundaries separating different knowledge communities and producers. The demarcation lines of the positivist era have been permeated and transformed into border-crossings that require no passports or visas. To avoid failure in an effort to foster interdisciplinary and transdisciplinary research, the differences between the sciences and disciplines and between scientists and societal stakeholders must be *acknowledged, accepted* and *respected* right from the outset. Tolerance among the knowledge providers cannot be based on indifference, but must be based on insight and mutual respect.

However, in no way would it be accurate to say that post-positivism has done away with all the theoretical hurdles that need to be overcome. The business of theory formation in interdisciplinarity is still unfinished and in need of further development and refinement to secure holistic research maturity and sound standards of evaluation.

NOTES

1. Wilson (99), p. 12.
2. Hubbard, Kitchin,Bartly et al. (02), p. 4.
3. Bechtel (98), p. xi.
4. Klemke, Hollinger, Rudge et.al. (98), p. 20.
5. Pippin (06), p. 191.
6. Kjørup (99), p. 23.
7. Thompson Klein (90), p. 19.
8. Wilson (99), p. 227.
9. Sardar (00), p. 8.
10. Snow (98), p. 57.
11. Sardar (00), p. 62.

12. Kjørup (99), p. 288.
13. Hess (97), p. 1.
14. Heisenberg (99), p. 83.
15. Wallerstein et al. (97), p. 17.
16. Cited from Latour (99), p. 3.
17. Burke Johnston (97), p. 282.
18. The concept: *brain-in-a-vat* is used by Latour (99), p. 113.
19. Northrop (99), p. 3.
20. Machlup (99), pp. 135–153.
21. Bechtel (88), pp. 71–93.
22. Bechtel (88), p. 72.
23. Hess (95), 14–19.
24. Hollinger (98), p. 105.
25. Pearson (37), p. 333.
26. Pearson (37), p. 16.
27. Pearson (37), p. 15.
28. Cited from Wilson (99), p. 24.
29. Wilson (99), p. 53.
30. Cited from Davies (83), p. 222.
31. Cited from May (04), p. 79.
32. Bechtel (88), p. 93.
33. Bechtel (88), pp. 72–73.
34. Weinberg (93), p. 12.
35. Bechtel (88), p. 76.
36. Collini (98), p. xlvii.
37. Wilson (99), p. 53.
38. Trout (99), Carnap (99), Oppenheimer (99) and Garfinkel (99). pp. 387–462.
39. Livingstone (05), pp. 305–346.
40. Harvey (84), p. 5.
41. Hollinger (98), p. 105.
42. Hess (95), p. 17.
43. Durbin (88), p. 334
44. Latour (99), p. 299.
45. Wallerstein et al. (97), pp. 47–52.
46. Capra (99), p. 340.
47. Heisenberg (99), p. 199.
48. Searle (97), p. 75.
49. Wallen (81), p. 19.
50. Gunn, (92, p 255.
51. Cited from Hollinger (98), p.109.
52. Heisenberg (99), pp. 79–80.
53. Davies (83), p. 85.
54. Heisenberg (99), p.81.
55. Heisenberg (99), p. 80.
56. Wilson (99), p. 205.

57. Kuhn (70), p. xx.
58. Bechtel (88), p. 51
59. Blyth (02), p. 295.
60. Funtowicz and Ravetz (08), p. 367.
61. Krohn (08), pp. 369–383.
62. Bechtel (88), p. 17.
63. Hollinger (98), p. 107.
64. Wilson (99), p.5.
65. Wilson (99), p. 367.
66. Ziman (00), pp. 321–322.
67. Wilber (00), p. xii.
68. Weinberg (93), p. ix.
69. Weinberg (93), pp. 230–240.
70. Weinberg (93), p. 61.
71. Wilber (00), p. xii.
72. Chu, Strand and Fjelland (03), p. 25
73. Høyrup (00), p. xx.
74. Caldwell (82), p. 18.
75. Ziman (00), pp. 26–28.
76. Collini (98), p. xliv.
77. Sørensen (99), p. 87.
78. Taylor (98), p. 124.
79. Taylor (98), p. 125.
80. Wilson (99), p. 304.
81. Coveney and Highfield (95), p. 13.
82. Fay and Moon (98), p. 172.
83. Taylor (98), pp. 110–127.
84. Rudge (98), pp. 197–205.
85. Greve (07), pp. 97–99.
86. King, Keohane and Verba (94), p.5.
87. King, Keohane and Verba (94), p. 1.
88. Fay and Moon (98), pp. 171–187.
89. Weinberg (93), p.56.
90. Cited from Bellomo, Li and Maini (08), p. 593
91. Weinberg (93), p. 43.
92. Schuum (98), p. 95.
93. Schuum (98), p. 27.
94. Schuum (98), p. 5–96.
95. Schuum (98), p. 119.
96. Hubbard, Kitchin, Bartley et al (02), pp. 12–13.
97. Peet (98), p. 292.
98. King, Keohane and Verba (94), p. 35.
99. Goonatilake (98).
100. Hughes and Sharrock (97), pp. 196–197.
101. Lambert and Britten (98), p. 225.

102. Stace (98), p. 352.
103. Stace (98), p. 355.
104. Stace (98), p. 352.
105. Latour (99), p. 153.
106. Wilson (99), p. 65.
107. Carnap (98), pp. 316–332.
108. Stace (98), p. 352.
109. Kjørup (99), pp. 290–291.
110. Zeilinger(05), p. 743.
111. Aalen (07), pp. 79–82.
112. Cartwright (98), p. 224.
113. Pecseli (05), p. 159.
114. Hawking (02), p. XIII
115. Cartwright (98), p. 237.
116. Collini (98), p. xiv.
117. Bronowski (72), p. 57.
118. Hjort (20), p. 161.
119. Hjort (20), p. 173.
120. Lepenies (89), p. 64.
121. Hollinger (98), p. 489.
122. Hubbard, Kitchin, Bartley et al. (02), pp. 75–76.
123. Hollinger (98), pp. 489–490.
124. Wilber (00), p. ix.
125. McAnulla (02), pp. 281–282.
126. Funtowicz and Ravetz (98), p. 364.
127. Smolin (07), p. 298.
128. Cited from Weinberg (93), p. 189.
129. Kjørup (99), p. 183.
130. Capra (99), p. 331.
131. Wilson (99), p. 49.
132. Bronowski (72), p. 37.
133. Commission of the European Communities (04), p. 3.
134. Wallerstein et al. (97), p. 80.
135. Zeilinger (05), p. 743.
136. Heisenberg (99), p. 85.
137. McMullin (98), p. 517.
138. McMullin (98), p. 531.
139. Skirbekk (94), p. 3.
140. McMullin (98), pp. 515–538.
141. Ziman (00), p. 51.
142. Hollinger (98), p. 485.
143. Cited from Hess (95), p. 20.
144. Cartwright (98), p. 234.
145. Hess (95), p. 20.
146. Wilson (99), p. 57.

147. Rudner (98), p. 493.
148. Rudner (98), p. 497.
149. Gribbin (99), p. 1.
150. Wilson (99), p. 230.
151. Killeen (03), p. 1.
152. McMullin (98), p. 516.
153. Latour (99), p. 8.
154. Searle (97), pp. 84–85.
155. Collini (98), p. xlviii.
156. Johnston (97), p. 284.
157. Davis and Gribbin (92).
158. Gribbin (99), p. 15.
159. Jørgensen and Lilleholt (93), pp. 27–39.
160. Coveney and Highfield (95), p. 330.
161. Fay and Moon (98), p. 171.
162. Fay and Moon (98), pp. 186–187.
163. Ziman (00), p. 27.
164. Capra (99), p. 131.
165. Wallerstein et al.(97), p. 70.
166. Wallerstein et al. (97), p. 58.
167. See Selden, Widdowson and Brooker (97) and Phelan and Rabinowitz (05). See also Hillis (07).
168. Kjørup (99), p. 91, and pp. 288–307.
169. Faarlund (03), pp. 281–285.
170. Whitehead (78), p. 3.
171. Wallerstein et al. (97), pp. 56–57.
172. Kjørup (99), p. 302.
173. Cited from Sardar (00), p. 4. The proceedings of the New York Conference were published by Gross, Levitt` and Lewis (96).
174. Kuhn (70), pp. 19–20.
175. Sardar (00), p. 29.
176. Capra (99), pp. 328–336.
177. Hubbard, Kitchin, Bartley et al. (02), pp. 58–62.
178. Marcus (00), p. 14.
179. Hubbard, Kitchin, Bartley et al. (02), p. 239.
180. Shapiro (00), p. A14.
181. See discussion in the Introduction to this Chapter.
182. Ziman (00), p. 49.
183. Cited from Weinberg (93), p. 167.
184. In this respect, mention should be made of Thompson Klein's two books (90) and (96). Another important contribution is Salter and Hearns (96).
185. Wallerstein et al. (97), p. 40.
186. Thompson Klein (90), p. 30.
187. Wallerstein et al. (97), pp. 38–56.

188. Kennedy (89) and Knutsen (99). Paul Kennedy is a historian whereas Torbjørn Knutsen is a political scientist, but they both apply an ideographic approach with the aim of achieving nomographic purposes.

189. Wallerstein et al. (97), p. 46.

190. Gallagher and Greenblatt (00).

191. Cited from Sandberg (07), pp. 36–39.

192. Lothe and Storeide (06).

193. Sandberg (07), p. 37.

194. De Mey (92), pp. 78–79.

195. Cited from Heilbron (04), p. 36.

196. Thompson Klein (90), p. 12.

197. Dogan and Pahre (90), p. 87.

198. Thompson Klein (96), p. 79.

199. Schuum (98), p. 3.

200. Dogan (01), p. 14855.

201. Thompson Klein (90), p. 45.

202. Interview with Jan Trulsen and Hans L. Pecseli, in *CAS Newsletter*. No. 2 (04), p. 5.

203. Thompson Klein (90), p. 105.

204. Coveney and Highfield (95), p. 190.

205. Coveney and Highfield (95), pp. 191–192.

206. Kjørup (99), p. 288.

207. Thompson Klein (90), p. 127.

208. Caron, Chapin III, Donoghue, et al. (94), p. 189.

209. Kaje (99), p. 14.

210. Brock, et al. (), p. 77.

211. Wilson (99), p. 292.

212. Wilson (75).

213. Hessen (03), pp. 191–193.

214. Hubbard, Kitchin, Bartley et al. (02), p. 62.

215. Wallerstein et al. (97), p. 62.

216. Snow (98), pp. 70–71.

217. Coveney and Highfield (95), pp. 335–336.

218. Wilson (99) p. 219.

210. Searle (97), pp. 83–84.

220. Rosenberg (98), pp. 154–170.

221. Brock, et al. (86), p. 99.

222. Heffernan (04), pp. 3–22.

223. See Richards (04), Johnston (04) and Blunt (04), pp. 23–95.

224. Jones (01), p. 226.

225. Thompson Klein (96), p. 41.

226. Peet (98), p. 1.

227. Hubbard, Kitchin, Bartley et al. (02), p. 10.

228. Peet (98), pp. 1–2.

229. Jones (01), p. 226.

230. Knox and Marston (01), p. 2.

231. This point was put forward by Professor Graham Chapman in commenting on this mss.

232. Berge and Powel (97), p. 19.

233. Wallerstein et al. (97), p. 29.

234. Thompson Klein (96), p. 139.

235. Chomsky (02), p. 4.

236. Chomsky (02), p. 1.

237. Faarlund (05).

238. Interview, Faarlund in *CAS-Newsletter*, no. 2, (04), p. 6.

239. Hagelberg and Faarlund (04), p. 1.

240. Interview with Faarlund (04) p. 6.

241. Jones (92), p. 84.

242. Livingstone (05), p. 304.

243. Livingstone (05), pp .5–6.

244. Jones (92), p. 85.

245. Jones (92), p. 85.

246. Jones (01), p. 226.

247. Seyfang and Jordan (02–03), pp. 19–26.

Chapter Five

From Theory to Practice: The Case of INSROP

"To understand science is to understand (the) complex web of connections (politics, social, cultural etc) without imagining in advance that there exists a given state of society and a given state of science."

Bruno Latour, 1999[1]

The purpose of this chapter is to test the theoretical discussion of the previous chapters against the reality of interdisciplinary research as manifested in practice. The main case of study and reference used is the *International Northern Sea Route Programme* (INSROP)—a six-year collaborative effort initiated in 1993 by Russia, Japan and Norway. The overall objective is to discuss how, to what extent, and in what way the *history, politics, concepts, design, philosophy and methodology* of INSROP (and other cases) affected the integrative and synthetic results of the research.

To enhance the generality of the conclusions, a search was initiated to trace and identify cases using a similar analytical approach to that used by INSROP. The outcome was disappointment: Not a single case could be found that featured all or most of the variables applied in INSROP, which turned out to be unique in multiple and special ways. This is not to say that comparisons with other cases have not been made. Whenever such cases can shed light on one or more of the variables addressed in INSROP they are included in the discussion, enhancing the generality of conclusions.

To fulfil the objective of this chapter, it has been subdivided into six parts: The first provides a brief introduction to the INSROP programme in terms of the study's objective, purpose, goals, thematic approach and organizational set-up. The second part discusses the political reality and history of programme initiation and implementation. The conceptual and synthetic significance of the philosophy and theory of science are discussed in the third section. The fourth part assesses the integrative impact of the research design. The fifth section deals with the choice of methodological tools, and the concluding section sketches the integrative potential of the INSROP reality as a whole with a view to the interconnectedness and pattern of interaction between project history, politics, philosophy, theory and methodology in general.

THE INSROP PROGRAMME

The name assigned to INSROP clearly identifies the object of study: the Northern Sea Route (NSR), a series of shallow, ice infested shipping lanes running through the shallow, ice infested waters of the peripheral sea territories north of Siberia, stretching from Novaya Zemlja in the west to the Bering Strait in the east. Depending on the course set, the route varies in length from 2200 to 2900 nautical miles. Basically, the route runs through the Russian economic zone and Russian territorial and internal waters. It is claimed to be an internal route of transportation developed, managed and controlled by Russia.

For generations, scientific knowledge of the sailing conditions along this waterway has been incomplete and unsatisfactory. In the perception of non-Russians, the NSR has remained a peculiarity far off the beaten track of international trade and shipping, located in an inhospitable, harsh environment on the geographical outskirts of the civilized world. Severe ice conditions, shallow waters, Cold War politics, legal peculiarities, nationality issues, military-security interests, the fragile environment, inadequate technology, polar darkness, the lack of insurance coverage and a host of other obstacles to navigation caused western interests to keep their distance and concentrate their trade on the blue, less hostile waters of more southerly latitudes. The emphasis on the obstacles to navigation made international shipping overlook, downplay and/or even neglect the unique geopolitical and geoeconomic potential offered by the northern route, both as a transport artery passing close to the abundant mineral resources on and offshore in Siberia, and as a geographical shortcut for cargo transport between the most industrialized and economically prosperous regions in the world—Western Europe, North America and East Asia. In the late 20th century, the NSR became an object of collective oblivion for foreign economic interests, i.e. an issue of collective illiteracy.

Purpose and Goals

The *Central Marine Research and Design Institute* in St. Petersburg (Russia), the *Nippon Foundation/Ship and Ocean Foundation* in Tokyo (Japan) and the

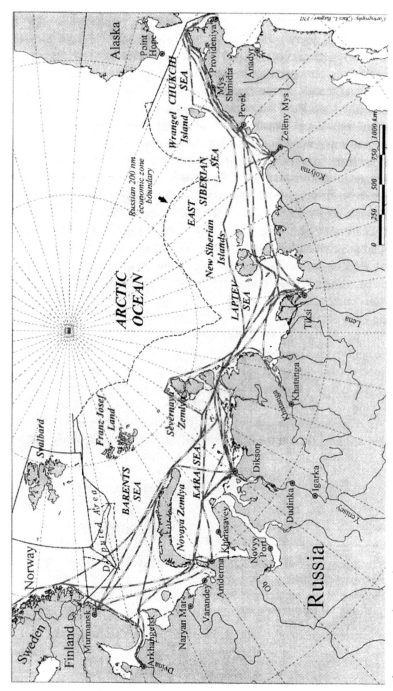

Figure 5.1. The Northern Sea Route. *Source: Lovås and Brude: INSROP-GIS, 1999.*

Fridtjof Nansen Institute at Lysaker (Norway) decided to change this state of affairs and to organize international research cooperation to examine the route. On 26 May 1993, the three parties signed an *Agreement on Research Cooperation*, in which they agreed to make the *navigational obstacles* and the *geopolitical potentials* of the route the two main pillars on which to base their joint research efforts.[2] The purpose was to build up a scientifically based foundation of knowledge, disciplinary as well as interdisciplinary, encompassing all relevant variables involved in NSR navigation with a view to enabling the public authorities, private interests and international organizations to make rational decisions based upon scientific insight rather than upon myths and inadequate knowledge. The prime motive of the three principal parties was thus to indulge themselves in practical-instrumental research to test the overall navigability of this route so long neglected by foreign interests.

Thematic Composition

To fulfil its objective, INSROP was organized as a five-year interdisciplinary and multinational research programme, split into four main sub-programmes: I. *Natural Conditions and Ice Navigation*, II. *Environmental Factors and Challenges*, III. *Trade and Commercial Shipping*, and IV. *Political, Legal, Cultural, Organizational and Strategic Factors* (see Figure 5.2). These sub-programmes were rendered more concrete when broken down into 52 separate sub-projects, all commencing in 1993. Upon completion of the programme in 1999, there were a total of 104 individual, but interrelated sub-projects.

The programme embraced a large number of individual disciplines from all the branches of learning: the *humanities*, the *social science*, the *natural sci-*

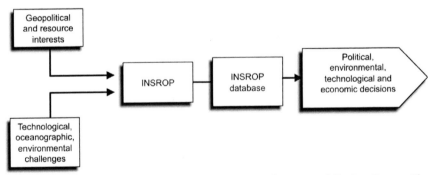

Figure 5.2.   Thematic Orientation, Objective, User Relevance and Design. *Source: The INSROP Newsletter, no.3, vol. 5, December 1997.*

ences and *technical sciences*. Within the humanities, historians from several countries took part, and the social sciences included economists, anthropologists, political scientists, international lawyers and geographers. The natural sciences brought in biologists, oceanographers, geologists, ecologists, geographers and ice specialists, and the technical sciences leaned on the expertise of a variety of marine engineers, and satellite and logistics specialists.

The programme also involved practitioners from many fields, e.g. insurers, sea captains, representatives of businesses, bureaucrats, etc. In other words, INSROP was designed as an *extended peer community*, also called *hybrid community*[3], i.e. a mix of researchers and practitioners working together across disciplinary boundaries and across the gap between practical life and academe. Thus, the integration objective of the programme was transdisciplinary—to achieve synthesis across three kinds of boundaries: *between disciplines within the individual branches of sciences, between the four branches of sciences, and between practice and academe.*

Thematic Approach

The programme had two priorities: to provide disciplinary depth and juxtapositional breadth. The work was divided into two phases; the first, lasting for three years, aimed at filling in some of the more important disciplinary gaps in the existing knowledge on the navigability of the NSR, whereas phase II was by and large devoted to integrating the principal findings of phase I. Phasing was necessary because science has to be aggregated before it can be integrated, and "until it is integrated it can seldom be useful in policymaking."[4] In between phases I and II (1996), the results of the programme were to be evaluated by an independent *International Evaluation Committee* (IEC) to check the quality, scope, multidisciplinarity and user relevance of the research work completed in phase I, and to come up with recommendations for priorities in phase II. In the evaluation submitted by the IEC, the original plan was modified to include some more disciplinary work in phase II to meet the overall integrative goal of the programme (see Figure 5.5). A large number of specialist studies (112 reports) were conducted in the first three years of INSROP. On the basis of these reports, links were forged between the sub-programmes to integrate the results across knowledge boundaries.

The Organizational Set-Up of the Study

Apart from what has already been said about the size of INSROP in terms of hybridity, multidisciplinarity, number of sub-programmes and projects, the following facts should be noted to comprehend the scope of organization and management needed to tie the programme together: At the close of INSROP on 31 December 1998, the total number of scientists that had taken part in the execution of the programme amounted to some 390 individuals from 14 countries on three continents. Of those numbers, 318 researchers from 69 different institutions in 10 countries took part in the actual research and submitted scientific reports. These quantities and the geographical scope and distribution of activities gave rise to an elaborate and detailed organizational set-up.

As illustrated in Figure 5.3, the supreme governing body was the *Steering Committee of Sponsors* (SCS), comprising six representative members

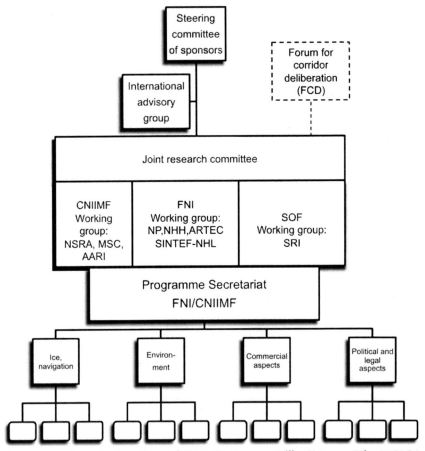

Figure 5.3.   Organization Chart of INSROP. *Source: Willy Østreng: "The INSROP SWEAT: What was it all about and how was it handled?" in Willy Østreng (ed.): The Natural and Societal Challenges of the Northern Sea Route. A Reference Work, Dordrecht, Boston, London, 1999, p. xli.*

elected from the sponsors in Russia, Japan and Norway. Under the Steering Committee was the *Joint Research Committee* (JRC), consisting of representatives from the three principal partners, including scientific experts from the Working Groups established on an individual basis by the same parties. The JRC—the head, heart and soul of the organizational scheme—drew up the guidelines for INSROP, suggested what projects should be implemented, and continuously monitored the programme from a scientific, organizational and financial point of view. The recommendations of JRC were then submitted to the SCS for discussion and final approval.

The interdisciplinarity of the programme also required an *International Advisory Group* (IAG), consisting of internationally acknowledged experts in all the polar fields employed by INSROP. The final entity in the formal structure was the *Secretariat,* located at the Fridtjof Nansen Institute, which was entrusted with responsibility for preparing for the meetings of SCS and JRC and for implementing their decisions. Interaction between these bodies was ensured by organizing regular meetings—10 gatherings in all. The SCS and JRC met with the Secretariat an average of twice a year to discuss scientific priorities, budgets, accounts, organizational matters, plans, personnel questions, etc.

In the main, scientific discussions, initiatives, coordination and integration took place within and between the four sub-programmes and the 104 individual projects. This was the level of programme fulfilment, whereas the levels above were primarily designed to define the framework and working conditions for the research.

Communication within the whole of the programme was to be ensured in four ways: First, between the programme coordinators (4 sub-programmes) and the individual project supervisors (104 projects) whenever a need arose for contact, clarification and/or coordination. This resulted in a series of workshops between and within the sub-programmes. Second, through an active secretariat that functioned all year round, communicating in actual practice with all the levels of programme implementation on a daily basis. Third, through a continuous flow of faxes, e-mails and telephone calls between all units and levels of the structure. Fourth and finally, through the 16 *INSROP Newsletters* published quarterly to present updates on research progress and other relevant administrative and scientific information.

THE HISTORICAL AND POLITICAL REALITY OF INSROP

This section will address the societal context of the programme, as expressed in the politics of research. To illustrate the dynamics of the societal context in which the programme worked, let us make a distinction between the *politics of initiation* and *politics of implementation* with a view to INSROP.

The Politics of Initiation

INSROP was a direct consequence of an initiative taken by the then Secretary General of the Soviet Communist Party, Mikhail Gorbachev. On 1 October 1987, he gave a speech that set the wheels in motion. Delivered in the Arctic city of Murmansk, the speech signalled a will to distinguish more sharply between military and non-military issues in the highly militarized Arctic. He opted to promote international civil cooperation in a region where military interests had for decades had the upper hand in Soviet priorities. The 'yield sign' had, as a rule, been in the lane occupied by the civil sector if and when the two interests were perceived as being on a collision course.[5] The aim of the Murmansk programme was to bring about ". . . a radical lowering of the confrontation level in the area," and to "let the northern part of the globe . . . become a zone of peace" and "the North Pole become a pole of peace." Gorbachev then presented a plan of implementation which has come to be known as the Murmansk programme. In it, he identified five non-military issue areas as particularly suitable for international cooperation to foster multinational security in the North:[6]

- Instituting a joint energy programme for the northern region;
- Instituting cooperation on environmental protection with the aim of establishing a joint environmental plan for the region;
- Establishing an international forum for Arctic scientific research;
- Establishing energy cooperation between Canada, Norway and the Soviet Union, and last but not least,
- Opening up the NSR for international shipping (the route had in practice, if not pursuant to legislation, been closed to foreign vessels since the late 1950s).[7]

In identifying the areas suitable for international civil cooperation, Gorbachev also indirectly made a distinction between *military* and *civil security*. The NSR was thus deemed a suitable object of international cooperation to secure the civil component of Soviet security interests in the Arctic. The break with former Soviet policy could not have been more radical or far-reaching.[8] Gorbachev had formulated a brand new political foundation for co-operation in the High North and the other Arctic states were actually invited to react to and follow up on it.

In most western governments, this historical shift in regional Soviet policy either went unnoticed, was played down or was categorized as one of many deceptive actions inspired by the logic of the Cold War. At the time, governmental reluctance to take it at face value was widespread and close to unanimous.[9] Ignorance and/or misconceptions ruled the day in governments for no less than three years. It was not until the beginning of the 1990s that the Murmansk programme unleashed, for the first time in Arctic history, a series of lasting cooperative initiatives among the Arctic states.[10] In the meantime, the Soviet government went on implementing part of the programme.

In November 1988, the Soviet Ministry of the Merchant Marine acted upon the implications of Gorbachev's speech and made contact with the director of the Fridtjof Nansen Institute (FNI) in Norway.[11] Following that meeting, an understanding was reached under which the director of FNI was to look into the possibility of establishing and organizing an international research programme focused on the NSR, on the condition that it would be open for participation for scientists from many countries and that access to Russian environmental data was guaranteed. After a complicated negotiating process that dragged out for five years, the three institutions finally reached agreement on 26 May 1993. The protracted period of negotiations and planning nearly forced some of the participants to give up any hope of succeeding.[12] In retrospect, however, it seems as though the time invested was well spent for sorting out some of the collaborative challenges encountered at a later stage of programme implementation (see below).

During the negotiations, politics continued to shape and define people's faith in INSROP. Three political incidents deserve special attention: The role played by the Norwegian Foreign Minister Thorvald Stoltenberg and his successor Bjørn Tore Godal, and that of the President of the Nippon Foundation/ Ship and Ocean Foundation of Japan, Yohei Sasakawa. Together, these three gentlemen brought the project to fruition in Norway and Japan despite strong opposition from parts of the business community, the bureaucracy and even the scientific community.

Foreign Minister Thorvald Stoltenberg gradually became convinced that Norwegian financial and political support to INSROP, and consequently to the Murmansk programme, would be an element in his own political architecture for building post-Cold War international cooperation with Russia in the North. He believed that it might also strengthen Norway's bilateral relations with Russia. He wanted to combine Norway's European policy in the South with its Russian policy in the North to involve Russia more directly in European cooperation.[13] In the latter part of 1993, he received a letter from President Boris Yeltsin's Special Advisor Alexander Granberg confirming that INSROP held high priority within the new regime. Stoltenberg subsequently decided to follow his political instinct and support the programme. As the founding father of the *Barents Euro-Arctic Region* (BEAR), established in

1993, in close collaboration with his opposite numbers in Russia, Foreign Minister Andrey Kosyrev and in Finland, Foreign Minister Vayrynen, Stoltenberg made the NSR an issue of cooperation within the BEAR structure as well.[14] At the same time, he decided that Norway's Ministry of Foreign Affairs should fund part of the research activity on INSROP to secure a dynamic source of information to be fed back into BEAR's NSR working group. In so doing, Stoltenberg put INSROP in the context of the overall efforts of international and Norwegian politics to do away with the political vestiges of the Cold War, and to follow up on the intentions of the Murmansk Programme. The same path was chosen by the new regime in Russia, the successor to the Soviet Union.[15] In this way, both Mikhail Gorbachev, Thorvald Stoltenberg and Boris Yeltsin acted to set up a *symbolic-instrumental* research programme, in that the prime reason for supporting research was political rather than scientific. INSROP was defined primarily as a means to achieve the political objective of dismantling the former East/West divide. Without strong political will to see the programme through, INSROP would most likely have been laid to rest in the cemetery of futile research proposals. The political expectation was that as long as research did not interfere with the attainment of political objectives, research could proceed without external intervention. It soon turned out that this precondition was harder to fulfil than what is usually the case for symbolic-instrumental research.

After three years of research (Phase I), the results were subjected to an international review by an independent International Evaluation Committee.[16] The Committee was, among other things, entrusted to advise the three cooperative partners on the necessity of going on with research and eventually launching Phase II. "After a scrupulous consideration of Phase I," the Committee unanimously recommended that INSROP proceed to Phase II.[17] This was so "because the issues addressed by INSROP (were) themselves cross-functional"[18] and "integration (would) be a key objective of central management of INSROP."[19] The Committee strongly reaffirmed that any study of the NSR would have to be conducted on a multidisciplinary basis.[20] It also indicated that more disciplinary data had to be gathered and several gaps in the data coverage had to be filled before the mission of the programme was completed (see Figure 5.5).[21] In the collective mind of the Committee, the job of INSROP was still to be finished. Phase II was mandatory to finish what had been begun. What Phase II had revealed to the Committee was that as the levels of aggregative generalizations increased during Phase I, important pieces of disciplinary information that had not yet been produced were identified. Since the strength of a generalization can hardly be stronger than its weakest link, Phase II was advised to come up with fresh, supplementary items of disciplinary information. As such, both knowledge- and practical-instrumental reasons were presented to support the original symbolic-instrumental backdrop of the programme. To put it differently: All scientific arguments, be they practical, symbolic or knowledge-instrumental, indicated the prudence of initiating Phase II.

Despite this, since the bureaucracy of the Norwegian Ministry of Foreign Affairs found itself squeezed between many financial obligations and still harboured some doubts about the viability of the programme, it proposed a substantial reduction in the Ministry's funding for the next phase of INSROP. Had it been implemented, this proposal could have unleashed a chain reaction of falling dominos among the sponsors, undercutting the financial basis of the programme and undermining the original political objective of research. Suddenly and unexpectedly, much was at stake: at worst, the programme could have been cut short halfway through its period of implementation. To avoid such an outcome, the head of the Secretariat and JRC requested a meeting with then Foreign Minister Bjørn Tore Godal. He argued that the symbolic-instrumental (post-Cold War) significance of INSROP was still a most important reason for continuing to fund the programme. Ultimately, Godal became convinced of the politics involved and raised the funding to a level acceptable for the fulfilment of Phase II. The arguments of politics became the crutch for making INSROP complete its designated course. Once again, financial constraints gave way to the needs of politics. A reminder of the politics of initiation was all that was needed for the foreign minister to re-establish the political foundation of INSROP.

The president of the Nippon Foundation, Yohei Sasakawa, also acted in support of the Murmansk Programme, but on a different pretence from that of Stoltenberg and Godal. His motivation was dual, both *practical- and symbolic-instrumental.* Concerning the former, he wanted to know the conditions under which the geopolitical aspects of the NSR could be brought to fruition, and the approximate time span in which this was feasible. In his mind, applied research in terms of transdisciplinarity was needed (see below).

Contrary to what might have been expected from a supporter of a practical-instrumental research programme, Sasakawa did not insist on keeping the results of the research confidential within the exclusive group of the principal cooperative partners (see Chapter 2). He wholeheartedly supported the programme's objective to make the results accessible to all parties interested in the NSR. This was the case because Yohei Sasakawa mixed his business inclination with a political vision to contribute to international cooperation and peace-building by actively supporting projects designed to bring people from different parts of the world together. He expressed the hope that the INSROP network "among experts and researchers (would) be enlarged, and that with their wisdom, the Northern Sea Route (would) be opened as a new sea lane linking East and West in the 21s century."[22] In his mind, INSROP hosted this potential twice over, in that it promoted international cooperation in research *par excellence* and, at the same time, was "a highly significant step in terms of future global development."[23]

From Yohei Sasakawa's vantage point, INSROP was a blend of practical and symbolic-instrumental research, the product of a fusion between different research categories with partially different objectives. In this respect the practical-instrumental aspects of INSROP concurred, coincided and complied with the symbolic-instrumental objectives of politics. Business and politics went hand in hand. Sasakawa thus became instrumental in the *politics of business and international cooperation.*

After having been presented with the content of the programme, Yohei Sasakawa acted with swiftness and determination. In a short meeting (less than one hour) with the director of the Fridtjof Nansen Institute[24] and the Norwegian Embassy in Tokyo in late 1992, Sasakawa decided on the spot to support the programme by providing a substantial sum of money. This decision unleashed a chain reaction in the shipping and business community, not least in Norway. The scepticism aired against the programme earlier had now mellowed and gradually turned to support, albeit lukewarm for some. Sasakawa's swift and positive reaction obviously raised doubts about the wisdom of not providing support for INSROP. As expressed by a representative of the Norwegian business interests: "If the Japanese believe in the programme, there must be something to it!"

The thumb originally turned down for INSROP by a host of bureaucrats, business people and even academics, was twice, upon its initiation and continuation, reversed by politics. However, it turned out that INSROP, to a large extent launched in symbolic-instrumental support of world politics, soon became an object of increasing practical-instrumental politics between some of the stakeholders in the programme. The implementation of the programme also revolved around politics.

The Politics of Implementation

After its inception, INSROP became an object of disagreement among the Russian security elite. Those supporting the Murmansk initiative urged that INSROP would strengthen Russian security for a host of different reasons: First, because opening the NSR would revive domestic Arctic shipping and help reverse the dwindling of Russian sea power. Second, because improved infrastructure along the route would ameliorate the Russian military capability to control the coastal zones in the Arctic and third, that the opening of the route would have a tremendous confidence-building potential between former adversaries and thus reduce the chances of military confrontation.[25] This analysis was derived from the policy of *Perestroika* that gave birth to the Murmansk programme and consequently to INSROP in the first place.

Certain circles among the Russian security elite challenged this position, claiming that the programme might potentially be detrimental to Russia's military security because of the prospect of massive infiltration of foreigners into strategically important and sensitive parts of the Russian Arctic.[26] They also contended that a growing influx of foreign capital into the High North, Siberia and the Far Eastern regions would tend to aggravate the trend towards these regions' autonomization and 'sovereignization.'[27] The sensitivity of this issue never seems to fade. In late February 2000, more than a year after the termination of INSROP, in an article in *Polyarnaya Pravda,* the Chief Navigator of the NSR Valery Kondratyev warned against the possibility that a foreign presence in NSR waters could be used for military surveillance and that it was mandatory for the Russians "to defend their national route."[28] This apprehension is deeply rooted in the minds of parts of the Russian security establishment. The Cold War was over, but its perceptions still seem to be looming in the air.[29] Those reverting to this line of argument seemingly kept INSROP under supervision from start to finish, and made their mark on the internal discussions between the cooperative partners.

The sensitivity of the matter was aired time and again by representatives of the Russian delegation to INSROP meetings. At almost all gatherings of the JRC and SCS, the Russians took the head of the Joint Research Committee and the Secretariat aside and reminded him of the thin and fragile security line on which parts of the INSROP programme balanced. The statements made in this unofficial and exclusive 'forum of corridor deliberations' (FCD) (see Figure 5.3)[30] were of two kinds: The first was presented as a 'general concern' that military and surveillance activities on the part of NATO countries in northern waters, in particular the Barents Sea, could backfire and hurt the programme beyond repair, whereas the second, was issued as a 'precautionary warning' to the foreign participants to exercise due care and prudence when asking for access to specific types of ice data from Russian archives. In both cases, the implicit message was that INSROP was under supervision and scrutiny by certain Russian authorities, and that Russian scientists were subject to political attention (bordering on pressure) to keep INSROP from harming Russian national interests. The answer provided by the head of the secretariat and JRC to these 'warnings' was equally repetitive: Due prudence would be exercised by INSROP when asking for access to Russian archives, and military dispositions in northern waters extended far beyond the influence of anyone involved in the programme. When these security concerns had been delivered and answered, the informal meetings of the FCD, which never lasted more than 15 or 20 minutes, adjourned and the formal deliberations of the JRC resumed.

The deliberations within the JRC also reflected the security concerns of the Russians, but they were expressed indirectly and camouflaged in scientific discussions about the quality of Russian ice data that would be required. The Russian delegation urged for moderation and caution when asking for access to such data, whereas the Japanese and Norwegian delegations asked for specific high-quality data in optimal quantities. This difference in approach grew more complicated in that differences in viewpoints on one aspect of research correlated with differences in others. To illustrate the point at hand: For scientists to chart the best course of navigation, data on the historic distribution of *polynyas* (open leads in the sheet ice), ice thickness, ice strength and ice age are mandatory. In this respect, the Norwegian and Japanese scientists wanted access to the raw data of different historical periods and regions, whereas Russian scientists accepted that they make their analyses on the basis of aggregated data. Those that did not approve of aggregated data also required detailed scientific references, whereas the Russians questioned the necessity of such practices since all involved were highly esteemed researchers in their respective fields; The group asking for original data and detailed references in research also approved of peer reviewing their draft reports, whereas Russian scientists, although reluctantly accepting the practice, felt peer reviewing to be an expression of mistrust, undue control, etc., etc. These differences were explicitly noted by the International Evaluation Committee which, among the reports it reviewed, "found methodological differences, a range of depths of analysis, and wide variations in the extent of referencing." Based on these differences, the quality of research, according to the Committee, ranged "from the excellent to the less rigorous."[31]

These differences alone cannot be explained as differences of scientific cultures, i.e. *sets of correlated requirements (values) defining what constitutes reliable and valid science.* The explanation is rather to be found in differences in political exposure between the three delegations; Norwegian and Japanese scientists had no restrictions placed on their scientific targets and culture, whereas Russian scientists were constrained and partly handcuffed by political considerations. For them, scientific interests encountered political interests and had to be reconciled to avoid collisions between potentially conflicting needs. The Russian dilemma was obvious.

Raw data on the distribution of ice thickness and *polynyas* are important not only from the perspective of civil cargo transportation, but also from the national security point of view. Strategic submarines (SSBNs) operating beneath the ice need to know the quality of ice conditions in any one area and season in order, if need be, to effectively launch their missiles or to surface through the ice for emergency or other reasons. Revealing detailed information on the state of ice conditions in areas perceived as strategically important implied that privileged knowledge would be released for anyone to use for military counter-measures.[32] Political reluctance to come up with classified data upon request from foreigners is therefore both understandable and reasonable. At the same time, these data are mandatory for science to identify the areas most suitable for surface transportation of cargo. Which interest should prevail—science or security? As a rule of thumb, states are, generally speaking, strongly inclined to give priority to the interest of high politics (security) over those of low politics (non-security issues). For governments, this is a 'zero sum' situation which has no simple solution. It is an ultimate dilemma that poses several crucial questions: How much data can be released without compromising national security interests? Is there a point of balance in which both interests can be sufficiently satisfied? Everybody knows that severe risks are involved in acting on these questions and that someone will suffer the consequences if something goes wrong. As discussed in Chapter 2, the proclivity of the political authorities is to keep their guard high and ready if they suspect—rightly or wrongly— that symbolic-instrumental research will harm rather than benefit national interests. This was the dilemma in which Russian INSROP scientists found themselves. They were assigned a most uncomfortable position, squeezed in between the scientific requests of their foreign colleagues and the 'off limits' signs posted by certain military quarters.

In situations like these, the goals of science and politics are easily frustrated. However, according to Bruno Latour, this frustration may be untangled by the *process of translation*, i.e. by the players taking detours through the goals of the others, resulting in a general drift in which the language of one player is substituted by the language of another.[33] In Latour's conception the "idea of translation provides the two teams of scholars, one coming from the side of politics and going toward the sciences and the other coming from the side of the sciences and following the circulating references, with the system of guidance and alignment that gives them some chance of meeting in the middle rather than missing each other."[34] To call the first ambition 'purely political' and the second 'purely scientific' is, according to Latour, completely pointless, because it is the 'impurity' alone that will allow both goals to be attained.[35] Thus, the process of translation is to combine different interests to form a single composite goal agreeable to all interests involved. Thus the process of translation is a middle-ground compromise reflecting the interconnectedness of societal spheres, in this context, between science and politics, between scientists and societal stakeholders.

The INSROP partners applied the translation process and used two measures of convergence to achieve resolution of their conflicting interests. They benefited from:

- the split of the Russian security elite, and
- the parties' willingness to take "detours through the goals of the others."

The likelihood is that the split of the Russian security elite worked to the benefit of and gave leeway for the Russian scientists, enabling them to comply with the requests of their foreign colleagues. The undisputable fact is that, despite security supervision, the Russians managed to get previously unpublished Russian ice data released for INSROP research is an indication that those supportive of the Murmansk Programme within the defence establishment provided some space for Russian researchers to move towards the interests of science. Indications are that Russian scientists had allies (stakeholders) in high places to moderate and restrict the most uncompromising positions of the sceptics to INSROP.[36] This is also suggested by the fact that the Russian Federation supported INSROP activities with a steadily increasing amount of roubles. Although modest in relation to the overall costs of the programme, public support increased from 15 million roubles in 1993 to 40 million (old) roubles in 1998.[37]

Along with the internal tug-of-war situation within the Russian defence establishment, intricate negotiations were conducted among the INSROP partners to reach middle ground. Give and take through translation became the strategy deployed by all. By way of compromise, good office, mediation and a substantial dose of stubbornness and patience, the parties finally reached an agreement that was scientifically acceptable to all and politically agreeable to the Russian security opposition. The translation process made the parties meet by way of "detours through the goals" of each other. *The Russian efforts to untangle the Gordian knot of politics and science within Russia without compromising national interests, is, in particular, to be commended.* The Japanese and Norwegian delegations assumed the important, but much easier role of defining what the quality of data ought to be to make the programme a worthwhile knowledge- and practical-instrumental endeavour. To optimise these standards, the Russians agreed, but opted to strike a balance between the potentially conflicting interests of politics and science. Once again, Bruno Latour's words come to mind: "No one said being a scientist (is) a simple job. To be intelligent, as the word's etymology indicates, is to be able to hold all these connections at once. To understand science is to understand this complex web of connections without imagining in advance that there exists a given state of society and a given state of science."[38] This is not, according to

Latour, a question of truthful scientists who have broken away from society and liars who are influenced by the vagaries of passion and politics, but one of highly connected scientists and sparsely connected scientists limited only to words.[39] "Instead of cutting the Gordian knot—on the one hand pure science, on the other pure politics—science studies struggle to follow the gestures of those who tie it together."[40] Based on experience, Lynton K. Caldwell claims that in these respects scientific accommodation need not mean capitulation or distortion of objective science. To the contrary, what is required is an interactive process among science analysts and programme planners (translation process) that will draw from scientific investigations the range of details and evidence that will enable policy-makers to arrive at decisions consistent with their mandate.[41]

The quality of the outcome of the INSROP translation process was recognized by the IEC which stated that the materials released already contribute substantially to what must be an informed debate on the future of the Northern Sea Route, and that INSROP's early success had been marked by close cooperation among representatives from Japan, Russia and Norway."[42] In the mind of the Committee, the aim of assembling and creating an extensive knowledge base about the NSR had thus been generally achieved.[43] In particular, the Committee recognized the success of Sub-programme I, which had "produced an enormous amount of valuable information that (was) both general and specific" and, on the whole, it had "fulfilled its main objective of data gathering, especially in the area of ice conditions along the NSR."[44] In the experience of INSROP, the implementation of the Environmental Impact Assessment (EIA) concept "has proven to be a key in dialogues across legislative, political and cultural borders. We highly recommend it for use in future development projects concerning the Northern Sea Route."[45]

This success notwithstanding, it stems from the logic of translation that the politics of implementation also had some of its concerns met, i.e. in restricting the optimality of the data releases. When it came to data on historical ice conditions, hydro-meteorological conditions and the development of a database to store, access and display the data to serve as a basis of forecasting methods, the IEC noted that most, if not all, sources had been identified, but that the "question remains whether all these data sources can be made fully available to all INSROP participants since the latest Discussion Paper of Project 1.3.4 states (p. 8) that '. . . many kinds of primary information . . . cannot be transmitted for general use to the participants of INSROP.' This serious issue requires resolution."[46] Here it should be noted that this conclusion is valid only for Phase I, whereas the final and crucial breakthrough in INSROP's access to Russian ice data came late in Phase II. In other words, the translation process of INSROP was gradual and protracted, covering both phases and close to five years of negotiations. Thus resolution did not come easy, but finally, with patience, perseverance and commitment to the goal of succeeding on the part of all parties involved, it was ultimately achieved. This underscores the graveness of the matter as seen from the point of view of both science and politics. This, however, poses the question of whether the process of translation compromised the reliability and validity of the research results: Was the intermingling of politics in the scientific dealings of INSROP of such a nature, magnitude and scope that it compromised the scientific integrity of INSROP?

The Political Impact on Scientific Results

In research programmes that deal with the interests of high politics, a complete and absolute separation of the realms of science and politics is hard to achieve insofar as the results are to be published and made available to all, friend as well as foe. By its sheer existence, politics may have an adverse impact on the conditions for this kind of research in two ways: either *directly,* by blunt proactive action to stop, curb or distort research in any way possible, or *indirectly*, by assuming the position of a passive bystander declining cooperation and refusing to release first-hand primary data. Either way, science is hurt in that it is forced to accept politically motivated restrictions on its own activity. Science becomes the compliant party. Politics decides, science accommodates. In instances like these, a reasonable scientific reaction would not be to insist on separate realms, as desired by one of the parties, but to assess whether the impact of politics is in relative agreement with acceptable standards of science. As stated by Roger Guillemin: "Science is not a self-cleaning oven so there is nothing you can do about the layers of artifacts incrusted on its walls"[47] The ones to decide whether the 'layers of politics' are agreeable to scientific standards or not are the researchers themselves. In line with the definition of scientific knowledge, *a compliance exists between the two elements when it has been scrutinized and universally accepted by the consensus of the rational opinion of the scientists involved in a particular programme*. In the case of INSROP, the question is then why the politics of implementation was ultimately conceived to be agreeable and acceptable by scientific standards?

As discussed above, politics made itself felt in INSROP by curbing the optimality of a *restricted*, although important portion of the total amount of data used for research in all four sub-programmes. The bulk of research data was produced by and provided by the Russian side without any interference from politics, either directly or indirectly. *The Northern Sea Route Dynamic Environmental Atlas* (DEA) may serve as an example. Most of the data for this atlas were supplied by Russian institutions and experts. If they had failed, the atlas would probably not have seen the light of day in the comprehensive form it enjoys today. At the time of publication, DEA had grown into a substantial base of systematized information, containing more than 4 000 individual georeferenced registrations on the temporal and spatial distribution of selected Value Ecosystem Components.[48] A substantial part of this information was released and published for the first time outside Russia, which made "the DEA a unique product."[49] In other words: It was the submission of a specific type of ice data that was the politicized bottleneck of INSROP, not of data delivery in general. The politics of implementation was narrowed and focused, not all-embracing. Was the quality of the ice data perceived by the ice researchers to meet the criteria of acceptable scientific standards?

The IEC rightly observed that in Sub-programme I, one issue that had not been addressed "in any of the projects.... (was) the assessment of the 'quality' of available data such as by associating a level of uncertainty to each data or data-set."[50] Due to the political sensitivity of the matter, INSROP never followed up on the Committee's tacit request to produce such an analysis in writing. Indirectly, however, the matter got top priority, and is why the process of translation lasted throughout the whole lifespan of the programme. This process was all about securing the delivery of sufficiently high quality of data where there was "a translation of political terms into science and vice versa."[51] It concerned the meeting of political and scientific requirements and their translation. Ultimately, the resolution was accepted by the rational opinion of a multinational team of independent, highly esteemed ice researchers. Their collective judgment was that the data sets released were acceptable and workable in the light of the objective of INSROP, either because the data deliveries were deemed to be the best possible foundation for the research that could realistically to be achieved, and/or because the reduced optimality of Russian data could be compensated by using supplementary data from other sources (remote sensing, foreign archives, etc.), and/or because doing something was better than doing nothing, and/or because fresh data had been presented in an acceptable amount and quality. Either way, none of the researchers declined to undertake INSROP research on the basis of the data received from Russian sources, nor did any of them declare the data unsatisfactory, or for that matter, distorted or fake. This implies that the translation process produced a resolution that was unanimously accepted by the extended peer community of INSROP—scientists as well as practitioners. Stated differently: the data sets were not scientifically *optimal,* but they were scientifically *acceptable.* In this vein, the collective acceptance of the available data was an expression of the unavoidable presence of *epistemic* and *pragmatic values* in scientific processes, INSROP included. As discussed in Chapter 4, these values restrict the (positivist) objectivity of research, but are neverthe-less intrinsic to all scientific processes, deriving from the finiteness of the time and resources available to conduct research. This being the case, in the collective mind of the IEC, Sub-programme I had "fulfilled its main objective of data gathering, especially in the area of ice conditions. . . ."[52] The final breakthrough in this respect came late in Phase II, several months after the Committee had delivered its evaluation. All in all, the 'political layers of sci-ence' mattered to the content of research, but not to the exclusion of accept-able standards of post-positivist research.

INSROP never experienced having any restrictions placed on the topics to address, nor any lack of relevant, acceptable data, or attempts to impose cen-sorship or actions to halt publications, or any other move that could be inter-preted as a violation of any of the three rights inherent in the freedom of re-search principle. There were not even attempts to exclude studies of the military-strategic utility of the NSR.[53] In principle, all parties, with the reluc-tant inclusion of the Russian delegation, accepted the viewpoint expressed by the IEC: If the political and security aspects of the NSR "had been ignored (in research), then all the technical and economic reports could mean very little in practical terms for opening of the route. What potential users need to know is the 'political risk' of involvement in NSR projects."[54] Here, the preoccu-pation of post-normal science with risk is brought to the fore and addressed in a setting of transdisciplinarity.

This is not to say that the political opposition to INSROP inside of Russia was altogether happy with every aspect of such studies or did not breath down the necks of their own scientists. Some of these studies undoubtedly bordered on the policy of implementation and ran the risk of being included. The deli-cacy of the matter came to the fore when the editor-in-chief of the INSROP integration book, *The Natural and Societal Challenges of the Northern Sea Route. A Reference Work* (REF), presented the results of the project in Chap-ter 5, dealing with the military, political, legal and human affairs of the route, to his editorial advisors from Japan and Russia. This flared a discussion that led the editor-in-chief to suggest the inclusion of a disclaimer in his foreword, stating that there was disagreement between the Russian and Norwegian members of the Advisory Group on certain specific highly sensitive issues discussed in the chapter.[55] Here, the conflicting parties practiced *adversarial collaboration* in which disagreements were accepted, put to fore and made public in writing. This procedure contributed effectively to resolving the mat-ter, and no further action was initiated by any of the parties to restrict publi-cation, alter conclusions or in any other way violate the freedom of research principle. Nor did anyone ever question the right of the author/editor to have a final say in all professional matters. What ultimately emerged from this benign and short-lived controversy was just another INSROP publication equipped with a somewhat longer foreword than originally planned. Nothing of substance was altered, only a disclaimer was added.

Nevertheless, the episode left the editor-in-chief with the impression that the concerns expressed indicated that the chapter balanced dangerously close to the edge of the unacceptable, and that corrective measures should be taken to keep it out of the active reach of the politics of implementation. Several factors substantiated this feeling. First, the chapter was not based on Russian data, but on available material from many open sources. Since the data was not criticized as being insufficient or distorted in any way, the cause of concern was not related to the ownership and/or quality of data, but more likely to the sensitivity of the conclusions. Second, the Russian member of the Advisory Group was not a co-author of the chapter in question or, for that matter, of any of the other chapters in the book. He could therefore not reasonably be held accountable for the content of a text written by others. Third, his involvement in the book was restricted to that of an advisor and not to that of an editor. Since it is common experience that the counsel of advisors is not always followed by decision-makers, he could not reasonably be held responsible for decisions made by others, in this respect a foreign editor-in-chief. In the light of these factors, the disclaimer could serve only one purpose: to show interested parties outside INSROP that the message had been delivered and that the Russians had washed their hands. Outsiders could only know for sure that the submission had taken place if confirmed in publicly available documentation.

Issues with the inherent potential to harm national interests are the more likely candidates of politicization. In the INSROP experience, high sensitivity implied cautious manoeuvring to avoid such topics from becoming part of the politics of implementation. Lingering doubts about the reality of a matter embodied a potential to escalate to the level of politics. By manoeuvring to restrict the number of issue areas in the political domain, the process to reach middle ground was eased for the one politicized area regarding ice data that originally prompted politics to become part of the implementation process. Thus, the Russian disclaimer contained in the lengthy foreword may have served the useful purpose of making the last phase of the translation process smoother and easier. In this perspective, the disclaimer served the overall interest of all sub-programmes.

The politicization of INSROP came in two forms. The politics of initiation was inclusive, embracing all the sub-programmes and their issues, while the politics of implementation was exclusive, focusing narrowly on just one single issue (see Figure 5.4). In spite of this, it was the latter—the less inclusive—that caused the most trouble for research. The function of

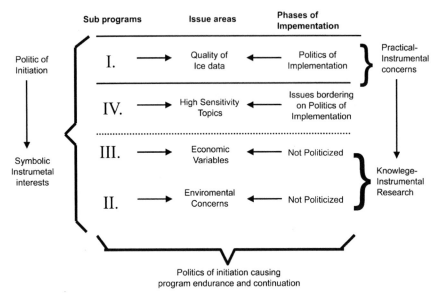

Figure 5.4.   The Political Backdrop and Infiltration of INSROP.

the all-encompassing politics of initiation was basically to render a helping
hand in the labour process in the run-up to the birth of the programme. This
was pre-research, whereas the politics of implementation was part of research. As long as research stays symbolic-instrumental, politics are most
likely to stay out of the realm of science implementation. Only when the
overriding goal of symbolic-instrumental research is perceived to be threatened by the research activity itself, may practical-instrumental reactions
produce a policy of implementation that will eventually interfere with the
conduct of science.

Apart from the politics of ice data, INSROP enjoyed all the privileges usually granted to knowledge-instrumental research, i.e. no interference by
sponsors or other interested parties in any stage of the research process. This
may be explained by four factors: First, the programme was part of an overall political scheme of multi- and bilateral diplomacy to build a new international order, both regionally and globally. Accordingly, the prominence of
the political commitments and prestige invested raised the threshold for programme cancellation and/or severe political interference to a threshold hard,
if not impossible, to cross. Thus, the symbolic-instrumental reasons originally evoked for establishing the programme in the first place made a lasting
contribution to keeping the programme going throughout all phases of implementation with a minimum of political interference. Here, the two cate-

gories of politics of INSROP levelled out: The politics of implementation caused problems for research, indirectly threatening to cut the programme short, whereas the politics of initiation contributed to its endurance and continuation (see Figure 5.4). Second, the process of translation was sufficiently successful to make the sceptics in INSROP keep their heads cool and moderate their actions. Third, freedom of research was a value subscribed to by all the three principal cooperative partners. It would have been hard, not least for the Russians striving to build a fresh democracy after Communism, to interfere too bluntly with the conduct of research. Here, INSROP was one of many litmus tests of the sincerity of their democratic commitments. If undertaken, any kind of interference in the research had to be low-key and channelled through intermediaries, causing as little publicity and disturbance as possible. In this respect, the Russians were most successful. Fourth, as INSROP unfolded, it gradually became clear that the research activity would not harm Russian national interests and that the process of translation would correct for such an outcome.

In Chapter 2, we indicated that experience suggests there is a close link between politics, societal needs and interdisciplinary research. The INSROP experience underscores the closeness of this interrelationship. Politics not only gave birth to the programme, but it never completely left it. In general, symbolic-instrumental research is sensitive to interference from two quarters: Either from the political opposition to the political objective of research and/or from the supporters of research if the scientific activity counteracts the goal of politics. Either way, symbolic-instrumental research is put on the spot by politics, and can, under certain circumstances, be supervised and even curbed by political measures. Serious curbing did not occur in the case of INSROP. Politics appeared mostly as a benign watchdog, howling only to restrict data delivery to foreigners. The overall conclusion is therefore that politics mattered both to the initiation and implementation of research, but never to an extent that was unacceptable to the standards of science. The fact of the matter is that INSROP produced fresh, new, coordinated and integrated insights on a range of interrelated topics in a manner and on a scope that made the joint effort worthwhile to all involved.

The overall goal of INSROP was to produce fresh disciplinary as well as interdisciplinary knowledge on the part of a large number of nationalities, disciplines, sciences, sub-programmes and projects, filtered through a four-layered decision-making organization, an extended peer community, politics and the interests of three principal partners. The complexity of this structure strongly suggests that the degree of interdisciplinary synthesis had to be affected. In particular, the politics of implementation contributed to reduce the level of integration.

This is so because the more coarse the data, the less subtlety of information, and the less subtlety of information, the less there is to integrate. This ties in with the assumption put forward in Chapters 2 and 4, that the slimmer, deeper and narrower the unit of topical specialization, the better the conditions for synthesis. The layers of artefacts encrusted on the walls of data have to be removed to locate the trading zones of disciplines—the hot spots of information exchange, transmission and integration. The smaller the items of information, the fewer artefacts are attached to its walls. Thus, increasing specialization is a process of data wall cleansing, artefact disposal, trading zone disclosure and information transmission. The more refined the data become, the easier they are to integrate, and the coarser they get, the less they integrate. Mismatch in the subtlety of data will produce uneven and less rewarding syntheses. To integrate successfully, it seems to be a prerequisite that there be a match in the subtlety of the data and information transmitted between disciplines; coarse data will provide raw integration, subtle data, subtle integration. The fact that a large portion of the ice data used in INSROP was presented in cumulative form increased the rawness of information, and reduced the ability to achieve the desired subtlety in syntheses. Raw ice data would have increased the integration potential manifold. Since all sub-programmes and projects depended on using ice data in their analyses, the integration potential of the whole of INSROP was affected by the decision to use only cumulative data. The two programmes hurt the most, however, were Sub-programmes I and II, whose primary focus was on the existence of ice.

THE PHILOSOPHY, THEORIES OF SCIENCE AND INTERDISCIPLINARITY

Politics of Implementation and Synthesis of Knowledge

In the planning and implementation of INSROP, the debate between positivists and post-positivists was never an issue. The same was true of the unity of science thesis, which was not invoked by anyone as a useful means of integration. What the INSROP partners leaned on when doing their work was their own disciplinary experience and interdisciplinary curiosity, making scientific practice and curiosity their common point of reference rather than philosophical speculation. Right from the outset, philosophy of science was considered a non-issue and separated from practice. This poses the question of whether INSROP was an exception to the rule, a philosophical anomaly among research programmes, or if the INSROP experience is commensurate with that of others?

From the science war we know that the heated debate among philosophers on what constitutes science only had a marginal impact on the conduct of research. To a certain extent, the two levels existed in parallel and went about their respective businesses without paying much attention to each other. As has been pointed out, just as the actual practices of research scientists have been little affected by philosophers' various re-descriptions of their activities, the popular understanding of the identity of scientists has not been greatly troubled by these developments either.[56] This should come as no surprise, since disciplinary debates have a long tradition of being internal to the discipline—to avoid crossing boundaries. Thus, INSROP shared in the experience of many scientists. The fundamental question, however, is not whether experiences are shared, but whether good science depends on the guidance of philosophy. In other words, is the relationship between the practice and philosophy of science a construct of a self-appointed supra-discipline (philosophy) or a necessity to foster high-quality research (in all disciplines)? Many and partly conflicting answers are provided.

The reasoning of John Hughes and Wes Sharrock is predicated on the idea that science is the purview of researchers with different convictions about the philosophy of science, just as it is conducted by confirmed atheists as well as by those who are devoutly religious. The problem as they see it is "not with science but with *scientism*." That is, with those philosophies such as positivism that seek to present themselves as having a close affiliation with the sciences and to speak in their name, and which then go on to fetishize the so-called scientific standpoint, arguing in favour of its unlimited and universal applicability.[57] The assumption that the sciences are put together 'from the top down,' and seemingly constructed according to a predestined master plan, they find to be strange, impractical[58] and even wrong since the natural sciences originated by separating themselves from philosophy.[59] Separation, not unification is the catchword used by Hughes and Sharrock.

Steven Weinberg concurs, claiming that knowledge of philosophy does not seem to be of much use to physicists. He goes on to argue that he has tried to read current work on the philosophy of science, "but only rarely did it seem to me to have anything to do with the work of science as I knew it."[60] He concludes that he is not aiming to play the role of a philosopher, but rather the role of a unregenerated working scientist who finds no help in professional philosophy, and he continues: "I know of *no one* who has participated in the advance of physics in the post-war period whose research has been significantly helped by the work of philosophers."[61] On the contrary, Weinberg often finds the service of philosophy to be a negative one, freeing science only from the constraints of philosophy itself.[62]

Stanley A. Schuum applies the same mode of thought, claiming that most earth scientists do not find philosophical discussions of their field very interesting, and that many of them treat the philosophy of science with 'exasperated contempt.' In his view, geologists tend to go about their scientific endeavours without giving much thought to the manner in which they proceed.[63] Ludwig Wittgenstein follow suit purporting that philosophy is more or less completely independent of science, and that philosophers from time to time are incoherent, confused and even talk nonsense.[64] How then can philosophy service the needs of science?

Most scientists will agree that speculation and theorization in one way or another play a role in empirical science. Lee Smolin admits that his discussions with Paul Feyerabend, has led him to believe that scientific progress "sometimes requires deep philosophical thinking, but most often it does not."[65] Even Steven Weinberg, a vocal critic of the usefulness of philosophy in physics, does not deny all value of philosophy. He admits that philosophers occasionally have benefited physicists, although generally by protecting them from the preconceptions of other philosophers. This, in his mind, is not to say that physics is best done without philosophical preconceptions. "At any one moment there are so many things that might be done, so many accepted principles that might be challenged, that without some guidance from our preconceptions one could do nothing at all."[66] The problem, as he sees it, is not with philosophy *per se,* as a guiding discipline of science, but that philosophers have not always provided scientists with the right notions. He is critical to parts of positivism, but supportive of philosophy's ability to come up with useful notions.[67] As pointed out by Phil Hubbard et al., "even if it is not explicitly articulated all research is guided by a set of philosophical beliefs. These beliefs influence or motivate the selection of topics for research, the selection of methods for research and the manner in which completed projects are subjected to evaluation."[68] For this reason, post-normal scientists call for the construction of an appropriate new philosophical synthesis, which will enable the practitioners in this area, including those in transdisciplinary research, to . . . produce a new synthesis of theory and practice. . . ."[69] In short, philosophical issues permeate most decision of research. No one escapes the philosophical culture of one's birthplace and educational setting.

This is to say that philosophy in many respects is a *second-order discipline.* The practice of research is to a large, if not absolute extent self-sufficient and self-propelled. As observed by Phil Hubbard, et al.: "Regardless of how the personal salience of each approach is explored, it should be appreciated that there is no right or wrong way of thinking (scientifically). Each approach has its critics and each has moral and political effects by informing teaching, research, actions and policy. To put it bluntly, there is no pure thought, no pure theory."[70] In this way of reasoning, the sciences do not need to turn to philosophy in order to solve the tasks that they set themselves.

Richard Peet reasons that social theory is preferable to philosophy as the guiding hand of science in part because philosophy often seems remote from anything resembling practical concern. To some extent, this is because disciplinary boundaries melt away, leaving social theory as the common ground for generalized yet practical ideas. Philosophy is best, according to Peet, when formed through empirical and theoretical practice rather than abstract speculation, reading philosophy in general, or reading theoreticians through the lenses of other theoreticians.[71] The contribution of philosophy to the sciences "lies rather in the possibility of a deeper, or reflective, understanding of the content of the claims advanced in scientific theories."[72]

In practical research, philosophy and theory fill in the lacunae between empirical facts, providing a coherent perspective on the functioning of reality. The combination of empirical facts and speculation are thus the building blocks of theories to be further tested against reality at later dates. In this way, philosophy is the speculative merger, i.e. the vision, of how the facts most likely interconnect with that part of reality that has not yet been unveiled empirically by scientific methods. Or, in the words of F.C.S. Northrop, the object of scientific knowledge is never known directly by observation or experimentation, but is only known by speculatively proposed theoretical construction.[73] Therefore, theories and philosophies will be in flux, subjected to constant adjustments and changes as the sciences produce more empirical facts. Whether we chose to designate what we use to fill the blanks between the facts as philosophy, social theory, seminal or interfield theories or some other conceptual construction is less important than acknowledging that scientific practice allows for a separation of theorization from empery, without sacrificing scientific quality.

As pointed out by Julie Thompson Klein, imposing a particular conceptual framework at the beginning of a project may mean basing the entire project on something inappropriate for the problem at hand.[74] De Wachter argues that interdisciplinary projects should ideally begin with the temporary suspension of all known methods (theories and models), so that the overall question is formulated in an unbiased, collaborative manner.[75] If a discipline, theory or methodology is allowed to dominate, it may inhibit role negotiations, delay communal work and create uncritical social and cognitive dependence.[76] To avoid such pitfalls, staying neutral in choice of scientific integrative measures at the outset of research may help avoid a dogfight that may ultimately have been resolved naturally by peaceful means among the disciplinarians as they walk their designated road together. The process of de-education from the confinement of the disciplines and the effect of synergetic discourse will in most cases be helpful for identifying integrative measures of consensus among the participants. Through reciprocal learning over time, team members will come to know each others' basic theoretical formulations and methods, building platform based on a common denominator.[77] In this vein, interdisciplinary research is nothing but specialties in a concerted search for bridge-building methodologies and theories on each others' territories to achieving the synthesis of knowledge.

The absence of explicit philosophical debate within INSROP probably made the execution of the programme less complicated. No one felt that their particular philosophical leaning or sympathies, if they ever had one, was set aside so that someone else could fulfil the objective of synthesis. The programme relied on its own tools, theories of many disciplinary origins and methodology that provided juxtapositional depth and breadth. However, this is not to say that the programme was totally deprived of guidance from philosophy. The fact of the matter is that the practical orientation of INSROP mirrored the looming preconceptions of post-positivism, holding a unique potential for interdisciplinarity. The disciplines were ranked as being equal in scientific value, and respect was granted for differences in methods and procedures. The borrowing of tools was encouraged across disciplinary boundaries, and synergetic discussions were expected and actively promoted through workshops between sub-programmes. The social and natural parameters were regarded as interrelated and interacting, interpretation was seen as a means of scientific methodology, values were defined as objects of study, and qualitative and quantitative methods were regarded as supplementary and complementary means of science. The aim was to achieve reasonably balanced representation of the natural and social sciences,[78] and between disciplines that lend themselves to quantitative and qualitative methodology.[79] The assumption was that these measures would facilitate constructive cooperation also in the form of adversarial collaboration across boundaries. Thus, in terms of the usefulness of the right preconceptions, the philosophy of science indirectly provided guidance to the INSROP without anyone making an issue out of it.

THE IMPACT OF DESIGN ON INTERDISCIPLINARITY

What we will do here is to assess how and to what extent, *cultural diversity*, *hybridity* and *organizational set-up* became an integral part of the research design. How was the design influenced by the cultural setting, hybridity and organizational set-up of INSROP?

Cultural Diversity

Culture is a complex system of interconnected parts stemming from a nation's history, language, convergence of ideas, relationship to nature, to fellow Man, to man's place in society, etc. The concept of culture is elusive, interpretive and hard to define in conceptual terms. On one side, there is the image a culture has of itself, on the other, there is the reality of how culture is perceived by others. These two images are seldom identical, the truth lies somewhere in between, and they change over time. Both are shaped by a complex set of forces largely beyond our control.[80] This being said, all communities produce a linguistic, literary and artistic genre, as well as beliefs and practices that characterize social life and our ways of thinking, feeling and behaving.[81] As a phenomenon, culture can help us understand why humans act in the way they do, and what similarities and differences exist amongst them. A fairly consensual assumption is that culture is important in shaping our mindsets and in uniting peoples in such a way as to distinguish mindsets and nations from each other. Clifford Geertz sees culture as the *webs of significance* that peoples creates for themselves—a context within which social events, behaviour and ways of thinking can be intelligibly described.[82] To be properly understood, the phenomenon of culture requires actions, statements and/or the behaviour of others to be interpreted within the context of related actions, as parts of a larger mosaic. Culture is thus an inherited framework of references, which may explain how we think and prioritize. The world is then defined as the home of all 'webs of significance.' Thus, the multiplicity of cultures provides for pluralism of behaviours and thoughts, and affects our priorities, professional as well as individual. Cultures are important shapers of thought processes, and thoughts interface where cultures meet. This does not mean that a person within a culture losses his individuality and free will. It does, however, mean that, "without knowing it, we are more than we think children of our time, place and training."[83] Culture means that without thinking about it, people respond alike.

Against this backdrop, defining precisely what culture is, gauging its impact on performance, distinguishing it from other behavioural factors, extracting it from the joint values of epistemic communities, discerning it from various interests and deciding its influence on outcome, is a most demanding social-psychological and anthropological task far beyond the competence of this author. Even anthropological writers on culture admit that using culture "as an analytical tool can be problematic."[84] For that reason alone, this section probably should not have been written, but left out and hushed up. Two facts counteract this inclination, however: First, the cultural diversity of the programme is beyond dispute, stemming from the geographical dispersion of participation—3 continents and 14 countries. Second, the lingering sensation within INSROP throughout the whole of the programme period ended in cultural differences mattering and that they should be addressed and accounted for, one way or another. The account provided here is based on a mix of personal observations embraced and buttressed by scattered anthropological statements, entangling many interacting factors, culture just being one of them.

Jacob Bronowski once wrote that anyone who has worked in the Far East knows how hard it is to get an answer to a question of fact there. In his mind, a person from the Far East: "wants to do what is fitting, (and) is not unwilling to be candid, but at bottom he does not know the facts because they are not his language. These cultures of the East . . . lack the language and the very habit of fact."[85] This assessment was offered in 1956. The INSROP experience of the 1990s both deviates from and complies with this description. The Japanese participants certainly proved that they know the language of facts, but certain languages are more fact-oriented than others. For example, the Japanese languages of natural science and technology are firmly rooted in facts and spoken as a distinct and clear 'mother tongue.' Through INSROP, the Japanese indicated that their culture is one of science facts. Hard fact languages have been the movers of civilization in Japan as they have in the West. However, when it came to the social sciences, i.e. the soft-fact languages, the Japanese turned out to be less clear in their expressions. As expressed by Bronowski, "he (the Japanese) does not understand that one wants to know." There is, in his conception, an "indifference to the world of the senses."[86] In the sphere of the soft sciences, this may have materialized in their reluctance to never actively to take part in discussions on the broad politics of the NSR. The Japanese focused their energy on the research that addressed the requirements for year-round sailing on the NSR. The interaction of sea ice and shipbuilding technology predominated their professional interest. To the Japanese: "Greater use of the Arctic Ocean could become feasible with the development of marine technology to make possible safe, economic voyages in ice infested waters, as well as exploitation of natural resources and advances in oceanography, marine geology and earth science as well."[87] Topics of environmental and policy interests attracted no Japanese research efforts and, in discussing priorities in these topics, the Japanese did not practice "a language of facts." They actually did not take part in any of these discussions in a substantive way, leaving them to the Russians and the Norwegians. At the same time, they did not explicitly oppose the soft sciences playing a role in INSROP. However, their body language expressed a certain feeling of discomfort when JRC discussions concerned questions of political controversies pertaining to the route. In other words, within the context of INSROP, the language of social facts did not come across as the preferred language of the Japanese. Their scientific culture seems firmly rooted in the Newtonian tradition: Hard facts are objective and consensus-building whereas soft facts are subjective and potentially conflict prone. By not practicing a soft-fact language, the Japanese assumed a tacit position, assuming a sort of neutrality between the Norwegians and Russians, which actually eased agreement on social science priorities in the deliberations of the JRC.

The Norwegian social anthropologist, Tord Larsen, once described Norway as "the homeland of bye-laws and object clauses." In his mind, the idea of playing for the sake of playing is alien to Norwegians. Life is too serious to waste time in pointless play. In the spirit of egalitarianism, Norwegians also love package solutions and like to simplify reality by stringing together opinions in long, solid arrays: if you are familiar with one link in the chain, you know the rest.[88]

Although, this author never consulted with his cooperating partners about the aptness of this description, it would come as no surprise if they were to concur. The stated objective of INSROP, i.e. to achieve overall integration of all relevant aspects of NSR navigation, became the lodestar—a navigational beacon—for the Norwegians. This was reflected in the emphasis of their research portfolio. The Norwegian side asked for and assumed coordinating responsibility for studies aiming at integration. In four out of a total of five integrative projects, the Norwegians took a leading role. Forming a whole out of particulars, sticking to the programme 'bye-laws' and 'stringing opinions together in long, solid arrays' showed in the Norwegian demeanour. Thus, Norwegian researchers took part in 19 per cent of the studies of natural conditions and ship technology, 60 per cent in environmental studies, 70 per cent in studies on trade and commercial shipping and 73 per cent in political, legal, military and cultural studies. In quantitative as well as in qualitative terms, the Norwegians emphasised the environment and the societal/human aspects of navigation, rather than the role of technology in sea ice navigation.[89] They applied a perspective which advocated that the diversity of values and interests—societal, human and natural—should be treated in a sustainable and integrated manner to preserve what was termed the *socio-biodiversity of navigation* (see below). Totality, not segments, was what should be addressed.[90] Here, the inclusiveness of Norwegian culture acted as a supplement to the exclusiveness of the hard-fact language of the Japanese. In Tord Larsen's view, all aspects should be linked together and included in a package solution to demonstrate the interconnectedness of the multiplicity of reality's variables. In Jacob Bronowski's imagery, the Norwegians tried to speak the language of all Sub-programmes, both the soft and hard fact languages, and they practiced them in an egalitarian manner. So did the Russians, but for a different reason and a different purpose.

According to Petter Norman Waage, Russian culture has been formed at the junction between the East and the West. As such, it may be regarded as a bridge, facilitator or merger between two distant cultures: "An intermediary between the Asian and Western cultures."[91] Russia is the cultural 'Middle Realm' of Asia and Europe. According to Waage, there is probably no nation which appears so mysterious to the West as Russia. She is close enough to be confused with Europe, while at the same time so remote that all domestic resemblance disappears.[92] Her culture is both akin and alien to the Norwegian and Japanese cultures. In Bronowski's terminology, the Russians speak the fact languages of a mixed scientific culture blended together in a unique manner. As has been observed, Russian preoccupation with the natural sciences in education and research is well known.[93] Even within social science disciplines, Russians are strongly influenced by the philosophical thinking relating to the natural sciences. Social systems are, for example, to a large extent studied on the basis of biological models, whereas studies of international relations often take their inspiration from economic theory. The predominance of this orientation has actually made western social scientists advise their Russian colleagues to rely less on mathematics and 'simple economic theory' in their research, and to make more use of the scientific achievements of modern social science.[94] This friendly advice aside, the Russian demeanour is thus both different from and similar to the Norwegian and Japanese positions.

In INSROP, the Russians not only copied the Norwegians in getting involved in all four Sub-programmes, but they made their participation profile even more equal and egalitarian than that of the Norwegians. Russian scientists and experts took part in 44 per cent of the projects relating to ice navigation, 37 per cent of the studies concerning environmental protection, 61 per cent of the trade and commercial studies, and in 45 per cent of the social science projects. Their participation profile indicates that they commanded fluency in all relevant languages of facts, soft as well as hard. This is not, however, entirely the truth.

As stated by the head of the Russian delegation, Vsevolod Peresypkin: "Russian investigations on the International INSROP Programme proceeded from the fact that the Northern Sea Route is Russia's main national transport line in the Arctic."[95] In other words: The holistic approach seen in the participation profile of the Russians was a reflection of the Russian claim to ownership of the route: Any proprietor will by inclination take a comprehensive interest in his own property. But there was more to it than this; it also reflected the politics of implementation. As discussed above, Russian scientists were 'watched' by certain official security circles that harboured concerns about the possible negative impact of INSROP on national interests. To reduce the risk of controversial outcomes, Russian scientists therefore had no alternative but to assume a prevalent and visible posture in the most controversial sub-programmes, i.e. Sub-programme IV, dealing with indigenous peoples, military security, legal affairs, etc., and to a lesser extent with Sub-programme 2 dealing with environmental matters. The sensitivity inherent in these programmes was expressed in many ways, indirectly (JRC) and directly (FCD), but also in the foreword of the book *The Natural and Societal Challenges of the Northern Sea Route. A Reference Work*, edited by Willy Østreng. By taking an active posture in these two Sub-programmes, Russian scientists could claim a legitimate right of review regarding the work done by foreign researchers and they could also consume a fair chunk of the funding appropriated to the two sub-programmes. Both elements reduced, if not eliminated, the risk of 'undesirable conclusions' and gave them latitude to wash their hands. The written statements provided undisputable documentation of their professional disagreements with their foreign colleagues.

The core interests of the Russians in INSROP were expressed in the deliberations of the JRC. Here, the inclusiveness of their participation profile was replaced by the exclusiveness of their primary interest. Again, Vsevolod Peresypkin is a most prominent source: The INSROP programme, as he saw it, "was intended to attract the attention of business circles, shippers and shipowners to the Northern Sea Route, to demonstrate economic efficiency of the use of this transport line as compared with the southern alternative for cargo, and in this connection to retain and keep in service the Russian Arctic Fleet and icebreakers, preserving the infrastructure of the NSR, providing jobs for the population of Arctic towns, settlements and polar stations, and to secure a profitable operation of the route."[96] That is to say, the overall purpose of INSROP was hinged on *economy* in all its multiple manifestations and ramifications. The listing of economic factors presented by Peresypkin designated top priority to the topics of Sub-programmes I (ice navigation) and III (trade), whereas Sub-programme II (environment) was mentioned only in passing, referring to the fact that an assessment had been made of "the potential effects of higher intensity of navigation on the Arctic environment."[97] The topics of politics and military affairs were left out when he commented on the overarching intention of INSROP. It is worth mentioning that the intentions identified by the head of the Russian delegation deviated from the officially stated objective of INSROP (see above). The Russians had their own national reason and agenda for taking part. This showed in their writings. According to the IEC, all papers coordinated by the CNIMF on the commercial potential of the route had "a rather technical approach, uncritical of market potential and environmental problems, and lacking sufficient non-Russian references."[98] The Russian Minister of Transportation Sergey O. Frank reiterated Peresypkin's viewpoint at an international User Conference on the NSR in 1999, stating that: "It is quite possible to ensure an all-year navigation in the eastern zone of the Arctic area (the western part has been utilized for year round navigation ever since 1978), as well as all-year round Arctic transit, provided that there is a sufficient demand."[99] The Russian attitude was that the experience they have gained over more than 60 years of ice navigation would suffice to assure year-round transportation, and that if the volume of international shipping was sufficiently large, it could be done on good commercial terms in a fairly short-term perspective. In arguing their case, the Russians played down the challenges associated with sea ice navigation in order to get commercial operations started as soon as possible.

The primary language of the Russians was that of hard facts, with a special proclivity for economic profitability without market reasoning. Here, they were in line with the Japanese position. In taking a keen interest in all Sub-programmes, they were also in line with the Norwegian position, but from a different perspective than that of the Norwegians. In this way the Russians, in Waage's conception, apparently assumed a 'Middle Realm' position between their cooperative partners. Their heart, however, proved to be with the hard languages, feeling a certain kind of discomfort with the soft languages. In IN-SROP, the Russians were in actual practice closer to the Japanese than to the Norwegians.

The differences in priorities and behaviours observed between the three parties are at least partly a reflection of the differences in cultures which affected the research process in three ways: first, they added to the richness of interdisciplinary research in that they provided different veins of thought, focuses and value priorities, forcing the participants to take a second and fresh look at their own preferences and approaches. Inherent in these differences lay the roots of invisible collegial and cultural synergies. Thus, cultural plurality may cause intellectual dynamics and interactions, introducing beneficial new and fresh perspectives to the formulation of research objectives. One example: At an early stage of INSROP, doubts were expressed about the prominence attributed to environmental studies in the programme. A year later, all parties agreed to the high profile given to this issue. Second, cultural differences can also inhibit communication and make the threshold of meaningful interaction so high that it will inevitably give rise to misunderstandings and unnecessary squabbling. INSROP was no exception. At important junctions in programme implementation, the meetings of the JRC often dragged on beyond the point of collective exhaustion before relief was offered by adjournment. This bred frustration and a feeling of hopelessness, forcing some people to succumb to the effort required for success. Patience and perseverance are thus elements which cross-cultural programmes cannot do without. Third, to keep different mindsets moving in a concerted direction over a long period of time, mutual diplomacy is required for progress and success. IN-SROP had its sad moments of failed diplomacy and its happy moments of success where mediation and good office rekindled the torch for all to see and find their ways back to the same collective path. By and large, the programme muddled through on the basis of the successful diplomatic efforts provided by all parties. As pointed out in the *Artic Marine Shipping Assessment*—study (AMSA) of the Arctic Council in 2009 "Creating and carrying out INSROP was a challenge from beginning to end with respect to administration. It took years of networking, negotiations and lobbying to shape the program and to obtain funding, and its eventual initiation was a witness to the strong wills and personalities of the key figures involved."[100] In this connection, leadership and diplomacy were prime facilitators of collective progress. Here the INSROP experience complies with that of multiple transdisciplinary projects.[101]

Extended Peer Community

The extended peer community, also called hybridity, of INSROP involved three categories of participants: University professors bearing a special responsibility for securing the knowledge-instrumental quality of the programme; practitioners (insurers, sea captains, business representatives, shipping experts, bureaucrats, etc.) to keep an eye on the applicability of research; and third, researchers from applied research institutions spanning all sciences to act as 'mediators,' and bridging the assumed gap between the two former categories. Thus naturally some representatives of the latter organizations assumed the position of bridge scientists or polyvalent specialists, making the communication across the various divides a smoother enterprise. Of the 468 people[102] involved in the implementation of all aspects of the programme, some 90 were practitioners, 93 came from universities and 283 earned their livelihood at applied research institutions.[103] This pattern was also reflected in the diversity of the participating institutions. Of the 97 different organizations and sub-departments engaged in INSROP, 31 were university related, 27 practical and 39 applied research institutes.[104] The hybridity of the INSROP team was designed to mirror the overall programme objective to build a scientifically based foundation of knowledge, disciplinary as well as interdisciplinary and knowledge-instrumental as well as practical-instrumental, to serve the navigational needs of real life interests, including the public authorities, private/commercial interests and national/international organizations. Thus, the production of knowledge and insight was designed to achieve synthesis across three kinds of boundaries: between disciplines within the individual branches of learning, between the four branches of sciences and between the academe and practice. To harvest these gains from hybridity, representatives of all three groups were represented in all corners of the INSROP organization and at all levels of implementation in the decision-making bodies and in the research and evaluation processes (see Figure 5.5).

Notwithstanding this planning process, transdisciplinary teams must deal with the existence of a strong element of tension between heterogeneity and effectiveness. On the one hand, heterogeneity is the primary characteristic of transdisciplinary teams; on the other, it represents the greatest threat of conflict between team members, not least when it comes to the validity of knowledge. This is what has been labelled the *transdisciplinary paradox*.[105] The combination of this paradox with the cultural diversity of the INSROP team could have posed a serious threat to the team's productivity. In general, the working relationship and conditions between the three groups, four sciences and multiple disciplines of INSROP were surprisingly good and constructive. None of the participants was caught up in any preconceived notions that their respective field of expertise held scientific supremacy over the others. A postpositivist attitude of equality in scientific value was characteristic of the working atmosphere. The willingness to listen and to include 'foreign' contributions across boundaries was widespread, and the programme never experienced any kind of condescending attitudes from the hard sciences, for instance, which as could have been expected on the basis of the Newtonian inclination and cultural leanings of the cooperating parties. The consensual sense was that hybridity mattered to support and realize programme objectives and that the bridges of communication had been erected for all to cross. The IEC noted this, referring to "INSROP's early success" as the "close collaboration among representatives from Japan, Russia and Norway," whereas the Norwegian Shipowner's Association found reason to congratulate the research team for having carried out its complicated task in such a "professional and excellent way."[106] The IEC also commended INSROP for its ability to meet its multidisciplinary objectives, which had "been substantially achieved,"[107] and for completing useful applicable research within all four Sub-programmes.[108] The Norwegian Shipowner's Association followed suit, concluding that[109] the four sub-programmes "present results of great interest — results that are helpful for us in judging the business opportunity in the area."[110] In particular, the data base and the Geographical Information System (GIS) were deemed by the IEC as "significant achievements" that would be useful to many sectors external to the programme.[111] The AMSA-study concluded in 2009 that "INSROP's results and impact were considerable given the starting point. A wealth of new and unique knowledge on the Russian Arctic was produced, as well as making available to the international community information that was previously only known internally in Russia. This information and results are still being sought and utilized today, 10 years later."[112] This result was not, however, due to hybridity alone.

For extended peer community to breed convergence, the individual members must harbour the urge, curiosity and will to transcend boundaries, the skill of interdisciplinary sensitivity and competence, an holistic problem orientation and the practice of Bruno Latour's *process of translation*. When these measures and abilities come together, the members of the hybrid community will take detours through the goals of the others to achieve a translation of political terms into science and vice versa (see above). Thus, hybridity is only one among a host of interacting elements to foster interdisciplinary communication and the convergence of knowledge. The primary role of extended peer community in synthesis is thus to secure representation of all involved voices, expertise and interests—to provide a team composition reflecting the objectives of research and the complexity of reality.

The IEC's overall positive evaluation of INSROP does not mean that the programme did not experience 'clashes' across its various boundaries. Such clashes are to be expected since *tug-of-war* and *give-and-take* constellations are integral parts of synergetic discourse. What is required is interaction and a 'bending of minds' through disagreements and discussions. The borders of disciplines are the commotion zones of science. This being said, there were two outcomes stemming from hybridity that were not anticipated.

First, not all the participants were equally skilled in dealing with the translation process. At an early stage, there were clear signs that some disciplinarians claimed enhanced significance for their own discipline at the expense of the others. The problem did not rest with the natural sciences, but rather with the social sciences, in particular economy, indirectly defining the other fields as feeder disciplines to the core discipline of economics. The claim was that the focus of research should be on the economic features of the NSR, i.e. if the analysis proved the route did not meet the operational requirements of market economies, then the other disciplines had no independent role to play. The market economy features of the route were defined by some to be the crux of the matter. This was a deviation from the official objective of INSROP, which was to build a scientifically based foundation of knowledge "encompassing all relevant variables involved in NSR navigation" for the purpose of testing the overall navigability of the route." By shifting their focus to the latter aspect, some of the economists achieved a perfect fit with the confinements of their own discipline, claiming centre stage and the right to feed on the contributions of subordinate disciplines. This viewpoint took on a fairly prolonged life because it gained some support within the programme by the inclination of some of the collaborative parties and commercial practitioners to focus

their personal emphasis and interest on the economic attributes of the route
rather than on <u>all</u> aspects of navigation. Although, in the heat of discussions,
some of the economists threatened to leave the programme if they did not get
it their way, and one did actually leave, these skirmishes were sorted out in
a fairly cooperative manner by reiterating the original objective as the
lodestar from which to set the course for the programme. One important reason for this benign outcome was that the disagreement was not between a
united front of economists on the one hand and a joint coalition of the other
disciplines on the other. Both interpretations received varying support for
their principal stand across disciplinary and group boundaries. The misconception was not a product of traditional disciplinary in-fighting so it lent itself to being sorted out by way of reason. That Sub-programme III, dealing
with the commercial aspect of navigation, did not receive any special attention in relation to the other programmes was noted by the IEC.

The problem of Sub-programme III actually turned out to be the very opposite of what some of its representatives had argued. The IEC set the record
straight: "It is evident from the review of papers in Sub-programme III that
there has been insufficient cooperation between researchers from different
disciplines working in the same and similar projects, and a lack of integration."[113] Not only was the programme criticized for its lack of integration with
other disciplines, but also for a lack of integration between "two distinct
schools of thought: the business economics approach of Nordic/Anglo-Saxon
researchers, and the technical engineering approach of Russian researchers."[114] This prompted the Committee to recommend that more disciplines, approaches and research communities be included in the research
done under Sub-programme III since it stood out as the least integrated of all
the four sub-programmes once phase I was completed.[115] In terms of the number of reports produced and pages written in the course of the programme period, Sub-programme III individually produced fewer reports than Sub-programmes I (on ice and navigation) and IV (on politics, security and legal
matters); Sub-programme III produced 40 reports (3044 pages), as against 50
reports (4148 pages) emanating from Sub-programme I and 43 reports (3592
pages) from the two latter sub-programmes.[116] In retrospect, a lot of valuable
time was wasted discussing a non-existent problem. In actual practice, Sub-
programme III received no special or superior attention. It was just another
important sub-programme, like all the others.

Two lessons can be derived from this experience: First, avoid discipline-
shaped interpretations of interdisciplinary objectives. If derailing takes
place, valuable time is wasted and creative leadership and cautious diplomacy are required to get the programme back on track. It is imperative that
project participants think in terms of 'we' rather than 'I.' In a world organ-

ized in discipline-shaped blocks, hybridity harbours a rich potential for disciplinary take-over. This is not to say that objectives are holy cows never to be touched or discussed in the course of a programme. To the contrary, the hallmark of a dynamic and creative research process is that objectives can be adjusted when the process indicates that a different path will be more rewarding than the one defined at the commencement of the project. However, to challenge an objective with no reference to fresh research results usually entails more obstruction than construction.

The second challenge of extended peer community was that of money allocations between the four sub-programmes, disciplines and groups. It is no secret that those most inclined to give priority to the hard sciences were less understanding of the need to make historic studies of the 16th century, and those that saw military strategic aspects as the main stumbling block to navigation were less incline to get enthusiastic about studies of the love life of benthic communities on the seabed of the marginal seas.[117] The question of who should get what opened up a Pandora's box of surprising divisions, not only between but also within sub-programmes. It became a free-for-all, setting sciences and disciplines alike at loggerheads. The reason is as simple as it is trivial: Resources were finite. No one could have everything, and they all knew they would have to settle for less than what would be optimal. Thus, some of the darkest moments of INSROP were when budget discussions had to be dealt with once a year for six years. Discussions often dragged on for hours, even into the night, without any visible progress being made. Disciplinary rationality on the one side was pitted against disciplinary rationality on the other. All appeared rational within their own constraints, but it was hard to find consensual rationality that benefited all. Resolution ultimately came, either after long negotiated compromises or by mutual fatigue and exhaustion, or through a combination of the two. Prolonged diplomacy, mediation, aggressiveness, patient and forthcoming leadership became measures of resolution. As has been generally noted, a "laissez-faire type of leadership which hopes that the different parts of the work of transdisciplinary teams will grow together organically has not proven successful.[118] Thus, hybridity does not always, under all circumstances, bring synthesis. It may as well produce turf battles between veterans of disciplinary wars. In the throes of funding allocations, money seems to make disciplines close their ranks and consolidate their boundaries. Without bridge scientists working full time to find acceptable solutions, the appropriation processes of INSROP might have stalemated beyond repair.

There are, of course, alternatives to annual appropriations of money. Appropriations can be made by the leadership for the whole period upon the initiation of a programme, bearing a clear-cut message: *take it or leave it, accept or resign.* Such an authoritarian solution would, however, have deprived the programme of the flexibility to make the necessary adjustments to deal with unexpected needs that popped up in the course of the research, and it would most certainly cause conflicts to flare up. Variations of this method may also be considered, for example, offering some degree of flexibility by having two periods of appropriations instead of six. However, considering the alternatives, it may be that the method employed by INSROP, despite the infighting it caused, provided the flexibility, hope of future compensation and trust necessary to carry the programme through from start to end. Interestingly enough, when all the appropriation issues had been resolved within the programme, the parties pulled their act together, gathered momentum, and pooled resources to fulfil their *joint* obligations as if the battles never had taken place. Everyone was aware that disagreements about funding at one stage could be compensated at a later stage. In other words, hope never faded. You could lose a battle but still win the war. However, it should be noted that such battles were energy consuming and tiring.

The Organizational Set-Up

As stated above, the mere size of INSROP required a fairly elaborate organizational set-up which required a division of labour between the scientific and the administrative parts of the programme. It soon proved that this fairly clear-cut structure did not work as anticipated. In actual practice, the levels and the procedures blurred and intermingled, suggesting that administrative problems cannot be separated from scientific ones or vice versa. Thus constant adjustments had to be made between the levels and activities as the programme implementation progressed. The division of labour became interactive, dynamic and even blurred. To leave the scientific part of the programme totally and independently in the hands of disciplinarians and experts would most likely also have enhanced the risk of losing sight of the interdisciplinary objective of research. Disciplinarians and experts are by training and inclination specialists, not generalists. The programme leadership had to keep a close eye on the scientific developments and to correct the course of science if need be. Equally important, a structure that adheres slavishly to a certain division of labour may increase the conflict potential between scientists and administrators and reduce the overall effectiveness of the programme because the functions are not sufficiently integrated and related to all the needs involved. To facilitate smoothness, the INSROP experience suggests that the management of large-scale programmes should involve scientists as well as administrators. In due course, their interactions will have to respond to a blend of administrative and scientific needs. Ideally, they should breed a mutual understanding of the totality of needs involved in the implementation of the programme. Should the two branches of implementation fail to reach common ground, science needs must be given priority since it is the craft and skills of the scientists and experts that will ultimately produce the convergence of knowledge.

As has already been discussed, where politics enters the arena of programme implementation, the significance of mixed leadership will multiply many times over. Politics has to be understood at all levels of implementation, so that the process of translation can find a suitable expression without compromising the integrity of research. In such situations, a *science politician* is invaluable. He or she will act as a sieve, filtering the various components of science, politics and management in measured doses to allow the translation process to make its way towards a mutually agreeable resolution. In the context of INSROP, this was one of the main tasks resting with the head of the Secretariat and the JRC, and the mission turned out to be one of sensitivity, delicacy and complications. The complexity of the situation was appreciated by the Norwegian Shipowner's Association: "Certainly it has not been an easy task, it has asked for both technical skills and for flexibility in cooperation with so many different countries, experiences and cultures involved. It is our impression that you (INSROP) have solved this in an excellent way."[119] Stakeholder statements concerning the quality and utility of transdisciplinary research are a measure of its failure or success.

The JRC and the SCS were the venues where bridges had to be built, i.e. where the hybridity of INSROP had to converge. In this respect, the organizational set-up proved to be most instrumental in bringing INSROP to a successful conclusion, and to making interdisciplinarity possible in large-scale research.

THE CHOICE OF METHODOLOGICAL TOOLS

The INSROP partners never engaged in any positivist discussion about what type of methods to apply. This was a non-issue, an odd topic of no relevance. The overall acknowledgement was that INSROP contained many sciences and types of expertise, and a rich stock of methodological devices, quantitative as well as qualitative, disciplinary as well as interdisciplinary. It stems from what has been said so far, that what united INSROP researchers across disciplinary boundaries and different theoretical convictions was *the scientific approach* as defined by Stanley Schuum. He contends that science must abandon the concept of a single scientific method and instead consider a scientific approach which is less involved with procedure and method and more with *the state of the mind of the investigator*.[120] Of the five requirements involved in Schuum's approach, four were adopted by INSROP researchers.[121]

The first requirement was that the research was to be *systematic and orderly*. The five years used patiently by INSROP sub-programmes to negotiate the release of relevant, if not optimal Russian ice data, imply compliance with this requirement. The second requirement that research shall be characterized by a *lack of biases* was complied with by peer reviewing all Working Papers. Here, the internal justice of the programme was that the comments of the reviewer were to be included in the text of all the reports, and if the author disagreed with the comments of his or her peer reviewer, the latter's viewpoints were represented explicitly in the final version of the report. Thus the Working Papers assumed a format of *adversarial collaboration*, and disagreements were made public for all to see. This instrument also acted as a guarantee to ensure the fourth requirement, namely, that the research process was to be characterized by *absolute intellectual honesty* and that *metaphysical explanations were to be avoided*. The last requirement was that researchers were to *develop broad generalizations where appropriate and possible*. This was complied with in the most synthetic of the Working Papers (see Figure 5.7).

Four synthetic goals were set for INSROP:

- Figure 5.5 depicts the two main integrative arrangements applied throughout the INSROP period. In Box A, the programme aimed at *overall integration* of all applied disciplines, embracing all principal findings from phase I and box C (of phase II). Box B aimed at what was called *theme integration*, i.e. linking and integrating knowledge from multiple disciplines to address realities of restricted complexity. One such example was the study: *Simulation of Ship Traffic along the Route* (SIM)
- Sub-programmes I and II collaborated to develop an *INSROP Geographical Information System: Software and Database* (INSROP-GIS), employing data from all four sub-programmes. This system was designed as an integrative tool for organization and storage of INSROP data and for project-related analysis work.[122]
- Collaboration was also to take place between individual projects within and between the four sub-programmes.
- Early on, Sub-programme II accepted the responsibility for making the first environmental atlas ever spanning the whole stretch of the NSR from Novaya Zemlya to the Bering Strait. This resulted in *The Northern Sea Route—Dynamic Environmental Atlas (INSROP-DEA)*,[123] derived from INSROP-GIS.

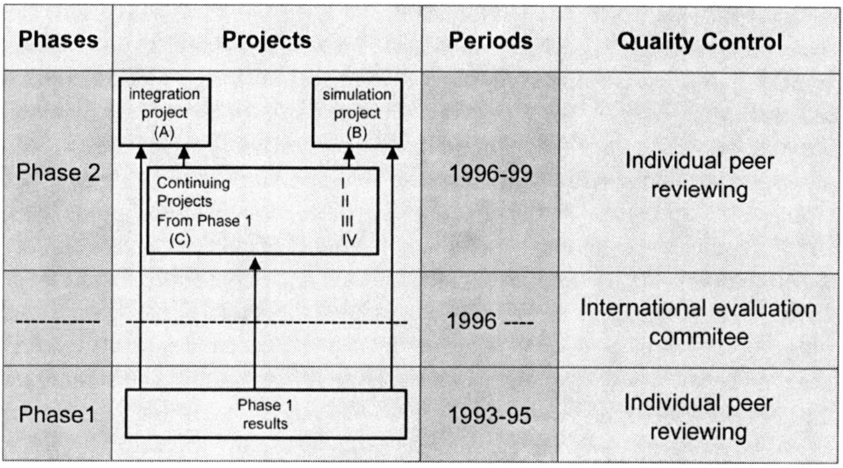

Phases	Projects		Periods	Quality Control
Phase 2	integration project (A)	simulation project (B)	1996-99	Individual peer reviewing
	Continuing Projects From Phase 1 (C)	I II III IV		
----------	--		1996 ----	International evaluation commitee
Phase1	Phase 1 results		1993-95	Individual peer reviewing

Figure 5.5. *The Integration Design of INSROP. Source: The INSROP Newsletter, no.3, vol. 5, December 1997.*

The geographical dispersion of the INSROP participants across three continents,[124] 10 countries[125] and 69 different institutions[126] entailed a very real potential for making integrative science a difficult and cumbersome challenge. Restricted budgets added to this by not allowing for frequent travel and lengthy stays between researchers of different countries and continents. In a certain, if not an absolute sense, geographical distance and restricted travel funds kept the majority of the researchers secluded from each other either permanently or for extended periods of time. The fact is that the majority of researchers never saw their team partners face to face for consultations and discussions during the phase of integration. Thus, the team cohesion of INSROP lagged far behind the optimal, a frequent but sad characteristic of large interdisciplinary groups.[127]

In actual reality, the majority of INSROP researchers were commissioned to do disciplinary work and were never invited to cross or transverse any boundaries. This is reflected in the fact that phase I was basically designated for disciplinary work, and that the bulk of working papers turned out to stay firmly within the confines of identifiable disciplines.[128] Most of these papers served the purpose of interdisciplinarity only indirectly by filling in the disciplinary gaps in our knowledge. As such, they represented the first mandatory step in the process of converging disciplines (see Chapter 3).

Comparative studies of synthetic teamwork indicate that the ideal number of people on interdisciplinary teams should not be more than in between five to nine, and the representation of disciplines restricted to four or five. The problem facing large teams is that they are more difficult to coordinate, and responsibility becomes more complicated to allocate and delegate. It has also been experienced that large size can inhibit creativity and provoke a tendency to work at the level of the 'smallest common denominator.'[129] This is not to say that large teams will never succeed. Under certain circumstances, there are examples that productivity has been successfully maintained in large teams with a stable membership and leaders with more than 14 years of experience.[130] The overall conclusion drawn, however, is that small groups with stable memberships appear to be the most integrative, and that cohesion tends to fall dramatically in large groups that are poorly organized.[131] Small teams have the opportunity to overcome issues of translation and language over time, and to develop coherent, integrative conceptual models such as inter-field theories through synergetic discourse, as collaboration increases across disciplines.[132] Although it is not directly specified what constitutes a large team, the number of persons mentioned "to flow through" an interdisciplinary team before a handful of individuals "settle out," is 50.[133] If this number is large, INSROP was gigantic, counting more than 300 researchers coming from all four branches of learning and some 30 disciplines and societal sectors.[134] At the outset, this comprehensiveness, diversity and geographical span strongly suggested that the synergy to be hoped for in INSROP was to be at the level of the smallest common denominator.

From the point of view of mere size, it would be impossible and even impractical to invite all INSROP researchers to take part directly in the act of synthesis. In consequence, the researchers specifically asked to engage in interdisciplinary research were selected on the basis of three criteria: *The groups were to be small, the members were to be able to interact easily*, and *each group was to host at least one person with some or all of the abilities of a bridge scientist.*

In terms of number, the integrative studies were conducted in groups made up of 2 to 12 individuals. Most of these groups stayed within the limits of the synthesisers recommended in the general literature. In terms of easy interaction between the synthesisers, the groups were basically composed of members from the three principal cooperating partners living and working in close proximity to each other. This made it possible to compensate for those features of INSROP that would otherwise have reduced the scale and level of integration. Among other things, transdisciplinary, bridge-building concepts were developed between disciplines referring to the interrelationship between societal and natural conditions along the NSR and different modes of navigation applied to sustain what was denominated the *socio-biodiversity* of these waters. *Aggregated hot spots* is one such concept, depicting areas where societal and natural parameters multiply, mix, interact and cluster, making the passage of ships a complex, multi-faceted challenge. *Issue-specific hot spots,* hosting just one parameter with a particularly high societal or natural value, is another. *Cool spots*, describing areas that harbour just a single parameter to be overcome, is a third concept. Each and every one of these spots was subsequently broken down further into sub-spots on the basis of the accessibility or inaccessibility of the respective areas to ships.[135] These concepts, which were applied in the AMSA-study[136], are examples of what Michael Jones has labelled *chaotic concepts* and *cherished concepts*, straddling disciplinary boundaries and hosting multiple values from different sciences. Using concepts of this kind, a brand new navigational chart could be made, providing ideas for a host of different, more conscientious ways of navigation.

In relating these concepts to each other, the discussion became both *inter-scientific* (in bringing the natural and societal sciences together) and *conceptually cumulative* (in that the various spot concepts breed higher level concepts of navigation). This cumulativeness is an expression of the multi-level description of reality which, in philosophy, is gradually doing away with the old Cartesian dualism. Natural and societal values on the one hand and modes of navigation are not two distinct elements, but rather elements interlinked by the concept of *socio-biodiversity*.[137] They are dual-levelled, holistically interconnected and belong to the methodological inventory of interdisciplinarity. Werner Heisenberg's statement comes to mind: "Some concepts form an integral part of (the) scientific methods, since they represent for the time being the final result of the development of human thought in the past."[138]

Synergetic discourse was used between some of the sub-programmes, in particular between I and II, and to a certain extent also III. Lasting for just two to three days, these workshops were arranged to meet the synthetic objective of INSROP, in particular theme integration. In practice, they turned out to be more effective in negotiating releases of more and better quality ice data; they were most instrumental in extending the discussions of JRC (see above). If synergies emerged under the auspices of these workshops, it was basically in the lunch breaks and informal corridor meetings between co-authors, and not in the plenary sessions. In our context, the synergies were *project-specific*, rather than *programme-specific*, which is consonant with the assumption that most border crossings occur at the level of units of topical specialization and not on the boundaries between entire disciplines or programmes.

As concluded in Chapter 3, interdisciplinary research is process dependent and time consuming. The participants need a period of grace to socialize and internalize what interdisciplinary research has to offer; and they need the educational part of *extended synergetic discourse* to make its way in de-educating them from the most common constraints and perceptions of their own disciplines. Previously, we made the distinction between *short-term synergies*

TYPES OF INTERDISCIPLINARITY / METHOLOGICAL ELEMENTS	Monodisciplinarity	Extended Monodisciplinarity	Simple Multidisciplinarity	Synergetic Multidisciplinarity	Simple Crossdisciplinarity	Synergetic Crossdisciplinarity	All-encompassing Crossdisciplinarity
Scientific attitudes	Disciplinary curiosity	Disciplinary Orientation + Interdisciplinary preying	Interdisciplinary curiosity + Holistic problem-orientation + Aggregative approach	Interdisciplinary curiosity + Holistic problem-orientation + Aggregative approach	Interdisciplinary curiosity + Holistic problem-orientation + Aggregative approach	Interdisciplinary curiosity + Holistic problem-orientation + Aggregative approach	Interdisciplinary curiosity + Holistic problem-orientation + Aggregative approach
Scientific skills	Disciplinary guardian	Interdisciplinary sensitivity	Interdisciplinary practice	Interdisciplinary competence	Interdisciplinary identity + Dual competence + Bridge scientist / polyvalent specialist	Interdisciplinary identity + Triple competence + Bridge scientist / polyvalent specialist	Interdisciplinary community + Quadruple competence + Bridge community
Scientific tools	Mono depth	Mono depth	Juxtappositional depth and breadth + Interdisciplinary exchange and adjustment	Juxtappositional depth and breadth + Interdisciplinary exchange and adjustment + Synergetic discourse	Juxtappositional depth and breadth + Interdisciplinary exchange and adjustment + Synergetic discourse + Seminal theory	Juxtappositional depth and breadth + Interdisciplinary exchange and adjustment + Extended synergetic discourse + Synthetic theory + Interfield theories	Juxtappositional depth and breadth + Interdisciplinary exchange and adjustment + Extended synergetic discourse + Consilience theory + Unity of science thesis
Degree of INSROP integration	Some INSROP working papers	Majority of INSROP working papers	REF-Chapter V	INSROP-DEA, INSROP-GIS, REF-Chapters I,II,III and IV, SIM	REF-Chapter VI		

Figure 5.6.   The Degree of Integration of INSROP Publications.

(1 to 2 years), *medium-term synergies* (3 to 5 years) and *long-term synergies* (6 years or more). Although, many studies report that they are amazed by how much extra insight they have gained in a relatively short period of time, most studies suggest that 6 years of discourse or more is necessary to really harvest the most rewarding synergies by way of *invisible college*.[139] Synergetic effects do not come easily or with a snap of the fingers.

In the small teams of INSROP responsible for the integration of scientific attitudes, *interdisciplinary curiosity, holistic problem-orientation* and *the aggregative approach* ruled the day. After 6 years of cooperation, the members also gradually acquired the scientific skills necessary, including both *interdisciplinary sensitivity* and *interdisciplinary practice*. Some even developed some *interdisciplinary competence*. The tools applied were generally those of *juxtapositional depth and breadth, interdisciplinary exchange and adjustment,* and *synergetic discourse* of different types.

INSROP was a mixed programme in most respects. It applied disciplinary as well as interdisciplinary approaches. The former was undertaken to make the latter possible. In terms of integration, the studies varied in synthesis from extended monodisciplinarity via simple multidisciplinarity to synergetic multidisciplinarity with elements of simple cross-disciplinarity for one or several studies. The level of convergence of synergetic and all-embracing cross-disciplinarity was not within the reach of a programme with the size, complexity and diversity of INSROP. The cultural diversity, hybridity and organizational set-up of the programme did not in any serious way impact on the research design or consequently on the level of convergence.

THE INTERCONNECTEDNESS OF THE CONSTITUENT PARTS OF INSROP: A SUMMARY

The purpose of this section is to examine how the history, politics, philosophy, concept and methodology of interdisciplinarity interacted and ultimately affected the synthesis of INSROP.

INSROP was closely interconnected with politics and history. The politics of initiation undoubtedly gave birth to the programme. Without politics, the programme would never have got off the ground. In this respect politics served the interest of science in establishing research cooperation across national borders.[140] Politics actually permeated the programme from start to finish. The politics of implementation became more of a live wire, taking attention and energy away from research. It also created an informal excrescence on the organizational set-up of the programme. The FCD (the Forum for Corridor Deliberations) grew out of the Russian need to convey the qualms of that part of the security elite in Russia that harboured doubts about the security implications of INSROP. In that respect, it served a most important role in navigating the programme through the minefields of potentially hostile waters. That politics had its own exclusive forum allowed INSROP to fulfil its mission at a time when cooperation between Russia and Norway and Norwegian oceanographers and fisheries specialists in the High North was soured beyond recognition. In the later part of Phase I and throughout Phase II of INSROP, Norwegian oceanographic expeditions to the Barents and Kara Seas were not allowed to enter Russian waters despite prior Russian consent to and even participation in some of these expeditions. This puzzled the Norwegian authorities, who did not fully grasp the reason why a bilateral fishery relationship that had been productive and cooperative for years, should run into trouble all of a sudden. The Russian repugnance to cooperate was so prevalent that Norwegian authorities seriously contemplated discontinuing fishery cooperation with Russia in the Barents Sea.[141] Recently, the Norwegian press revealed that the Norwegian Ocean Research Institute in Bergen had long traditions of preferential exchanges of oceanographic data from northern waters with the Norwegian defence authorities.[142] These partly clandestine data exchanges took part throughout the Cold War, and the last one recorded took place in 1997, which coincides roughly with the year Russia changed its cooperative mood with a view to Norway. It is not known for sure whether the extraordinary souring of Russian-Norwegian relations in the late 1990s was caused by these data exchanges, but it certainly cannot be ruled out. Former Norwegian State Secretary of Foreign Affairs Aslaug Haga, who had no knowledge of these data exchanges prior to the revelations, claims that they may explain some of the Russian behaviour.[143] Whatever the reason and despite the generally bad Russian mood towards Norway during these years, the Russians never explicitly threatened to halt or sever their relations with its INSROP partners, who explicitly addressed waters perceived by Russia as having high security sensitivity for Russia. While Norwegian-Russian expeditions were cancelled time and again, INSROP was completed as planned. There may be a host of reasons for this, not least political in that so much political prestige had been invested in the programme, both bi-and multilaterally, and that the threshold was too high for cancellation even in the face of troubled relations. Thus politics not only initiated the programme and stayed with it throughout, but also contributed to preserving it in times of cool relations. In addition, there was the Forum of Corridor Deliberations which provided an outlet where the Russians could let off the steam generated by the political supervision with which they constantly had to deal. Equally important is the fact that the suspicions aired against INSROP by this faction of the Russian security elite were never based on any evidence in reality. INSROP stayed clean, politically speaking.

It is probably true that more external political attention and involvement was devoted to INSROP than what has been the case with most other interdisciplinary programmes. One significant reason for this is that INSROP was history in the making; it was an ingredient of the dramatic shift of international order from the Cold War to the Post-Cold War Era. As such, INSROP was a pawn in the game of nations and served the symbolic-instrumental interests of at least three countries, the Soviet Union, Russia and Norway. Finland and Sweden could also be added to this list since they belong to the inner circle of nations that established the *Barents Euro Arctic Region* (BEAR), which adopted the NSR as an object of cooperation to feed on the results of INSROP.

Politics were omnipresent in INSROP, but they affected the conduct of research only marginally when it came to the quality of sea ice data deliveries. At the outset, this was a disciplinary problem that turned interdisciplinary only when sea ice data were used for synthesis to converge with other disciplines and/or sub-programmes. The problem in this respect was that the data sets existed at two levels of refinement, i.e. one aggregated and one simplistic, making synthesis a bit of a challenge and probably less than thorough. Equal levels of refinement seem a prerequisite for achieving optimal synthesis. To put it differently: It is easier to identify the trading zones between items of information that maintain the same level of subtlety. This is just another reflection of the assumption put forward in Chapter 3 that the tinier and deeper the units of specialization, the more natural and easy integration becomes. The sea ice studies might have reached a higher level of integration if the refinement of sea ice data had matched the minutiae of the ship technology data applied. The politics of implementation undoubtedly made its marks on research, but in a way acceptable to the standards of science and the freedom of research principle. Politics were never anywhere near infringing seriously on the validity or reliability of the research.

INSROP became a three-headed troll. It not only served the symbolic-instrumental interests of contemporary politics, but also the practical-instrumental interests of economic life and the knowledge-instrumental interests of science. As such, the programme realized more objectives and fulfilled more tasks than originally planned. INSROP became a scientific hybrid doing everything and anything, integrating across three sets of boundaries—between disciplines, sciences and the members of the hybrid community. The process of translation, based on a widespread willingness to succeed, became a most valuable instrument in sorting matters out and resolving problems when need be and when strong wills pulled in opposite directions. The translation process resolved the tensions and interests of conflicts between politics and science, between the various disciplines, between the branches of learning, and between academe and practice. Thus translation seems a prerequisite for interdisciplinary teams to pull together. This is not least true for transdisciplinary teams. On the science side, it opened up a geographical area for international research that had been reserved for Soviet scientists alone for more than 50 years.[144] It also added fresh insights to the total knowledge of the Arctic region, basic as well as applied. On the practical side, it revealed that the transportation of oil and gas on the western side of the NSR would be possible in a short-term perspective, and that the technological, environmental and economic challenges are no longer insurmountable obstacles to navigation. It also showed that in the long haul the route may, under certain circumstances, be opened for transit sailings on a year-round basis.[145] This blend of science practices and forms is a viable indication that distinctions often made between basic and applied research are not essential to the progress of research. They move, if haltingly, together along the same path and in the same direction. In this respect, INSROP was executed in compliance with the basic characteristics of post-academic and post-normal science as depicted in the Circle of Science in which basic research merges with the techno science (applied science) and the strategic science needs of governments (see Chapter 2). The tacit social contract between the scientific community and society suggest that research in the post-academic era is supposed to be useful in both the short and long term. In retrospect, INSROP was useful to politics, in the short and long term to economy, and in all terms to science. Knowledge is knowledge, whether it is called basic, applied, disciplinary, interdisciplinary, transdisciplinary, practical-instrumental, symbolic-instrumental or knowledge-instrumental.

In terms of design, INSROP exceeded all reasonable limits suggested in the general literature on the optimal conditions for synthesis. Its comprehensiveness, number of participants, geographical dispersion, number of disciplines, sciences, cultures, hybridity and organization undoubtedly reduced the level of convergence, not least between the individual Working Papers. The sheer size of the programme invited disciplinary rather than interdisciplinary research. To manage this huge complicated body of scientific activity, the organizational set-up relied on complex scientific and political processes, demanding extraordinary patience, diplomacy, leadership, perseverance, stubbornness and endurance from its many members. The complexity of the organization added to the complexity of the object of study. Thus, INSROP was confronted with complexity on two levels, making it hard if not impossible to direct and manage research. From time to time, deep sighs among the participants indicated that the hard part of INSROP was not the scientific activity, but the continuous managerial and political dealings. Management became time-consuming and it stole time away from the integrative part of research, not least because some of those carrying the burden of the politics and management were also assigned responsibilities related to synthesis, i.e. the science politicians and the bridge scientists.

Basically, the integrative studies reached the level of synergetic multidisciplinarity, engaging the attitudes of interdisciplinary curiosity, holistic problem-orientation, aggregative approach, the skills of interdisciplinary competence, and the tools of juxtapositional depth and breadth, interdisciplinary exchange and adjustment and synergetic discourse (tools). One thematic study (REF Chapter 5) achieved less in terms of synthesis, only simple multidisciplinarity due to politics, and one study (REF Chapter 6) approached simple cross-disciplinarity, adding interdisciplinary identity and some virtual synergetic discourse.

History, politics and research design shaped INSROP in such a way as to affect the level of convergence between the sciences and disciplines. In so doing, these elements also directly and indirectly defined the methodology to be applied. Since the elements of the methodology involve the operationalization of the integrative mode, they also involve the concept of interdisciplinarity. History, politics, design, methodology and concept were thus strung together in an interrelated chain. One element depended on the other and vice versa. The philosophy of science is the only component that seems free-floating in relation to this interconnectedness, although the approach chosen for the research resembled that of post-positivism.

For all those who took part in INSROP, this interconnectedness became the beauty and the beast of interdisciplinarity at one and the same time. It proved beyond any reasonable doubt that inter- and transdisciplinarity is both practical and practicable. What is more, and according to the AMSA-study: "INSROP . . . pioneered cooperation between Russian and foreign researchers in Arctic-related fields, networks that live on. INSROP allowed a solid platform to be constructed upon which further Arctic multidisciplinary studies could be accomplished. Following the end of INSROP interest in the NSR subsided for some years, but as awareness increased of the realities of climate change and its impact on Arctic sea ice, a rise occurred in interest and demand for INSROP's results from the shipping industry, the authorities and the media. A number of new projects have been intitated."[146]

NOTES

1. Latour (99), p. 90.
2. Østreng (00), pp. 39–50.
3. Wisted and Mathisen (95), p. 21, See also Thompson Klein (90), p. 23.
4. Caldwell (82), p. 105.
5. Østreng (92), pp. 26–45.

6. Østreng (99/1), pp. 21–28.

7. Østreng (99/1), pp. 21–55.

8. Østreng (04), pp. 69–78.

9. The Finnish government, who initiated the Rovaniemi process, probably was the sole exception to the rule.

10. Østreng (99/1), pp. 21–51.

11. At the time of contact, the author of this book was the director of FNI. He was also director throughout the implementation period of the programme. During this period he was the Head of the Secretariat and the Joint Research Committee and met as Secretary to the Steering Committee of Sponsors.

12. The abundance of time consumed and its associated frustration was not, however, unique to INSROP. The planning stages of interdisciplinary projects are generally described as unstable periods of long-lasting duration. These periods are being "criticized as 'wallowing' periods," and, at the same time, praised as essential to "prevent wallowing later on."

13. Stoltenberg (94), p.x.

14. Stokke and Tunander (94).

15. Vartanov and Roginko (99/1), pp. 53–102.

16. The Committee was multidisciplinary and multinationally composed, and consisted of: *Professor Clive Archer*, Manchester Metropolitan University, United Kingdom, *Professor Jørgen Ole Bærenholdt*, Roskilde University, Denmark, *Captain Lawson Brigham (Chair)*, Scott Polar Research institute, United Kingdom, *Professor Tsuneo Nishiyama*, Hokkaido Tokai University, Japan, *Dr. David Stone*, Department of Indian Affairs and Northern Development, Canada, *Professor Tullio Scovazzi*, University of Milan, Italy, *Professor Michael Tamvakis*, City University Business School in London, United Kingdom, and *Professor Jean-Claude Tatinclaux*, U.S. Army Cold Regions Research and Engineering Laboratory, USA.

17. Report of the International Evaluation Committee (96), p. 3.

18. Report of the International Evaluation Committee (96), p. 3.

19. Report of the International Evaluation Committee (96), p. 4.

20. Report of the International Evaluation Committee (96), p. 2.

21. Report of the International Evaluation Committee (96) p. 4.

22. Sasakawa (96), p.1.

23. Sasakawa (99/2), p. xxvii.

24. This meeting was attended by the Norwegian Ambassador to Japan (Bjørneby) together with Henning Simonsen and Willy Østreng of the Fridtjof Nansen Institute.

25. Vartanov, Roginko and Kolossov (99/1), pp. 87–88.

26. Vartanov, Roginko and Kolossov (99/1), pp. 87–88.

27. Agranat (96), p. 5.

28. Polyarnaya Pravda, 29 February 2000.

29. For a different assessment of the security value of the NSR to Russia see: Brubaker and Østreng (99), pp. 299–331.

30. Usually only the Head of the JRC, an interpreter and the Head of the Russian delegation and the Deputy Head of the Administration for the Northern Sea Route took part in these discussions.

31. Report of the International Evaluation Committee (96), p. 3.

32. Østreng (99/1) pp. 21–102. The operational parameters of ice covered waters are thoroughly discussed in Østreng (87).

33. Latour (99), p. 89.

34. Latour (99), p. 87.

35. Latour (99), p. 88.

36. One indication of this is that retired Admiral A.N. Yakovlev, who was well connected in the military establishment of Russia, took an active part in the research of INSROP.

37. INSROP Programme Report 1993–1998 (99), p. 17.

38. Latour (99), p. 90.

39. Latour (99), p. 97.

40. Latour (99), p. 87.

41. Caldwell (82), pp. 83–84.

42. Report of the International Evaluation Committee (96), p. 1.

43. Report of the International Evaluation Committee (96), p. 2.

44. Report of the International Evaluation Committee (96), p. 7.

45. "Appendix 1: Russian EIA Procedures, Practice and Regulations" in Østreng (99/1), p. 430.

46. Report of the International Evaluation Committee (96), p. 5.

47. Cited from Latour (99), p. 300.

48. Value Ecosystem Components (VECs) refer to environmental features like the water-land border zone, indigenous peoples, benthic invertebrates, marine, estuarine and anadromous fish, birds and marine mammals.

49. Moe and Hansson (98), p. 8.

50. Report of the International Evaluation Committee (96), p. 6.

51. Latour (99), p. 87.

52. Report of the International Evaluation Committee (96), p. 7.

53. Examples of critical studies: Østreng (99/1), pp. 21–102, 239–266, and Østreng (99/2), pp. 281–426.

54. Report of the International Evaluation Committee (96), p. 23.

55. Østreng (99/2), p. xxxii.

56. Collini (98), p. xlvi.

57. Hughes and Sharrock (97), p. 208.

58. Hughes and Sharrock (97), p. 203.

59. Hughes and Sharrock (97), p. 1.

60. Weinberg (93), p. 168.

61. Weinberg (93), pp. 168–169.

62. Weinberg (93), p. 170.

63. Schumm (98), p. 2.

64. Cited from Alnes (03), pp. 79–82.

65. Smolin (07), p. 290.

66. Weinberg (93), pp. 166–167.

67. Weinberg (93), pp. 52–61.

68. Hubbard, Kitchin, Bartley et al. (02), p. 7.

69. Funtowicz and Ravetz (08), p. 367.
70. Hubbard, Kitchin, Bartley et al (02), p. 237.
71. Peet (98), p. 9.
72. Hansen (03), pp.97.
73. Northrop (99), p. 5.
74. Thompson Klein (90), p. 61.
75. Kaje (99), p. 15.
76. Kaje (99), p.13.
77. Thompson Klein (90), p. 135.
78. Brock, Comitas, Sigurd, et al. (86), p.79.
79. Kaje (99), p. 14.
80. Fladmark (98), p 219.
81. Murden (01), p. 456.
82. Geertz (73), p. 14.
83. Snow (98), p. 64.
84. Murden (01), p. 457.
85. Bronowski (72), p. 43.
86. Bronowski (72), p. 43.
87. Kozu (99), p. 5.
88. Larsen (84), pp.15–44.
89. INSROP Programme Report, 1993–1998 (99), pp. 21–24.
90. Østreng (99/2), pp. 365–426.
91. Waage (90), p. 281.
92. Waage (90), p. 12.
93. Hønneland (02), p. 97.
94. Hønneland (02), pp. 97–99.
95. Peresypkin (99), p. 9.
96. Peresypkin (99), p. 9.
97. Peresypkin (99), p. 9.
98. Report of the International Evaluation Committee (96), p. 17.
99. Frank (00), p. 9.
100. AMSA-Draft (08), p. 66.
101. Hollaender, Loibl, Wilts (08), pp. 385–397.
102. Summary and Statistics of all INSROP Products and Activities (99), pp. 25–34.
103. Summary and Statistics of all INSROP Products and Activities (99), pp. 25–34.
104. Summary and Statistics of all INSROP Products and Activities (99), pp. 35–46.
105. Hollaender, Loibl and Wilts (08), p. 386.
106. Saether (00), p. 36.
107. Report of the International Evaluation Committee (06), p. 3.
108. Report of the International Evaluation Committee (06), p. 4.
109. Cited from Livingstone (05), p. 325.
110. Saether (00), p. 36.

111. Report of the International Evaluation Committee (06), p. 11.

112. AMSA-Draft (08), p. 66

113. Report of the International Evaluation Committee (06), p. 19.

114. Report of the International Evaluation Committee (06), p. 20.

115. Report of the International Evaluation Committee (06), p. 20.

116. Summary of Statistics of all INSROP Products and Activities, (99), p. 20.

117. Moe and Semanov (99/2), pp. 121–220.

118. Hollaender, Loibl and Wilts (08), p. 387.

119. Saether (00), p. 36.

120. Schuum (98), p. 95.

121. Schuum (98), pp. 95–96.

122. Løvås and Brude ((98), p. 9.

123. Brude, Moe, Bakken, et al. (98), pp. 58.

124. USA, Europe and Asia.

125. Bulgaria, Canada, Finland, Japan, The Netherlands, Norway, Russia, Sweden, United Kingdom and USA.

126. See INSROP Programme Report, 1993–1998 (99), pp. 35–37.

127. See the 19 case studies discussed in Gertrude H. Hadorn et al (08), pp. 43–426.

128. See Østreng (99/2), pp. 431–444.

129. Thompson Klein (90), p. 129.

130. For a discussion of these aspects see Thompson Klein (90), pp. 126–133.

131. Thompson Klein (90), p. 129.

132. Kaje (99), p. 13.

133. For more on participation see Elzinga (08), pp. 345–360.

134. Østreng (99/2), pp. XI–XIi.

135. For a discussion of these concepts see Østreng (99/2), pp. 365–426.

136. AMSA-Draft (08)

137. See Østreng (99/2), pp. 365–366.

138. Heisenberg (99), p. 92.

139. Brock, Comitas, Sigurd et al. (86), p. 76.

140. Østreng (89).

141. See *Fiskaren*, 9th. November 2001.

142. For a discussion of these incidences, see the following volumes of *Fiskaren*, vol. 78, no. 130, week 45, 7 November 2001 and vol. 78, no. vol.193, week 45, 9 November 2001 and vol. 78, no. 194, week 46, 16 November vol. 2001 and 78, no. 196, week 47, 21 November 2001, and vol. 48, no. 138, week 48, 26 November 2001. *Fiskaren* is an independent weekly publication for coastal Norway, published in the municipality of Tromsø in northern Norway.

143. Interview with Aslaug Haga in *Fiskaren*, 30 November 2001, p. 13.

144. Østreng (92), pp. 26–45.

145. For a discussion of these aspects, see Østreng (99/2) and Ragner (00).

146. AMSA-Draft (08), p. 66.

Chapter Six

Post-Academic and Post-Normal Science, and the Preconditions of Interdisciplinarity

"It is my deep conviction that only by understanding the world in all its many aspects—reductionist and holist, mathematical and poetical, through forces, fields, and particles as well as through good and evil—that we will come to understand ourselves and the meaning behind this universe, our home."

Paul Davies, 1983[1]

In this Chapter the focus is on the interface of three influences shaping the preconditions of research in general and interdisciplinarity in particular, i.e. on post-academic and post-normal science as depicted in the Circle of Science. The first is *external* to the scientific community, reflecting the changing moods and elusiveness of history and politics. Here, the aim is to identify some of the historical preconditions that make polity to see whether it is in the national interest to rearrange the prevailing priorities of science. The second is *internal*, reflecting the never-ending conceptual and methodological discourse and competitiveness among practicing scientists, basic as well as applied. Here, the focus is on the ability of the epistemic community to accommodate societal needs without compromising the freedom of research principle. The third influence is *intermediate* in that philosophers—external to the scientific process, but internal to the scientific community—claim the authority, by means of theoretical speculation, to define the criteria of what constitutes good, valid and reliable science in practice. Here the aim is to assess the role of philosophy as a signpost for where science should move and how it should be conducted. Since the preconditions of complex system science to a large extent are forged at the point of convergence between these influences, the overall objective of this chapter is to discuss the interconnectedness and relative importance of

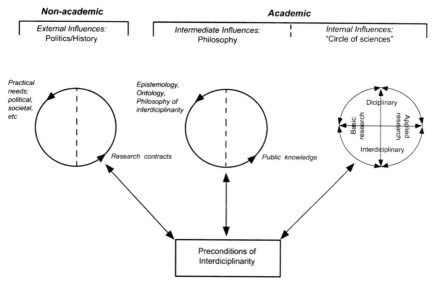

Figure 6.1.   The Shaping Influences of Interdisciplinarity.

these influences in providing incentives/disincentives for interdisciplinary undertakings (see Figure 6.1).

EXTERNAL INFLUENCES

External influences refer to a host of impacts, economic, political, cultural, religious, etc., coming from outside the scientific community and aiming at or forcing a sway in the direction and priority of research. Among the multiple influences, only two—historical circumstances and politics—will be addressed here by way of illustration.

Thomas Kuhn's main thesis is that the state, progress and changeability of science are history-dependent,[2] As argued in Chapter 2, that is to say that science depends on politics as a prominent shaper and mover of history. However, politics is not formed in a vacuum, but on the basis of historical circumstances. From this perspective, the two factors are two sides of the same coin. Undoubtedly, this observation has some merit and relates to the eternal question: What came first, the chicken or the egg, and can they be separated? Although the egg is an important step in the evolution of a chicken, the two expressions of the chicken may be studied separately in accordance with the reductionist tradition. An egg is an egg and a chicken is a chicken, but a mature chicken is definitely more complex than an egg. The same applies to history and politics. Although politics is an important shaper of history, history is the composite sum of many contributors and stems from multiple sources. In this sub-paragraph, we will try to reflect both the inter-relatedness and separateness of both influences.

Politics and Science

The Circle of Science confirms Bruno Latour's assertion that science is related to society as arteries are related to veins in the human body.[3] In the post-academic, post-normal era, all types of science (knowledge-, practical- and symbolic-instrumental) and all scientific practices (basic, applied, disciplinary or interdisciplinary research) serve as veins, supporting the arteries of politics and society. In principle, society is highly diverse and has no free-riding independent sectors. If basic science appears to be 'free-wheeling' and completely left to the unpredictable whims of the individual researcher, it is because society thinks of free-wheeling as a means to serve the collective interest. This integration forms a two-way street in which science and policy interact, the one influencing the conduct of the other. In Harvey Brooks words: "Science can both act upon policy and be acted upon by policy."[4] This poses the question of what impact policy may have on science?

To answer this question, an analytical distinction should be made between four notions: the *presence of politics in science, the presence of science in politics,* the *interference of politics in research and the interference of politics in scientific innovation.* The 'presence of politics in science' and the 'science presence in politics' are two juxtaposed expressions of mutual integration. The two latter notions—'interference of politics in research' and 'scientific innovation'—are two alternative outcomes of the presence of politics in science. Whereas the interference of politics in research is a breach of the freedom of research principle and an expression of fraud and conflict, 'scientific innovation' is the use of research for applied purposes and, as such, an expression of societal cooperation on the basis of the freedom of research principle. Given the purpose of this chapter, we will focus exclusively on scientific innovation here and leave out the 'interference of politics in research,' which is thoroughly discussed in Chapters 2 and 5.

Innovation is defined as a new product, service, production process and/or organizational form introduced into society for furthering applied purposes in the market and the public sector. Traditionally, innovation has partly been associated with practical-instrumental research conducted in applied research institutes, public and private organizations and consultancy firms using existing knowledge—societal and scientific—in *new* ways and combinations. However, the former Norwegian Minister of Research Kristin Clemet stated, contrary to this perception, that basic research is the first step in the innovation process, implying that *new* knowledge and not only new combinations of existing knowledge is an integral part of innovation. This is to acknowledge that all types of sciences, knowledge-instrumental, practical-instrumental and symbolic-instrumental research have a defined role in innovation, and that knowledge acquired from different sources and structured different ways serves a common purpose of utility. According to the Norwegian Government's recent White Paper on Research, innovation often happens at the interface of disciplines, trades and businesses. In other words, interdisciplinarity and inter-sector interactions involving academics and practitioners are means of innovation and tools for the resolution of practical problems and needs. Here, inter-disciplinarity—both academic interdisciplinarity and trandisciplinarity—has achieved the recognition of public authorities as a method for fostering innovation. Equally important is the distinction made between *radical innovation* and *incremental innovation.* Radical innovation is based on the production of new knowledge—both disciplinary and interdisciplinary—to make fresh breakthroughs in products, services, processes and organizations (knowledge-instrumental research), whereas *incremental innovation* involves improving existing products (practical-instrumental research). According to the White Paper, both types of innovation are necessary to get sufficient development of and readjustment in commercial life and the public sphere.[5] Interdisciplinarity is the new kid on the block when it comes to innovation. INSROP is illustrative in this respect

Politics and Science in INSROP

The INSROP experience confirms the 'presence of politics in science.' Politics created the programme which in turn mobilized parts of the Russian polity to monitor the programme throughout its lifespan. In this case, practical instrumental needs stirred political support as well as political opposition and apprehension in certain quarters of the Russian defence establishment. Politics never left INSROP. The veins and arteries of the programme were connected through the political thinking of the Cold War and the political uncertainties of the post-Cold War period. In this way, INSROP became a mix of politics and research, providing validity to the post-positivist and post-academic conceptions that science cannot free itself completely from the impacts of external influences. This became even more true in that the INSROP-partners wilfully applied transdisciplinarity as its basic approach transgressing the boundaries between society and academia. Here, politics, stakeholder interests and history in the making blended together.

In INSROP, all decisions concerning the choice of topics, conduct of research and publication of results were reached by consensus among the three
principal cooperating parties on purely scientific grounds. However, as
demonstrated in Chapter 5, there were disagreements, even heated ones,
among the same parties, and some of the arguments put forward appeared to
be influenced by political concerns. The sole incidence where politics may
have had a bearing on the conclusions concerned the availability and type of
ice data from Russian sources. Here, political restrictions prevailed and could
only be compensated for by drawing on alternative sources of data delivery.
This was to a certain extent done, and in the end none of the sea ice researchers found the final database to be scientifically unacceptable for their
own research. The ice data sample was restricted by politics, but at the same
time approved on scientific grounds. In this context, the presence of politics
in science came close to becoming a case of 'the interference of politics in research.' However, the interference was not intentionally aimed at harming the
research or creating fraud. On the contrary, the interference was motivated by
the feat that extended access to more Russian ice data could be detrimental to
Russian national security. In this respect, the interference on the part of Russian politics was a precautionary step rather than a deliberate action to distort
conclusions. It was an external element, an unintentional consequence originating from security considerations. What is more, the externality of the Russian restrictions was compensated for by the programme's access to alternative supplementary databases. Here, politics were to a certain extent
neutralized by scientific countermeasures.

INSROP contributed scientific innovations based both on new combinations of existing knowledge and by way of fresh basic research, disciplinary
as well as transdisciplinary. This production has been dealt with in some detail in Chapter 5 and will therefore only be reiterated here by using a few examples. Basically, the synthetic works that resulted from the programme
made applicable contributions in several respects: operational, technical, social and political. In terms of the overall integration some works produced
synthetic, integrative or chaotic concepts that straddled disciplines, and to a
certain extent also practical and scientific insights. Among those concepts
were *navigational hot spots, navigational cool spots, issue-specific hot spots,
aggregated hot spots, social-biodiversity, multi-value navigation,* etc. Concepts of this kind, which could only have been created by various combinations of multidisciplinary and stakeholder knowledge, have an operational as
well as political applicability.[6] Among other things, they seem to be applicable to meet the needs of the *Arctic Marine Strategic Plan,* approved by the
Arctic Council in its effort to base itself on an ecosystem-based management
approach.[7] In addition, these concepts could be adopted to form a novel basis

on which to forge a fresh navigational scheme for ice infested waters based on the environmental concerns of the 21st century.[8]

Another example of overall integrative work was the *INSROP Geographical Information System: Software and Database*, employing data from all four sub-programmes. Here, computerized quantifiable, measurable data and physical depiction were integrated across all sciences.[9] This was the first attempt in the history of the NSR to make a comprehensive, multidisciplinary tool of applicable utility.

An example of theme integration was *The Northern Sea Route Dynamic Environmental Atlas*,[10] derived from INSROP-GIS. Basically, this atlas integrates data from Sub-programmes I and II. This was the first environmental atlas ever made spanning the whole stretch of the NSR from Novaya Zemlya to the Bering Strait.

These are just a few of the many studies that applied interdisciplinarity as a tool of creativity to foster scientific innovation based on combining knowledge-instrumental and practical-instrumental research.

History and Interdisciplinarity

There is a strong linkage between interdisciplinarity and applied, mission-oriented research. In the 20th century, interdisciplinarity has been undertaken mostly at the request of societal institutions and governmental bodies. In times of heightened political tensions, politics seem to boost the demands for synthesis. The Manhattan Project to produce the atom bomb, the US military's need for a new turbo-engine during World War II, the US response to the launching of Sputnik in 1957, the US reaction to the alleged bomber and missile gaps in respect of the Soviet Union, and the area studies of the Cold War are just a few of the more prominent examples of hurriedly organized interdisciplinary teams being organized to serve the political needs of governments. These projects are benchmarks of successful convergences of scientific disciplines. All were induced by the drama of history. Julie Thompson Klein claims that from the standpoint of interdisciplinary history, area studies were to the social sciences and humanities what the Manhattan Project and operations research were to science and technology.[11] These programmes show that governments, in times of 'deadly' crises, are able to forge new transdisciplinary alliances and scientific convergence by overcoming the intellectual and institutional fragmentation of scientific communities and governments alike (see Chapter 2). The segmentation and specialization of society is thus not something unavoidable and forever blocking the emergence of additional and/or supplemental practices of science. In principle as well as in practice, governments hold the financial, political, economic and moral keys to change. In the above instances, the needs of contemporary history shaped and dictated the direction of science policy by way of funding power and government policy.

At one and the same time, the war-time projects advanced knowledge-instrumental research to an unprecedented level in the shortest period of time imaginable, satisfied practical-instrumental needs for getting the upper hand in international crises, and expressed a symbolic-instrumental determination to have a continuous say in world affairs. Even INSROP, a modest programme by any standard, played both a practical-, symbolic- and knowledge-instrumental role. This again illustrates the interconnectedness of the practices of research and how the practices are interchangeable horizontally, vertically and diagonally within the *Circle of Science*. They all serve the interests of politics and society, advancing the inventory of knowledge-instrumental research, both disciplinary and interdisciplinary. In these instances, the academic distinctions made between the practices of science proved to be uninteresting and to a large extent artificial. Science is science no matter how many boundaries researchers draw up between themselves and their activities. Against this backdrop, the logical assumption would be that the global environmental crises of the 21st century, e.g. the climate change, the possible rise in sea level, biodiversity, the need of sustainable development, etc. can be resolved by the same lessons learned from hot and cold war history.

The present historical dramas differ from those of World War II and the Cold War in that those conflicts originated in and were for a long time confined to Europe, then subsequently spread to involve parts of other continents. The threats of the new millennium are in the minds of many distinguished scientists, global in geographical scope, knowing no boundaries and concern all mankind. They are assumed to have potentially devastating consequences for life on Earth. These doomsday perspectives have been around for decades and have been argued by prominent scientists and politicians alike. The relevant question in our context is not who claims what or why, but rather whether the claims are based on public knowledge agreed upon among scientific peers. Any answer ventured here will probably be the object of continued fierce pro and con arguments, in line with the habit among researchers of using discussion to foster scientific progress. When it comes to climate change, there are strong indications that humankind is faced with challenges rooted in public knowledge. Drawing on the expertise of thousands of experts from all over the world involving multiple disciplines, the United Nations *Intergovernmental Panel on Climate Change* (IPCC) concludes unanimously that the likelihood is that at least parts of global warming are anthropogenic. IPCC's conclusions are commensurate with those drawn in the *Arctic Climate Impact*

Assessment Study (ACIA), urging that the attention of "decision makers and the public worldwide"[12] be devoted to environmental issues.

Never before has the assumed threat to humanity been more all-encompassing or non-discriminate. Humans find themselves at war with nature, the sole provider and sustainer of *homo sapiens*. The paradox is unprecedented in the history of mankind. There is a need for political action backed by complex system science to correct the paradox in a real universe of "chance and accident."[13] The call is for the whole Circle of Science to be commissioned in a Manhattan-like project to face up to the challenge and make amends with the planet. Or in the words of Silvio Funtowicz and Jerome Ravetz, ". . . along with the aggravating global ecological crises, the task of constructing an appropriate new philosophical synthesis takes on great urgency."[14] The need is for the emergence of "new cross-disciplinary alliances."[15] Notwithstanding, the political determination to step up research to the level of the assumed challenges lags far behind the assumed need and bears no resemblance to the efforts to counteract the threats of World War II and the Cold War. Why is this?

The explanations are multifarious, and only a few will be discussed here. When it comes to the organization of governments and universities, the past is no different from the present. Segmentation, fragmentation and specialization ruled and continue to rule the day in both spheres. The system still invites modern societies to suffer from the tyranny of small decisions,[16] based on specialized knowledge. As shown above, this 'tyranny' can be corrected if governments so desire. In times of war, cold or hot, governments have proved time and again their ability to overcome organizational constraints and to mobilize science for their own purposes. Fragmented organization is not generic or anything God-given that cannot be superseded, circumvented or changed. Where so required, the patronage of science holds the power of change. Thus the tyranny of small political decisions comes across as more of a peacetime than a wartime problem, implying that the notion of crisis has different meanings, depending on historical circumstances.

For crises to make governments mobilize research for practical-instrumental purposes, the challenge has to be perceived as *real, imminent and urgent.* These characteristics were the hallmarks of the wartime programmes. The needs of the two wars were to expand public knowledge to adequately cope with the threat of national socialism and communism. The needs of today are, in principle, no different. Complex system science is a necessary means to expand and provide the public knowledge needed to form a suitable basis for rational political decisions to be made and implemented. According to the IPCC, the public knowledge of the present strongly suggests that mankind is facing a man-made greenhouse effect expected to get worse as time passes. Former UN Secretary General Kofi Annan goes one step further: "These (climate change as an all-encompassing threat to humanity) are plausible scenarios, based on clear and rigorous scientific modelling. A few diehard sceptics continue to deny 'global warming' is taking place and try to sow doubt. They should be seen for what they are: out of step, out of arguments and out of time. In fact, the scientific consensus is becoming not only more complete, but also more alarming."[17] As could be expected, the reliability and validity of this conclusion is being questioned by individual climatologists, and the IPCC also admits that more research is required to increase the level of certainty. In the history of science, scientific progress is often, if not always, made by the few breaking away from conventional wisdom and the ruling paradigm. Some of them pave the ground for scientific revolutions and for new paradigms and public knowledge to emerge. The majority of peers, including those who define what the public knowledge is at any one time, is not always right in their consensus. This is why the truth of one particular era may be the misconception of another. In science, there is always a dose of uncertainty and incompleteness. Who should one believe—a handful of rebels or the majority of peers: both can be wrong and both can be right. On whose knowledge, rebel or public, should governments base their decisions?

The dilemma involved in this respect may explain why some governments, seem to drag their feet when it comes to confronting and counteracting today's crises. Granted, if governments devote substantial resources to coping with threats that turn out further down the road to be fictitious, they may end up as laughing stocks. By the same token, however, if today's public knowledge is correct, the risk is that the same governments may end up looking foolish for their lack of action. Vanity prone governments, and most of them are, do not find either of these outcomes particularly tempting, and will act accordingly. To cover their backs and not fall victim to the judgment of history, their most likely response will be to stay in the middle of the road, i.e. doing neither too much nor too little. The content of the *Kyoto Convention on Climate Change* is illustrative in that most governments consider it a step in the right direction, at the same time as the negotiating text is far too modest to meet the requirements of current public knowledge. The alleged filibustering of these negotiations by some governments actually suggests that the agreed-upon Convention will reflect the content *of the least ambitious programme,*[18] which is far less than a middle-of the road outcome. As a matter of fact, those governments that have not yet signed the Climate Convention are inclined to support their reluctance by resorting to research results that have not yet achieved the status of public knowledge, i.e. knowledge that competes with consensual knowledge to become public knowledge. They are applying a pick-and-choose strategy, without paying due regard to what existing public knowledge suggests.

In an ideal world, governments would act rationally within the parameters of public knowledge, i.e. within the assumed laws of nature, society and their interrelations. However, in the real world, public knowledge is not always an ally of politics. If the values of politics are in line with public knowledge, the uncertainties of science will most likely be ignored. If, on the other hand, they conflict with public knowledge, the uncertainties will likely be evoked. Either way, politics have a strong card to play due to the flexibility of support or opposition. Green parties often indulge in *worst case scenarios*, whereas parties focusing on economic growth are likely to subscribe to a strategy of *wait-and-see*. The compromise between worst case and wait-and-see is something in-between, hopefully an outcome closer to the middle of the road than to the 'law of the least ambitious programme.' Consequently, the preferred companion of politics is scientific uncertainty and incompleteness, rather than public knowledge. In a world of conflicting interests and values, certainty is far too inflexible to be treated equally by all political factions.

The relationship of politics to complex system science is thus one of ambiguity. On the one hand, governments have the power to overcome the fragmentation of science and to mobilize the brain power of researchers for their own synthetic purposes. This has been proved time and again in cold wars as well as 'hot' ones. On the other hand, governments do not always approve when public knowledge gives direction to policy. In peace time, they prefer to provide the direction suggested by their own political ideologies and the immediate wishes of their constituents. The lack of absolute certainty in science is therefore the preferred card to be played by some politicians, that is, either to disregard the uncertainties boldly or to embrace them warmly. Short-lived elected governments actually seem to find it difficult to cope effectively with long-term challenges that extend beyond their own time in office. The constituencies, which often prefer short-term economic growth to uncertain environmental consequences, are probably the only force that can change the political mood and proclivity of democratic governments. For this to happen, the electorate may have to suffer the consequences of global warming, at which time public knowledge may suggest that it is too late to turn developments around. Politics is not only a strange bedfellow, but also sets strange agendas.

In summing up, the presence of politics in research puts governments in a unique position to change and preserve the course of science in support of national interests. Owing to their appropriating powers, no single entity can do this as effectively as governments. It is also abundantly clear that they are not bound by public knowledge in forming their policies. Since there are always controversial research results fighting to become public knowledge, governments are in a position to pick and choose which research results support their national interests. In general, the lodestar of governments is national interests, not public knowledge. This is why governments do not always exercise their powers to change the direction of science. For governments, science is often, if not always, a vehicle of policy support. Consequently, governments are not reliable partners for redirecting the course of science.

Be political decision-making as it may, what has been demonstrated is that the external influences of politics and history may influence and change the course of science to serve national and international interests without compromising the integrity of research. In recent times, innovation is the political term explicitly applied to the provision of new knowledge that straddles disciplinary and other boundaries. Historically, imminent political crises have had the same effect many times over, producing knowledge-instrumental research of an interdisciplinary nature to meet pressing practical-instrumental needs. In these cases, scientists meet their societal obligation and fulfil their part of the tacit contract with society (see Chapter 4) by adjusting the way they conduct their trade.

INTERMEDIATE INFLUENCES

No synthetic theory can be more all-embracing than the unity of science thesis. This being the case, one would expect most 'interdisciplinarians' to invest their hearts, souls and minds in support of the realization of positivist ideas. Upon closer examination, however, such an expectation is built on loose, sandy ground that can easily be eroded.

The unity of science thesis implicated a hierarchical ordering of the sciences on the basis of their deterministic and predictive abilities. On top were the natural sciences and physics became the 'gold standard' against which weaker forms of disciplines could be measured. This hierarchy undercut the status of the discipline whose theory was reduced, placing the practitioners of those disciplines in the role of applied scientists serving the needs of higher ranking disciplines. The lower ranking branches of learning, in particular the humanities, counteracted and closed their ranks in opposition to the unity of science thesis. Thus, the irony of the matter was that an interdisciplinary theory suggesting perfect synthetic harmony between all the sciences caused a deep and to some extent hostile split between them. In this way, the unity of science thesis strengthened and preserved the very opposite of its own intention, disciplinarity. Dialogue between the disciplines ceased because the ruling philosophy of science was miles away from the reality of practicing scientists. The map of philosophy did not match the topography of science, and meta-theory was based on logical speculation rather than on the practice of science. Positivism alone provided no real life guidance. To many, the very trade of philosophers came into discredit.

The discrepancy between theory and practice opened up for post-positivist thinking which focused on the *actual practice of scientists,* as documented throughout history. Here, the message of pure logic is replaced by a message of practice, and the message of wishful thinking by a message of reality. Post-positivists maintain that judgments cannot be grounded on *a priori* principles, but must be established by pragmatic judgment based on what has been successful science. When faced with difficult philosophical questions, the only solution is to dive even deeper into the empiry to see how scientists can get out of this difficulty.[19]

At present, scientists have divergent views on the usefulness of philosophy to science in general. There are researchers who argue that science is not based on any philosophy at all, and that research does not depend upon the support of any philosophy of science, positivist or post-positivist. As they see it, the problem is not with science as such but with *scientism* claiming to have a close affiliation with the sciences and speaking in their name and arguing for philosophy's unlimited and universal applicability.[20] Others claim social theory to be more useful than philosophy to science because the former deals more directly with the social, political and cultural tendencies and characteristics of real societies.[21] Philosophy is best when formed through empirical and theoretical practice rather than abstract speculation.[22] Others underscore that the philosophy of science operates in the grey zone between philosophy and scientific practice, and that science theory is critical for understanding scientific practice.[23] In this vein of thought, philosophy is not regarded as a science in its own right, but as a reflection of the practical approaches applied to acquire reliable scientific knowledge and insight. At present, the focus is on *empiry* and *scientific practice.* Some researchers think this more than suffices and discard of any role for philosophy, whereas post-positivists encourage a merging of the aspects of reality with those of philosophy to provide for theoretical stringency and some additional guidance. The INSROP experience reflects both these positions.

Philosophy and the Practice of INSROP

In planning and executing INSROP, the ongoing debate between positivists and post-positivists was never an issue. Nor was the unity of science thesis suggested by anyone as being a useful means of integration. What the IN-SROP partners leaned on in the execution of their work, was their own disciplinary experience and interdisciplinary curiosity, making scientific practice and curiosity rather than philosophical speculation their common point of reference (see Chapter 3). At the outset, the philosophy of science was regarded as a non-issue and to a large extent separated from practice (see Chapter 5).

The INSROP experience affirms a common erudition, namely that interdisciplinary, and in particular transdisciplinary projects should begin with the temporary suspension of all known methods, theories and models, so that the overall questions are formulated in an unbiased, collaborative manner.[24] Allowing a discipline, theory or methodology to dominate may inhibit role negotiations, delay joint efforts and create uncritical social and cognitive dependence.[25] To avoid such pitfalls, staying neutral in the choice of scientific integrative measures at the outset may avoid a dogfight that could ultimately have been resolved naturally by peaceful means among the disciplinarians as they walk their designated road together.

According to this reasoning, interdisciplinary research is nothing but disciplines and knowledge expertise in a concerted search for bridge-building methodologies and theories moving into each others' territories with the aim of achieving a synthesis of knowledge. In the absence of a shared theoretical understanding, members will most likely maintain their expert roles, resorting to their disciplinary skills.[26]

Achieving the overall integration of research, which was the paramount objective of INSROP, required all relevant disciplines to stay the course and contribute throughout the life span of the programme. Consequently, the disciplines were considered equal in scientific worth, and respect was invoked for differences in methods and procedures. Borrowing and sharing tools was encouraged across disciplinary boundaries. Synergetic discussions were expected and actively promoted through workshops between the various subprogrammes. The social and natural parameters were regarded as interrelated and interacting. Interpretation was seen as a means of scientific methodology. Values were defined as objects of study and qualitative and quantitative methods were regarded as supplementary means of science. Last but not least, stakeholder knowledge was integrated on terms equal to that of scientific knowledge. The assumption was that these were measures to ease constructive cooperation also in the form of adversarial collaboration across boundaries. What was shared by all INSROP researchers was the *common scientific approach,* as depicted by Stanley Schuum.

In this way, the practical needs of the programme design made INSROP indirectly reflect the philosophical images of post-positivism.

Post-Positivism and Science

The criticism of positivism notwithstanding, most scientists will agree that theorization in one form or another can play a useful and supplementary role in empirical science, as long as the speculations of philosophers are based in the reality of science. The problem is not with philosophy *per se,* as a guiding influence on science, but that philosophers have not thus far provided scientists with the right concepts. The contribution of philosophy to the sciences lies rather in the possibility of a deeper, or reflective, understanding of the content of the claims advanced in scientific theories.[27]

To provide guidance to science, the challenge of post-positivism is to build a coherent meta-theory that blends the practices of scientists with the reflective understanding of philosophers. By building theory based on the experiences of practice, philosophers are narrowing the gap between themselves and the scientific community. Or, in our context, their common point of reference is putting them on speaking terms and opening a brand new chapter of communication. If post-positivism continues to evolve in this direction, the influence of philosophy on the course of science will most likely increase in future. This effect has to a certain extent already been demonstrated in relation to the freedom of research principle.

The joining of forces between scientific practice and philosophical rationality has for decennia proven effective in guarding science from negative external interference. As discussed in Chapter 2, external influences may, directly or indirectly, deliberately or not, violate the ideals and integrity of science, which by its uncompromising pursuit of the truth may pose a threat to group interests outside science. The freedom of research principle and its three expressions are forged at the interface of science and philosophy. In this context, philosophy offers rationale to meet the needs of science, and science provides content to the logic of philosophy. In this interrelationship, the freedom of research principle is a necessary working condition for scientists to produce quality, a fundamental value for philosophers to practice rationality and a necessity for societies to stimulate scientific innovation. This embodies practice, utility and theory, hand in hand. Although there are significant differences between positivists and post-positivists, they are all ardent proponents of the freedom of research principle. In this perspective, the freedom of research principle shows that philosophy based on the needs of scientific practice make philosophy useful to and a potential influencer of the course of science.

Although interdisciplinarity no longer has a consensual philosophy of its own, post-positivism is a most congenial platform upon which to conduct interdisciplinary research, academic as well as transdisciplinar. It tells us that values are present in all types of research; that elusive Man is just a little more elusive than complex Nature; that the total methodologies of the sciences are to a certain but restrictive degree interchangeable; that all sciences are both law- and fact-finding, particularistic and theory-building; that there is a scientific approach that applies to all sciences; and that the vein of thought of positivism is logically impossible. However, these common features call for a qualification. Post-positivists claim that the tenet of science progressing towards producing a complete and comprehensive scientific world picture, which will constitute the ultimate reality, is beyond reach. This is so because scientific knowledge excludes many human aspects, such as moral and aesthetic values, which are just as real as physical data and biological traits. Nor is it attainable in practical terms. Scientific knowledge is always expanding too rapidly to be grasped as a whole.[28] Wholeness will always lag behind disciplinary progress, and continuously have to be updated without ever being able to present the final unified picture. This poses the question of whether declining the possibility of all-embracing integration also disposes of interdisciplinarity as a viable practice of science?

The answer to this question is a resounding 'no.' Complexity arises at many levels in nature and society, which implies that a little bit of wholeness is better than none at all. Thus, post-positivism provides for interdisciplinarity in a restricted but doable manner. In post-positivism, the practice of science is the defining criteria on which philosophy is built. In this perspective, post-positivism and science blend together, making it hard to separate the role of philosophy from that of science. At present, they present a united front, enhancing each others' influence.

INTERNAL INFLUENCES

In post-academic and post-normal science, the border between all the types and practices of science is blurred and is, in certain respects, of moderate significance. Changes of paradigm may happen due to the internal dynamics of the Circle of Science. This makes it hard to establish a clear-cut distinction between internal and external influences of change. The sum of their interaction becomes the cause of change. Brian Robson and Elisabeth Shove take issue with this assumption, arguing that the focus and emphasis on the uses and users of research have fostered an increasingly complex, but also divided research culture with its own demands and requirements. Rather than supporting the assumption underlying the emergence of one new research environment, post-academic science, they argue that the focus on applicability in research has created two different scientific cultures with two distinct and separate ways of conducting research and influencing the course of science. The first is labelled *internal circuits of credibility* and revolves around conventional basic research and research councils, whereas the other connects applied research and non-academic needs and requirements and is called *external circuits of credibility* (see Figure 6.2).[29] Let us address this two-world

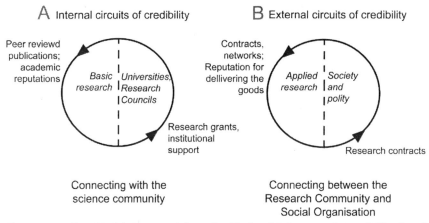

A   Internal circuits of credibility   B   External circuits of credibility

Peer reviewd publications; academic reputations

Basic research | *Universities, Research Councils*

Research grants, institutional support

Contracts, networks; Reputation for dellivering the goods

Applied research | *Society and polity*

Research contracts

Connecting with the science community

Connecting between the Research Community and Social Organisation

Figure 6.2.   Two Models Extracted from the Circle of Science: a. Internal Circuits of Credibility and b. External Circuits of Credibility. *Source: Adapted and developed from: Brian Robson and Elisabeth Shove: "Geography and Public Policy. A political turn," in John E. Matthews and David T. Herbert: Unifying Geography,...., Op. cit. p.* 357 *(published with the written permission of the authors).*

perspective on science and compare it with the outlook provided by post-academic and post-normal reasoning.

The Internal Circuit of Credibility

In the 'internal circuit of credibility,' publication and scholarship are essential ingredients. In other words, the point is to contribute to contemporary scholarly debate, to fuel and follow intellectual fashion, to make scientific marks on one's peers and to achieve academic recognition. In this community, academics act according to their self-determined convictions, conducting knowledge- and symbolic-instrumental research in the traditional manner. Scientific reputations are built through the circulation of ideas, cross-referencing and the peer-reviewing of publications, which ultimately attracts research funding from research councils in an ongoing, repetitive 'credibility circle.' Due to the importance of quality, research in this circuit is generally conducted on a more leisurely timetable than what is the case with research that serves the needs of outsiders. In this culture, conventions, trends and ways of working are determined by the academics involved—by those internal to the culture. As put by Robson and Shove: ". . . individual and group research agendas and priorities are oriented around the pursuit of respect and recognition. . . . What counts, and what is valued, is the advancement of academic debate. Prestige is attached to those "who star in the role" (see Figure 6.2.a)[30] In this depiction, the circuit is self-sufficient, self-governed, self-content and self-centred, existing on its own premises in seclusion from the users of scientific innovations. The reward system for basic researchers is still to a large extent inward looking and exercised according to self-defined conventions.

In this circuit, a reasonable expectation is that changeability exists in moderation, and that the internal credibility processes will sustain the tradition of basic disciplinary research. This circuit is to a large extent protected from the influence of external players and is consequently a stronghold of preservation.

The External Circuit of Credibility

What counts and what is valued in the 'external circuit of credibility' is the capacity "to contribute effectively to an externally defined suite of problems, to do so on time—that is according to a timetable dictated by external events—and with a language and style that are recognized by and accessible to the external user communities."[31] This is research commissioned from outside the scientific community. Successful completion of research contracts enhances researchers' reputations as being reliable, dependable and effective, and thereby worthy of new contracts. In this reputation–building process, the researcher broadens his or her networks of contacts outside the scientific community and creates credibility in the outside world. This circuit is one of communication and interaction between contrasting spheres, allowing scientists and practitioners to compare notes. To succeed, researchers have to develop skills other than those required for scientific publication. Their abilities to chair meetings, to organize events, to communicate effectively, to adhere to timetables and to engage multiple audiences are essential. What they are expected to do is to produce the evidentiary basis on which policy and/or societal action might be built (see Figure 6.2.b). This model mirrors what goes on in applied problem-oriented research and the reward system is to a large extent outward looking and exercised on the basis of externally defined conventions. Monetary power combines with knowledge power, and the two interact because both are in need of the other. In this circuit, the main players from the science side come from applied research institutes established to translate scientific findings into practical utility. To a certain but limited extent, research companies and consultancy firms also are part of this circuit (see Figure 1.4, Chapter 1).

In this circuit, the changeability of science is strongly influenced by the non-scientific community. In Chapters 1 and 3, we argued that society and polity are confronted with problems that may know no disciplinary boundaries. In this circuit, the scientific community is thus required by monetary power to respond to all kinds of problems—disciplinary as well as interdisciplinary, broad as well as narrow.

Two Separate Circuits

Robson and Shove argue that 'academic recognition' stemming from the internal circuit does not go well with 'user engagement' stemming from the external circuit. The credit, skills and work accumulated in one domain are acquired at the expense of time and energy in the other. The two circuits are bound together because they "run on different skills, revolve around different agendas, depend upon contrasting concepts of value and are rewarded through independent channels and mechanisms (see Figure 6.3)."[32] The difference is between scientific quality produced through basic high quality research in the internal circuit and user relevance required by society and polity produced by research institutes in the external circuit. As Robson and Shove see it, there are few obvious routes through which the internal and external circuits feed into each other.[33] The two circuits simply do not connect.[34] The assumption that the two cultures exist in seclusion from each other may in the long haul cause a widening of the gulf between basic and applied research and consequently between the two circuits,[35] and between society and academia. There will be one segment of the scientific community that forces a change in the direction of interdisciplinary problem-solving and one that

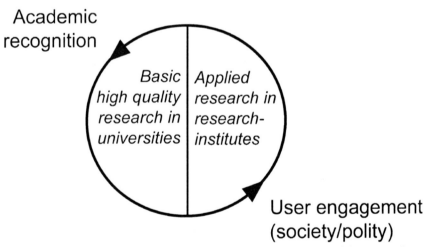

Figure 6.3. The Non-Existent Circuit. Source: *Adapted and developed from: B. Robson and E. Shove: "Geography and Public Policy," in J. E. Matthews and D. T. Herbert: Unifying Geography*, p. 357 (published with the written permission of the authors).

stays the course of basic disciplinary refinement. This development can, according to the two authors, be stopped through a reinvigoration of academic priorities based on first hand experience of grappling with shorter term challenges or a re-specification of the nature of policy agendas through the influx of new styles and strategies of enquiry. Against this backdrop, Robson and Shove argue that the key challenge of the present is to bridge the two circuits and reap the rewards of both.[36] The challenge is to find routes through which the two cultures can be bridged and ultimately feed into each other. At present, such bridges have not yet been established, according to Robson and Shove. This reasoning is in sharp contrast to the thoughts underpinning post-academic and post-normal science.

The Potential for Bridging External and Internal Circuits

Robson and Shove find geography, their own field of expertise, to be potentially well placed to exploit both circuits in an interdependent fashion. Two features are especially relevant in this respect. First, there is geography's long-standing interest in handling empirical evidence, which is also a prime concern of society and polity, and not least of post-positivists. The needs identified involve giving added centrality to the analysis and interpretation of empirical data, and assigning greater value to the interpretation of quantitative as well as qualitative evidence and to the formal analysis of the evidentiary base of the subject. Second, there is geography's essentially multidisciplinary nature, which fits well with the imperatives of much applied research. As has been demonstrated throughout this book, most of the questions that preoccupy users lie on the margins of traditional disciplines or call for multidisciplinary approaches.[37] This is just another way of saying that societies have problems, sciences have disciplines, and that overlapping between them is essential for making science even more relevant to external users. The means identified by Robson and Shove to correct and rearrange the monodisciplinary organization of the sciences is to accommodate the needs of society. In their minds, the fact that academic geographers operate alongside colleagues whose interests span the human and physical dimensions of the subject can sensitize them to the languages, concepts and concerns of multifarious sub-disciplines, and make geographers well placed to contribute to the external circuit of research. They conclude: "the evidence-based experience of the subject, its multidisciplinary nature and its capacity to configure academic expertise from a range of subjects could become key components in deliberately developing a distinctive and invaluable role for geographical research. Cast in this way, the tensions and tendencies we have outlined could appear in a new light."[38] That is to say that a linkage or bridge should be established between the practical-instrumental needs articulated in the external circuit and the knowledge-instrumental response effectuated in the internal circuit by way of a multi- and interdisciplinary approach. Here they are in agreement with a core premise in transdisciplinarian thinking. This reasoning is good as far as it goes, but it requires some critical evaluation and supplementation.

As shown in Chapter 4, geography is not the only multidisciplinary discipline featuring a bridge-building potential. The competing imperatives of academia and external user communities are placing a strain on other social and natural sciences as well, indicating that science is approaching a new age of synthesis[39] According to some, the new paradigm has already arrived and can be described in various ways; it can be called a holistic world view, seeing the world as an integrated whole, or it can be called an ecological view based in the concept of *deep ecology,* which refuses to see the world as a collection of isolated objects, but rather as a network of phenomena that are fundamentally interconnected and interdependent. This is a paradigm with a very profound sense of connectedness, of context, of relationship, of belonging.[40] Johan Heilbron states that ever since the emergence of the notion of interdisciplinarity and its diffusion through the academic system in the 1960s, the erosion of disciplines has been the dominant mode of academic organization. The growing importance of practical fields of knowledge, and the increasing heteronomy of academic institutions suggest, according to Heilbron, that we have entered an era in which the disciplines have not disappeared but tend to function as one mode of organization beside several others.[41] Some scholars argue that this is due to a troubling feeling within the scientific community that normal science is running out of creativity, which increasingly occurs at the junction between traditional disciplines and the interface of science and society. This in turn affects the alleged differences between the circuits.[42]

Interdisciplinary Bridges Between Circuits

Contrary to what is depicted in Figure 6.3, the above cited evidence suggests that there are connections, bridges and/or overlaps between the two circuits. In addition to the interdisciplinary bridge discussed above, at least three more can be identified: I. *Internal university bridges* formed by the research community itself, II. *External polity bridges* established by external users, and III. *Cooperative hybrid bridges* forged by external and internal users in combination. Let us briefly address each of them to illustrate the emerging pattern of circuit interconnections and the existence of a third culture, the *Intermediate circuits of credibility* (see Figures and 6.4 and 6.5).

Internal University Bridges

The university bridges are of three kinds: *university research centres, applied basic university research* and *transdisciplinarity university projects.*

Literally speaking, Figure 6.2 depicts a division of labour in which grant researchers do basic research in the internal circuit and contract researchers do applied research in the external circuit. This clear-cut image, which is the core of Robson/Shove model, is blurred when they conclude that university research centres are becoming more reliant on external funding "not just to sustain their contribution to academic debate, but as a defining condition of their very existence."[43] Here, the university centres are correctly designated as having two purposes referring to the two circuits, respectively. On the one hand, the centres contribute to the academic debate, implying that they do (grant) research or high quality research belonging in the internal circuit. On the other hand, they are increasingly dependent on external funding, implying that they do (contract) research, which is the assignment of the external circuit. This is just another way of saying that the centres act in accordance with the prerequisites of both circuits, and that scientific quality and user relevance are achieved by the work of these centres.

The ambiguity of their reasoning is further exacerbated when they state that "Since the currency of promotion and esteem within universities is still, predominantly, a matter of publishing in peer-reviewed journals, the result is a still-divided world."[44] This assumption gives rise to three additional comments. First, modern university centres and their staffs are increasingly becoming entities *of* the university and not, as previously, entities *at* the university. As such, they are subject to much the same promotion procedures and esteem criteria as those applying to the faculty of university departments. Second, if the faculty of university centres contributes to the academic debate, they are bound to publish in peer-review journals on an equal footing with disciplinary researchers. Third, university centres are in most cases established to deal with issues straddling disciplinary boundaries, representing an alternative to the disciplinary orientation of departments. In Sweden, there has been a marked tendency in recent years to multiply the number of university centres, many of which have been designated to serve interdisciplinary needs and purposes. What is more: Many of these establishments have resulted in extensive cooperation between and within universities, and the projection is that this trend will be further strengthened in the years to come.[45] In the light of the interchanges of personnel that often take place between university departments and centres, the researchers of the two circuits have a venue of interaction to even out whatever differences there might be in the quality of research between the two groups of faculty. Robson and Shove imply that the quality of research in the external circuit is 'grey' and not destined for peer-review publication.[46] At the same time, they admit that individual researchers may deal in both circuits, shuttling back and forth and capitalizing on the opportunities each domain affords, a quality especially valued by non-academic users.[47] This shuttling back and forth facilitates an interchange of ideas, skills and personnel between the two circuits. Thus, being *of* the university, contributing to the academic debate, receiving external funding, dealing in both circuits, doing contract and project research and providing a venue for disciplinarians and interdisciplinarians to meet, make the university centres bridging entities between the two cultures. They act as melting pots for the circuits.

'User relevance' is the key concept for understanding how credibility is built and exploiting the contracts achieved in the external circuit. Robson and Shove basically define the concept in terms of the ability to adhere to external timetables, deliver the goods, speak and write in a non-expert language, network, make presentations at practitioners' conferences, be known, etc. These are no doubt all important skills, but so is 'quality of research,' which is hardly mentioned and definitely played down in comparison with the other skills. Any product of research claiming to serve an applied purpose will do so only if it complies with accepted scientific standards and can withstand the scrutiny of peers. Prompt delivery of a failed product written in understandable language is of no use to anyone, neither to science nor to society. The time has long passed when basic research took place only in universities and was undertaken solely for its own sake. Well known examples of this are the research going on in applied research centres such as IBM, Microsoft, Bell Laboratories, the Max Planck Institute and others. Thus, quality of research is a prime element also in the concept of user relevance, and there is no a priori reason to assume that the quality of research is higher in the internal circuit than in the external circuit. Actually, although the research literature on the topic of university-industry relations is still in its infancy, and more results will emerge in the years to come, these studies so far conclude ". . . that there seems to be a positive relationship between commercialisation and academic quality, . . ."[48]

The applied purpose of the Manhattan Project involved basic research at the highest, most sophisticated level of science. It was funded by the US government, took place in the external circuit, was interdisciplinary in orientation and recruited most of its personnel from the internal circuit. In other words: Both circuits contributed to the success of the project in that academic recognition and user engagement and relevance developed into a higher unity. Thus science history contradicts the assumption that the two aspects do not go together in a third culture that draws on the characteristics of the two prime circuits. We call this culture the *Intermediate circuit of credibility*, connecting basic science with user utility (see Figure 6.4a). But there is also another version of this circuit revolving around *applied science* and *academic recognition* in which the former connects with the latter (see Figure 6.4.b).

The traditional division of labour between universities and applied research institutes is long gone. Thus the assertion that the circuits exist in perfect isolation from each other is a theoretical construct contradicted by the facts of history. The two circuits overlap and interact through the forces and needs of the market. Circuits are combined and recombined in different ways and versions to satisfy the needs and requirements of the users, scientists and companies alike. These recombinations build bridges between the circuits—bridges for all to cross over, depending on historical circumstances and what needs are to be fulfilled, practical as well as knowledge-instrumental. Today, basic research is both researcher and user-driven and brought about by curiosity as well as practical needs. It is therefore difficult to draw a sharp and absolute distinction between applied and basic research in the innovation process.

The corollary of this is that basic research is not always grant research. It may also take the form of contract research. In the latter instances, basic research is applied as response to needs in the market. Applied basic research often comes to expression in multidisciplinary disciplines such as medicine, technology, environmental studies, petroleum geology, polar research etc., and has been undertaken ever since World War II. In applied basic research programmes, scientific quality combines with user relevance, i.e. the two circuits overlap or may even merge. In such instances, 'academic recognition,' the hallmark of the internal circuit, combines with 'user relevance,' the hallmark of the external circuit. The Manhattan Project is just one of many examples.

Along with this development, science studies show that the volume of university-based interdisciplinary research is soaring. In the *Research Assessment Exercise* among 6000 university researchers undertaken in England in 1999, 80 per cent stated that they were involved in some sort of interdisciplinary research. On average, the respondents indicated that they devoted approximately 46 per cent of their research time to interdisciplinary research projects. Twenty per cent claimed that they spent all their time on such projects, whereas the majority of researchers alternated between disciplinary and interdisciplinary projects. Only 19 per cent spent all their time on disciplinary work. In this connection, it is interesting, and partly surprising to note that the social sciences and the humanistic disciplines were the less inclined to engage in interdisciplinary dealings. This notwithstanding, the conclusion was crystal clear: "The evidence is that interdisciplinary research is pervasive"[49] Similar trends have been identified in Swedish, Spanish, Finnish, Australian and other European studies,[50] and are assumed also to reflect the situation in countries where such studies have not yet been undertaken.[51] In the Spanish study, 80 per cent of the research groups interviewed claimed that they used knowledge from other disciplines in their own research and that they kept journals from a relatively broad range of disciplines.[52] The Swedish study showed that in the many applications received by the Research Council, approximately 40 per cent involved at least two disciplines as the basis of their research.[53] Of the projects funded by the Finnish Academy of Sciences in 1997, 2000 and 2004, 58 per cent were disciplinary and 42 per cent interdisciplinary in some sense. As was expected, narrow interdisciplinarity was more frequent than broad interdisciplinarity. About one-third of the interdisciplinary research undertaken under the auspices of the Finnish Academy was broad, making up 14 per cent of all research funded by the Academy, whereas more than 60 per cent was narrow (see Chapter 3).[54] As has been concluded: "Interdisciplinary research activity has increased dramatically internationally over the past few decades. At the moment (2005), research policy and funding agencies in many countries are developing policies and procedures for meeting the heightened societal demand and new epistemological drivers for interdisciplinary knowledge production. . . . The Academy of Finland, for example, is at the moment working actively to increase collaboration with research funding agencies in other countries, and there are already several internationally collaborative research programmes."[55]

This trend in university research was anticipated. In the early half of the 1990s, a couple of influential science analysts projected that interdisciplinary research would stand out in significance in the years to come, both as a means of defining the *direction of research* and as a *work form*.[56] This provides yet another internally erected university bridge between science and society. The Swedish study concludes that this change of direction was initiated from within universities and driven by practicing researchers, and not, as could have been expected, by external stakeholders. This is one more indication that the problem-orientation of interdisciplinary research attracts the interest of internal players, establishing bridges of interaction between the two spheres. Accordingly, researchers involved in interdisciplinary projects attracted more funding and interaction from societal institutions than did pure basic researchers.[57] The drive towards more interdisciplinary research is no longer stimulated by external forces alone, it has allies within universities as well. As has been noted: "Interdisciplinarity develops in a spontaneous manner and is a natural part of the internal dynamic of science."[58] One is tempted to add: An important part in this dynamic spontaneity is likely to rest on the rapidly growing interaction between university centres and university departments. This two-way interaction affects both the demarcations between the circuits, and the traditional demarcations between types of knowledge. Rather than dividing research into 'disciplinary' and 'interdisciplinary,' some researchers conclude that these terms only reflect different ways of categorizing the same phenomenon: the *generation of new knowledge*.[59] The likelihood is not that this new generation of knowledge is being created by circuits living in sovereign isolation from each other. Rather, it emerges in venues that invite a meeting of minds of all the stakeholders to the production and utilization of scientific and expert knowledge. This ability to produce new knowledge on the basis of different expertise is displayed convincingly in the *Handbook of Transdisciplinary Research*.[60] Of a total of 67 authors, 36 were university professors, 19 were affiliated with independent research institutes and 12 came from ministries and international organisations such as the UN and the EU. Of the 29 chapters contained in the book, 23 were written by two or more authors with different institutional affiliations and expertise. The jointly written chapters bear evidence that the authors did not give in to the challenges of their epistemological differences. The 19 individual cases discussed also show that transdisciplinary projects do not emerge on stakeholder initiatives alone. University professors and applied researchers are equally responsible. The circuits overlap and interact.

The fact that around 50 per cent of all research at Norwegian universities is based on external funding, public as well as private, shows that university centres, applied basic university research and transdisciplinarity university projects act as bridging devices between the two circuits. In countries more market-oriented than Norway, universities often depend even more on external funding than is the case for their Norwegian counterparts.

External Polity Bridges

Polity bridges are erected by governments and executive agencies predicated on recognition of the fact that basic research is the first step towards innovation as well as interdisciplinarity. In serving the needs of society, the question is not only one of making fresh combinations of existing knowledge, but also of producing new knowledge by way of basic disciplinary and interdisciplinary research. Either way, applied research is based on existing or new basic research. Western governments increasingly acknowledge that scientific progress through basic research is the lifeblood of applied research and that the former is needed to maintain the quality of the latter. In this perspective, basic research is the mandatory first step in making science applicable. Basic research is what the EU calls "investigator-driven frontier research" which is regarded as "a key driver of wealth and social progress, as it opens up new opportunities for scientific and technological advance, and is instrumental in

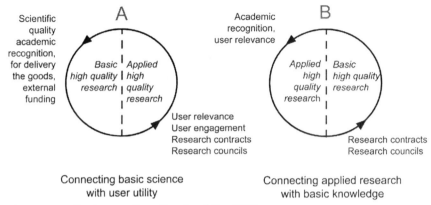

Figure 6.4. **The Intermediate Circuits of Credibility.**

producing new knowledge leading to future applications and markets."[61] This research is to be implemented by the European Research Council, whose autonomy and integrity is to be guaranteed by the European Commission.[62] What is more, funding institutions have been established with the explicit aim of serving and promoting the interconnectedness of basic and applied research. As mentioned above, the paramount objective of the Research Council of Norway (RCN) is to stimulate, among other things, interdisciplinary research with the aim of closing the gap between basic and applied research.[63] The main reasons for this include the need to provide a basis for producing useful research policy advice for the government "based on an holistic national perspective," to integrate high-quality basic and applied research and to meet the social and industrial needs of society.[64] This is so because in many areas the distinction between the fundamental and the applicable is being reduced, and the ability to work both within and between disciplines is becoming more important. "This means we need nimble researchers and agile research funders, able to bring together the powers of the most distant and dissimilar disciplines."[65]

As discussed in Chapters 1 and 3, interdisciplinarity in research is reality based, and not an artefact of scholarship. At the outset, a problem knows no other boundaries than those set by the problem itself. Its resolution will involve all those disciplines touched by the problem. In interdisciplinarity, reality, as perceived by humankind, is the point of departure when formulating a research problem. Mono-disciplines seek to resolve problems defined by the confines of disciplines, i.e. those that have been cut and shaped to fit their boundaries. Here, reality is being divided among disciplines and addressed in bits and pieces. The holism, diversity and adaptability of crossing boundaries make up an approach that is unique to interdisciplinarity. As pointed out by Bruno Latour, the difference between theory and practice ". . . is a divide that has been *made*. More exactly, it is a unity that has been fractured by the blow of a powerful hammer."[66] In line with this reasoning, some conceive of interdisciplinary research as a means of reuniting action and thought[67] at the juncture between pure theoretical research "and that of informed action, which emphasizes usefulness, efficiency, and practical results."[68] Post-normal science takes this the last mile, making societal stakeholders and their specific problem orientation an integral part of transdisciplinary research by doing away with the boundaries separating practical and scientific knowledge and calling for a new philosophy of science.

The secluded circuits of Robson and Shove are no longer separate in the science policy of an ever increasing number of western governments. In due time, the intermediate circuit of credibility may turn out to be a most preferred science culture of the executives as well. The first principle of the polity bridges is to make basic research, disciplinary as well as interdisciplinary, relevant for users, i.e. to combine academic recognition and user engagement in an intermediate circuit of credibility. The transdisciplinary approach, in particular, complies with the content of this circuit. In the light of the freedom of research principle, this is the optimal fulfilment of the unwritten utility-contact between the scientific community and society, and the ultimate consequence of the market economy.

Cooperative Hybrid Bridges

As shown in Chapter 5, the composition of the INSROP research staff abolished the usual distinction between applied and basic researchers and involved experts with relevant experience from outside of the scientific community. In this way, INSROP was formed as an extended peer community working together to integrate across all sorts of boundaries and to bridge the gap between practical life and academe.[69] INSROP served not only the symbolic-instrumental interests of contemporary politics, but also the practical-instrumental interests of economic life and the knowledge-instrumental interests of science. As such, the programme attained more objectives and fulfilled more tasks than originally envisaged. It was a typical example of how the Circle of Science works and how different types of knowledge can merge through the transdisciplinary approach.

The hybridity of the programme team was composed to mirror the overall programme objective to build up a scientifically and practically based foundation of knowledge encompassing all relevant variables involved in the navigation of the NSR, so as to enable public authorities, private interests and international organizations to make rational decisions based on scientific insight that also reflected practical experience. The team had three categories of participants: *University professors* with a special responsibility for securing the knowledge-instrumental quality of the programme, *practitioners* (insurers, sea captains, business representatives, shipping people, bureaucrats, etc.) to keep an eye on the applicability of research, and third, *researchers from applied research institutions* that spanned all the sciences to act as 'mediators,' bridging the asserted gap between the two former categories. Some representatives of the latter organizations naturally assumed the position of bridge scientists or polyvalent specialists, making communication across the various divides a smoother enterprise. Thus knowledge and insight production was designed to break down four kinds of boundaries: between disciplines within the individual branches of learning, between the four branches of sciences, between applied and basic researchers, and between academia and society. To harvest these gains from the extended peer community, representatives from all three groups were represented in every corner of the INSROP organization and at all levels of implementation, including the decision-making bodies and the research and evaluation processes.

The challenge facing members of transdisciplinary projects is that they do not speak the same professional language and thus lack a common frame of reference with all the inherent hazards this entails in terms of misunderstandings, ranking of priorities, frustrations and futile cross purposes. To overcome the pitfalls of the Tower of Babel, there is no way but to engage in the learning process of transdisciplinary (hybrid) projects that include facilitators. In INSROP, hybridity acted as a bridge between application and research as the team members became familiar with the basic theoretical formulations, methods and approach of each other, gradually building a platform of common denominators.

INSROP received financial support from multiple sources: research councils, governments, foundations and private industry. The project was applied, transdisciplinary and involved fresh basic research as well as new combinations of existing knowledge. INSROP was to large extent the sum of all the circuits: internal, external and intermediate.

It might be argued that the hybridity displayed by INSROP was not representative of the way in which most research teams are organized, and that the uniqueness of INSROP makes it less relevant as an example of bridge building. Although hybrid teams are fairly common in the implementation of major multi-and interdisciplinary programmes, this objection is to a certain extent valid, at least in the short term. In the longer term, however, the polity bridges carry the potential to stimulate hybridity for the sake of securing scientific quality, applicability and practicality in one and the same project. The participation of external non-scientific experts in research may in the longer haul be perceived by governments as a safeguard mechanism, protecting against unnecessary theorization on the part of scientists. As has been suggested, there is a need for an extended peer community, not least to bridge the gap between scientific expertise and public concerns. In this way, scientific evaluations become a dialogue among all the stakeholders in a problem, from the scientists themselves to the practitioners.[70] The independent International Evaluation Committee appointed to review the results of INSROP after three years of work complied with this requirement. In line with this, the EU Commission launched the idea of an *industry-academia pathway and partnership,* offering long-term cooperation programmes between academia and industry to enhance knowledge sharing by staff secondments and the organization of joint events.[71] Market players have already picked up on the idea of implementing an industry-academia pathway and partnership. The Internet company, InnoCentive is one extraordinary example. It has gathered a virtual workforce of 135 000 problem-solvers, including scientists, technicians, engineers and students alike, from 175 countries around the world. The overall strategy is to allow a massive number of experts from science and society to help companies or groups accomplish their goals by posting technical and theoretical challenges online, and then offering rewards ranging from USD 5000 to USD 1 million for solutions. The company's crowd-sourcing approach taps the global talent pool for societal problem-solving, reflecting a "broader trend of democratization of science."[72] Studies of the company conducted by researchers at the Harvard Business School have concluded "that the more diverse the pool of solvers, the greater the odds of a solution."[73] Of 166 challenges posted on the Internet by 26 companies, 49 were solved, a rate that the Harvard scientists considered impressive given that most of the problems had stumped well-funded research and development companies. What InnoCentive has done is to practice lock, stock and barrel the core idea underpinning post-normal science and transdisciplinarity as an approach. Here, segments of the market and segments of science apply the same philosophy and measures to resolve societal problems, bridging the internal and external circuits of science. This is a two-way street.

The Circuits and Post-Academic, Post-Normal Science

The three principal bridges identified show that the circuits intermingle, overlap, share common characteristics and recombine in different versions over time to suit the internal needs of science along with those of its external patronage. The two versions of the intermediate circuits of credibility, a blend of the internal and external cultures, strongly suggest that post-academic science, as expressed in the Circle of Science, is already a reality. In this respect, the two unconnected circuits depicted in Figure 6.2 appear to be a theoretical construct that lacks important aspects of the interconnected reality of science and society seen at present. John Ziman agrees: "There is no going back to a system of 'pure' academic science."[74] This is just another way of saying that the Robson/Shove model (see Figure 6.2) in its purest form is obsolete, of little validity and missing the crux of post-academic and post-normal developments.

As observed by Timothy O'Riordan, in this day and age of public-private partnerships, governmental control and private sponsorship, it may be that the independent scientists—the ones belonging to the internal circuits—are in the disappearing minority. This possibility notwithstanding, he contends that there is still an important place for such people.[75] Independence as prescribed by the freedom of research principle is vital to the creativity, self-criticism, self-consciousness, self-esteem, quality and, not least, user relevance of science. This raises the questions of whether post-academic science will allow independent researchers (belonging to the internal circuit of credibility) to practice their 'uncompromising scientific wonder and scrutiny,' and whether there is still room for them to function in seclusion from the disturbances of the market and the dynamics of the external circuit of credibility? John Ziman echoes the same queries from a different vantage point: "We now need a new 'contract' for basic science that will preserve its multifarious societal roles. That means serious thought about where it should be located institutionally, how it should be funded, what should be its relation with higher education and other scholarly pursuit, and, above all, what should be the conditions of employment of professional basic researchers."[76] As will be argued below, the differing stands of O'Riordan/Ziman on the one hand and Robson/Shove on the other may, under certain circumstances, converge, as each refines the perspective of the other.

THE ROBSON/SHOVE MODEL AND POST-ACADEMIC, POST-NORMAL SCIENCE

The overall objective of this chapter has been to discuss the relative and combined importance of the external, internal and intermediate influences swaying science in the direction of more interdisciplinary undertakings. It is not possible to weigh the relative ability of these influences up against each other to decide which is the more influential with any kind of quantitative preciseness. A qualitative assessment must do.

Of the three influences, the external influence exerted by governments and societal organizations seems to be the most potent. This is basically because of their monetary and funding power. On the other hand, this influence is only applied in support of national or company interests and, as such, is highly conditioned and linked to historical circumstances. However, when released, the impact of external influences may be significant, holding a clear sway over the two others. Since external users of science are problem-oriented and problems often cut across disciplinary boundaries, the external influence stimulates interdisciplinary research, both academic interdisciplinarity and transdisciplinarity.

The impact of philosophy is time-dependent. In the 1950s and 1960s, positivism was the navigational beacon of a fair amount of scientists, natural as well as behavioural. The impact of philosophy was so strong in certain quarters that it cut the scientific community in three: true believers stood up against true doubters on the validity of the unity of science thesis, whereas the majority of researchers took a leaned back position and stayed more or less indifferent to the raging debate. In this day and age, the doubters have gained ground due to the widespread feeling that philosophy, in terms of positivism, has contributed more conflict than substance to science. Thus, philosophy as a guiding light for science has generally lost credibility in recent decades. Only slowly, post-positivism is eroding some of the scepticism from the positivist era, opening the door to real life interdisciplinarity based on the practice of scientists. Against this backdrop, it seems reasonable to conclude that philosophy's stimulation of interdisciplinarity in general is, at best, less than moderate, ranging far below the stimulation provided by external users. The recent adoption of scientific practice in philosophy may eventually enhance the role of the latter, contributing to a coherent meta theory on the conduct of science.

The ability of science to alter its own course is a long-term project. No correction is slower than self-correction. A generation or so has been suggested as a timeline for turning science around and adjusting it to the needs of a new scientific era. Changes of paradigm are arising in response to inherent shortcomings of science itself, as happens when the ruling paradigm acts like a roadblock to scientific progress and depends on the retirement and death of the old guard. The paradigm of the present still leans on disciplinarity, depriving science both of the creativity engendered in the transition zones between disciplines and of the ability to address complex system challenges adequately. Science's ability to change its own direction is probably second only to that of the external influences, and above that of philosophy.

The question on the table, then, is how these three influences interact and how they influence each other to further the role of interdisciplinarity in science. Robson and Shove contend that the scientific community is split into two contrasting cultures with no interaction at all, whereas the theories of post-academic and post-normal science implies that scientific culture is one of close interaction, accommodating the needs of basic, applied, disciplinarity and interdisciplinarity research in the same melting pot. Johan Heilbron comments: "During the rapid expansion of higher education in the second half of the twentieth century, disciplinary structures were increasingly attacked as being too bureaucratic and too rigid. More flexible organizational forms were demanded from different quarters, from innovative researchers, and critical student movements, as well as from external pressure groups who opposed higher education for being too academic and not sufficiently responsive to the interests of public and private powers."[77] The new knowledge mode is "oriented and driven by problem-solving, in which the former is seen as the glue connecting the individual, the organization and the stakeholders together."[78]

Figure 6.5 illustrates the interconnectedness of the three influences (the internal is split in two) in shaping the course of science. They interact and merge in the *zone of pooling* without overlapping completely (see Figure 6.6). In this zone, politics asks for scientific innovations and philosophy provides

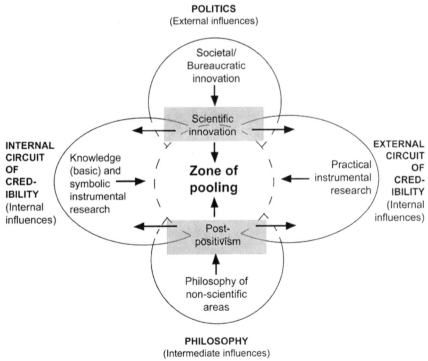

Figure 6.5.   The Interconnectedness of the Internal, External and Intermediate Influences in Shaping the Course of Science.

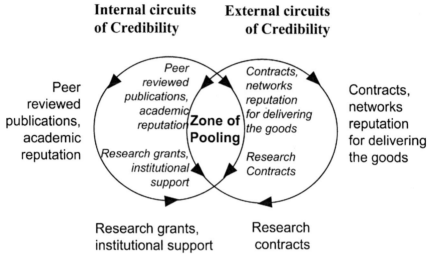

Figure 6.6. Two Interacting Models, or the Zone of Pooling, within the Circle of Sciences: "Internal Circuit of Credibility" and "External Circuit of Credibility."

post-positivism, whereas the two circuits conduct contract research, most often of an interdisciplinary nature. This is the zone in which the post-academic science in terms of the Circle of Science comes to full expression and where the two versions of the intermediate circuits of credibility operate. Thus it is in the zone where the three influences interact and strengthen each others' drive towards fostering both academic interdisciplinarity and transdisciplinarity.

Outside the zone of pooling, each of the three (four) influences has an *independent space* in which they can practice and cultivate their own professions and trades without the interference on the part of the players interacting within the zone. This space allows for external players to produce *societal and bureaucratic innovations* using their own experts, and by combining existing public knowledge in new ways. The internal circuit of credibility is located in the space where basic research takes place, whereas the external circuit of credibility is located in the space of applied research (see Figure 6.7). The internal and external circuits are thus still in operation on the basis of many of the characteristics described by Robson and Shove (see Figure 6.2). Thus, post-normal sciences do not support the contention of John Ziman that "there is no going back to a system of 'pure' academic science." Although basic research is an integral part the post-academic science, it also is located in a separate *space for cultivation* in accordance with a long tradition of independence and freewheeling. Here integration and separation act together.

The space reserved for the internal circuit is where the independent basic researcher can find refuge from all the noise and distractions of the market. The same applies to the external circuit located in the space of the applied sciences, whereas the space of philosophy provides an opportunity to practice critical inputs on the functionality of post-positivism. In this way, all influences can develop their own trades and interests based on the values and inputs intrinsic to themselves and different from those stemming from interactions within the zone. That is to say that the sciences, applied as well as basic, disciplinary as well as interdisciplinary, can enjoy the privilege of doing grant research, which is reserved for the spaces whereas contract research takes place in the zone (see Figure 6.7).

The implication is that there are basic researchers who are never subjected to the demands of the market and that there are basic researchers who take time out from the market, alternating between contract and grant research. Those researchers are active in keeping up a modified version of the internal circuit of credibility. The modification is in terms of their borders being punctuated and bridged by different versions of the other circuits. Parts of them are at one and

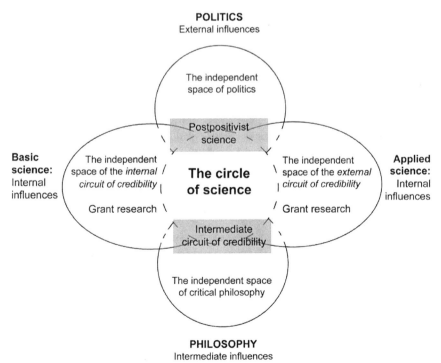

Figure 6.7. The Zone of Pooling Between the Influences and their Independent Spaces.

the same time both connected and unconnected (see Figure 6.7). In this way, the Robson-Shove model (Figure 6.2) fits in with the requirements of post-academic and post-normal science that find expression in the zone of pooling.

The three influences will never merge completely because that would deprive the scientific and user communities of the critical element secured by the independent scientists operating within the four spaces. To maintain the quality of post-academic and post-normal science, the spaces of applied and basic research have to be preserved and guarded. Full compliance between the influences could result in deflated quality of scientific outputs, harming applied, basic, disciplinary and interdisciplinary research. The confluence of the influences seems to have reached a level that satisfies all involved parties.

The freedom of research principle applies to the zone as well as to the spaces. In the latter case, the freedom is exercised in full without any compromise, whereas freedom of research in the zone may be somewhat affected, although without compromising non-negotiable scientific standards, e.g. the freedom of independent conduct. One acceptable way for the user community to influence the conduct of research is increasingly to ask for problem-related, transdisciplinary research in the zone of pooling. This may unleash the potential of post-positivist philosophy to develop the practice of interdisciplinarity further and to trigger a swifter change in the working paradigm of the scientific community. This invites some concluding remarks on the role of interdisciplinarity in the era of post-academic science.

As suggested in Figure 6.7, the battle between disciplinarians and 'interdisciplinarians' is long overdue, and should be brought to an end. The allegation that interdisciplinarity has little scientific legitimacy because it lacks an identifiable methodology and a theoretical foundation, while it had some validity at some point in time, is equally overdue. In practice, disciplinary science is the backbone of interdisciplinarity—both the academic and transdisciplinary versions—which is founded on *cooperation, corporation, mutual interdependence* and a *division of labour.* This makes interdisciplinarity an effort to be undertaken by the scientific community as a collective effort in the zone of pooling. Disciplinarity is the first step in interdisciplinarity. To move forward, all stakeholders need to sit down and discuss how to pool their resources more effectively, to divide their labour more rationally and to meet the needs of all involved, including the scientists, the funders and the users. This discussion will involve representatives from all circuits located in the zone of pooling and the four independent spaces, respectively (see Figures 6.5 and 6.7). As noted by Bino Catasus and Bengt Kristensson Uggla, "the university is facing one of its most critical periods, and *Re-Thinking Science* offers no other solution for the university than to follow society."[79] Flexibility in the organization of knowledge production has become a virtue in the context where the external conditions of universities " make it necessary to develop strategies to adopt to societal needs, to articulate substantial 'answers' to a society that has started to 'speak back'"[80] The distinction between science and society is slowly eroding.

Contrary to the assumption that disciplinarians isolate themselves *in absurdum* by their extreme specialization, the presupposition of this book is that specialization that takes place in the internal circuit is a gateway to synthesis, possibly to be further facilitated by a shift of paradigm. Specialization can help convert disciplinary curiosity into interdisciplinary curiosity and promote an interest for the problems as such. Since problems seldom come in discipline-shaped blocks, specialization will ultimately break away from the disciplinary confines and create an interest for problems in their entirety, rather than for disciplinary segments of problems, which is the first focus of units of topical specialization.

It is high time to denounce the myth that academic interdisciplinary research has no methodology of its own apart from those of the disciplines. It has methodologies and thereby qualifies as a legitimate practice of science on an equal footing with disciplinary research. The bridges between the sciences and disciplines have time and again been built through multiple projects over the years, and convergence has been accomplished by various methods and modes of synthesis. These methods and modes have been conceptualized, practiced and documented, and comply with sound scientific procedures as demonstrated in the cases discussed in Chapter 3 and 5. This is not to say that methodological improvements are not needed and possible. In particular, the practice of transdisciplinarity is in need of further methodological developments and epistemological clarifications. To comply with the definition of science, methodology must be rigorous, rational, logical, reliable and verifiable, and it has to be conceptualized so that every step in the research process is identified, argued and even made consensual. One cannot claim to be scientific without meeting the requirements defined as science. Fortunately, academic interdisciplinarity meets those requirements.

NOTES

1. Davies (83), p. 229
2. Kuhn (70), pp. 83–95
3. Latour (99), p. 80
4. Brooks (64), pp. 73–96.
5. Report No. 20 (2004–2005) to the Storting (White Paper) (04–05), p. 83
6. Østreng (99/2) pp. 365–425.
7. Østreng (06), pp. 71–81.
8. Østreng (05), pp. 80–98.

9. Løvås and Brude (98), p. 9.
10. Løvår and Brude (98), pp. 58.
11. Thompson Klein (96), p. 108.
12. Arctic Climate Impact Assessment (04), p. 125.
13. Weiner (90), p. 199.
14. Funtowicz and Ravets (08), p. 367.
15. Reisz http://www.timeshighereducation.co.uk/story.asp?sectioncode =26&story code=401125 (08), p. 2.
16. Caldwell (82), p. 114.
17. Annan (06).
18. Underdal (80).
19. Latour (99), p. 127.
20. Hughes and Sharrock (97), p. 208.
21. Peet (98), p. 6.
22. Peet (98), p. 9.
23. Hansen and Simonsen (04), p. 8.
24. Kaje (99), p.15.
25. Kaje (99), p.13.
26. Thompson Klein (90), p. 128.
27. Hansen (03), p. 97.
28. Ziman (00), p. 322.
29. Robson and Shove (04), pp. 353–365.
30. Robson and Shove (04), pp. 356–357.
31. Robson and Shove (04), p. 358.
32. Robson and Shove (04), p. 360.
33. Robson and Shove (04), p. 360.
34. Robson and Shove (04), p. 365.
35. Robson and Shove (04), p. 362.
36. Robson and Shove (04), p. 363.
37. Robson and Shove (04), p. 364.
38. Robson and Shove (04), p. 365.
39. Wilson (99), p. 12.
40. Capra (99), pp. 6 and 326.
41. Heilbron (04), pp. 38–39.
42. Pfirman (05), p. B15.
43. Robson and Shove (04), p. 358.
44. Robson and Shove (04), p. 362.
45. Sandstrøm, Friberg, Hyenstrand et al. (05), p. 127.
46. Robson and Shove (04), p. 358.
47. Robson and Shove (04), pp. 361–362.
48. Gulbrandsen (09), p. 59.
49. Evaluation Associates Ltd (99).
50. Sandstrøm, Friberg, Hyenstrand et al. (05), pp. 34–36.
51. Langfeldt (05), p. 174.
52. Sanz, Bordons, Zuleta (01), pp. 47–58.

53. Sandstrøm, Friberg, Hyenstrand et al. (05), p. 36.

54. Bruun, Hukkinen, Huutoniemi et al. (05), pp 99–1002.

55. Bruun, Hukkinen, Huutoniemi et al. (05), p.174.

56. Ziman (99) and Gibbons et al. (94).

57. Bruun, Hukkinen, Huutoniemi et al (05), pp. 34–38.

58. Bruun, Hukkinen, Huutoniemi et al (05). p. 128.

59. Bruun, Hukkinen, Huutoniemi et al (05), p.169.

60. Hadorn, Hoffman-Riem, Biber-Klemm (08).

61. Building the Europe of Knowledge (05), p. 36.

62. Building the Europe of Knowledge (05), pp. 36–37.

63. Hambro (02), p. 3.

64. Arnold, Kuhlman and Meulen (01), p. iii.

65. Arnold, Kuhlman and Meulen (01), p. 122.

66. Latour (99), p. 267.

67. Thompson Klein (90), p. 97.

68. Thompson Klein (90), p. 122.

69. Wisted and Mathisen (95), p. 21.

70. Sardar (00), pp. 64-65.

71. Building the Europe of Knowledge (05) , p. 38.

72. Travis (08) pp. 1750–1752.

73. Cited from Travis (08), p. 1752.

74. Ziman (04), p. 2.

75. O'Riordan (04), p. 129.

76. Ziman (04), p. 2.

77. Heilbron (04), p. 38.

78. Catasus and Uggla (07), p. 74.

79. Catasus and Uggla (07), p. 71.

80. Catasus and Uggla (07), p. 72.

Selected Bibliography

Aalen, O. O.: "Causality and Mechanisms: Between Statistics and Philosophy," W. Østreng (ed.): *Consilience*, See below.

Abbott, K. W.: "Modern International Relations Theory: A Dual Agenda" in *Yale Journal of International Law*, vol. 14, 1989.

Abrahamsen, A.: "Bridging Boundaries versus Breaking Boundaries: Psycholinguistics in Perspective," in *Synthesis*, 1987, vol. 72.

Activity Report 2006, Complex, NTNU, 2006.

Adler, E.: "Seizing the Middle Ground," *European Journal of International Relations*, vol. 3, no. 3, 1997.

Agenda 21: Programme of Action for Sustainable Development, United Nations Conference on Environment and Development (UNCED), Rio de Janeiro, Brazil, 3–14 June, 1992.

Agranat, Grigory: "Nuzhni li Rossii ee prostory?," *Segodina*, no. 3, 1996.

Alnes, J. H: "Philosophy and Science," W. Østreng (ed.): *Synergies*, See below.

Amundsen, B.: "Filosofen som reiste seg," *Forskning*, RCN, Oslo, no. 6/02, vol. 10.

Andersen, H: "On Turbulence and Language Change," W. Østreng (ed.), *Convergence*, See below.

Annan, Kofi: "Address," Climate Change Conference in Nairobi, 15 November 2006

Anton, T.: *Bold Sciences: Seven Scientists who are Changing our World*, W.H. Freeman and Company, New York, 2000.

Archer, M, R. Bhaskar, A. Collier, T. Lawson and A. Norrie (eds.): *Critical Realism. Essential Readings*, Routledge, London & New York, 1998.

Arctic Climate Impact Assessment: *Impacts of a Warming Arctic*, Cambridge University Press, Cambridge, 2004.

Arctic Marine Shipping Assessment 2009 Report, Draft 1, 14 November 2008.

Arnold, E., S. Kuhlman and B. Meulen: *Technopolis. A Singular Council. Evaluation of the Research Council of Norway*, Oslo, December 2001.

Bammer, G.: "Multidisciplinary Policy Research—An Australian Experience," *Prometheus*, vol. 15, no. 1, 1997.

Barnes, B, D. Bloor and J. Henry: *Scientific Knowledge. A Sociological Analysis,* Athlone, London, 1996.

Bauer, H.H.: "Barriers against Interdisciplinarity—Implications for the Studies of Science, Technology, and Society," *Science Technology and Human Values,* vol. 15, no. 1, 1990.

Baylis, J., S. Smith (eds.): *The Globalization of World Politics. An Introduction to International Relations,* Oxford University Press, Oxford, New York, 2001.

Becher, T.: "The Counter-Culture of Specialization," *European Journal of Education,* vol. 25, no. 2, 1990.

Bechtel, W.: *Philosophy of Science. An Overview for Cognitive Science,* Lawrence Erlbaum Associates, Publishers, Hillsdale, New Jersey, Hove, London, 1998.

Bellomo, N, N.K.Li and P.K. Maini: "On the Foundations of Cancer Modelling: Selected Topics, Speculations, and Perspectives," *Mathematical Models and Methods in Applied Science,* vol. 18, no. 4, 2008.

Berge, G. and N. Powell: "Reflections on Interdisciplinary Research. A synthesis of Experiences from Research in Development and the Environment," *Working Paper, 1997, 4,* CDE, University of Oslo.

Berger, G.: "Opinion and Facts," *Interdisciplinarity: Problems of Teaching and Research in Universities,* OECD, Paris, 1972.

Bioengineering Systems Research in the United States: An Overview, National Academy Press, Washington D.C., 1987.

Blumberg, B.: "Foreword," Coveney, P. and R. Highfield: *Frontiers of Complexity..,* See below.

Blunt, A.: "Geography and the Humanities Tradition," S. L. Holloway, et al. (eds.): *Key Concepts in Geography,* See below.

Blyth, M.: "Institutions and Ideas," Marsh, D. and Gerry Stoker (eds.) *Theory and Method in Political Science,* Palgrave MacMillan, London, New York, 2002.

Bohlin, I.: "Modern Polarforskning: Ett vitskapsteoretisk perspektiv på polarforskning" *Førsta årsrapport in FRN-prosjektet,* Institutionen før vitskapsteori, Gothenburg University, Gothenburg, 1988.

Boman, L.: "The Philosophy of Modern Physics" in W.Østreng (ed): *Confluence....,* See below.

Boyd, R., P. Gasper, J.D. Trout (eds.): *The Philosophy of Science,* Bradford Book, MIT Press, Cambridge and London, 1999.

Brock, T., C. L. Comitas, B. Sigurd, Å. O. F. Sundborg: *Interdisciplinary Research and Doctoral Training. A Study of the Linkoping University,* Swedish National Board of Universities and Colleges, Liber Tryck, Stockholm, 1986.

Brooks, H: "The Scientific Advisor," R. Gilpin and C. Wright (eds.): *Scientists and National Policy-Making,* Columbia University Press, 1964, New York.

Brubaker, D., W. Østreng: "The Northern Sea Route Regime: Exquisite Superpower Subterfuge?," *Ocean Development and International Law,* vol. 30, 1999.

Brude, O.W., K.A. Moe, V. Bakken et al.: *Northern Sea Route Dynamic Environmental Atlas,* FNI, INSROP Working Paper, no. 99–1998.

Bruun, H., J. Hukkinen, K. Huutoniemi et al.: *Promoting Interdisciplinary Research. The Case of the Academy of Finland*, Publication of the Academy of Finland 8/05, Helsinki, 2005.

Caldwell, L. K.: *Science and the National Environmental Policy Act: Redirecting Policy through Procedural Reform*, University of Alabama Press, Alabama, 1982.

Camic, C. & Hans J. (eds.): *The Dialogical Turn: New Roles for Sociology in the Post-Disciplinary Age*, Lanham, Rowman & Littlefield, 2004.

Campanario, J.M.: "Commentary on Influential Books and Articles Initially Rejected Because of Negative Referees' Evaluations," *Science Communications*, vol. 16, no. 3, 1995.

——: "Have Referees Rejected Some of the Most-Cited Articles of all Times?," *Journal of the American Society for Information Sciences*, vol. 47, Issue 4.

Capra, F.: *The TAO of Physics. An Exploration of the Parallels between Modern Physics and Eastern Mysticism*, 25th Anniversary Edition, Shambala, Boston, 2000.

Carnap, R.: *An Introduction to the Philosophy of Science*, Dover Publications INC, New York, 1995.

——: "Logical Foundations of the Unity of Science," in R. Boyd, et al (eds.): *The Philosophy of Science*, Op. cit.

——: "The Nature of Theories," Klemke, et al. (eds.): *Philosophy of Science..*, See below.

Caron, D. D., F. Stuart Chapin III et. Al. (eds.): *Ecological and Social Dimensions of Global Change*, Institute of International Studies, University of California, Berkeley, 1994.

Carson, Rachael: *Silent Spring*, Houghton Mifflin Company, Boston, New York, 1962/94.

Cartwright, N.: "The Truth Doesn't Explain Much," Klemke, et al. (eds.): *Philosophy of Science,....*, See below.

CAS Newsletter, no. 1, March 2004, CAS, Oslo.

Catasus, B., B. K. Uggla: "Reinventing the University as Driving Forces of Intellectual Capital" in C. Chaminade and B. Catasus (eds): *Intellectual Capital Revisited. Paradoxes in Knowledge Intensive Organizations*, Edward Elgar Publishing, Cheltenham, 2007.

Cetina, K. K.: *Epistemic Cultures. How the Sciences Make Knowledge*, Harvard University Press, Cambridge, London, 1999.

Cetto A.M. (ed.): *World Conference of Science, Science for the Twenty-first Century: A New Commitment*, UNESCO, Paris, 2000.

Chomsky, Noam: *On Nature and Language*, Cambridge University Press, Cambridge, 2002.

Christiansen, A. C.: "Technological Change and the Role of Public Policy: An Analytical Famework for Dynamic Efficiency Assessments," *FNI Report* 4/2001, Lysaker, 2001.

Chu, D., R. Strand, R. Fjelland: "Theories of Complexity. Common Denominators of Complex Systems," *Complexity*, Wiley Periodicals, Inc., 2003, vol. 8, no. 3.

Chubin, D.: "The Conceptualization of Scientific Specialties," *Sociological Quarterly*, vol. 17, 1976.

Clark, Burton R.: *The Higher Education System: Academic Organization in Cross-National Perspective*, University of California Press, Berkeley, 1983.

Clark, B. R.: *Places of Inquiry: Research and Advanced Education in Modern Universities*, University of California Press, Berkeley, 1995.

Clark, M. E. and S. A. Wawrytko (eds.): *Rethinking the Curriculum: Toward an Integrated Interdisciplinary College Education*, Study of Education 40, New York, 1990.

Clayton, K.: "Remarks on Interdisciplinarity Revisited," *Mimeographed paper*, University of Linkoping, 5 October, 1984.

Clemet, K.: "The Norwegian Government's Research Agenda," *International Conference on Science in a New Situation—The Role of Basic Research*, RCN, June 15–16, 2004, Oslo.

Clugston, M.J. (ed.): *The Penguin Dictionary of Science*, Penguin Books, London, New York, Toronto, 2004.

Collini, S.: "Introduction," C.P. Snow: *The Two Cultures.......*, See below.

Communication from the Commission: Europe and Basic Research, Commission of the European Community, Brussels, 14.1.2004, COM (2004) 9 final.

Cook, I., D. Grouch, S. Naylor and J. Ryan (eds.) *Cultural Turns/Geographical Turns*, Harlow, Prentice Hall, 2000.

Coveney, P. and R. Highfield: *Frontiers of Complexity. The Search for Order in a Chaotic World,* Fawcett Columbine, New York, 1995.

Davies, P.: *God and the New Physics.* A Touchstone book, Simon & Schuster, New York, London, Toronto, Sydney, Tokyo, Singapore, 1983.

Davis, P. and J. Gribbin: *The Matter Myth: Beyond Chaos and Complexity*, Harmondsworth, 1992.

De Mey, M.: *The Cognitive Paradigm. An Integrated Understanding of Scientific Development*, University of Chicago Press, Chicago, London, 1992.

Dogan, M.: Specialization and Recombination of Specialties in the Social Sciences," N. J. Smelser and P. B. Baltes (eds): *International Encyclopedia of the Social and Behavioral Sciences*, vol. 22, Elseviers Sciences, Oxford, 2001.

Dogan, M. and R. Phare: *Creative Marginality: Innovation at the Intersections of Social Sciences*, Boulder, Westview Press, 1990.

Dongen, J. van: "Fame, Philosophy, and Physics" in *Science*, 10 August 2007, vol. 317.

Durbin, P.: *Dictionary of Concepts in the Philosophy of Science*, Greenwood, New York, 1988.

Easton, D. and C. S. Schelling (eds.): *Divided Knowledge Across Disciplines, Across Cultures*, Sage, London.

Edwards, M. et al.: "I hver sin verden. Skogforvaltning og samfunnsutvikling på Madagaskar's høyland," K. Skretting, L. Olstad (eds): *Forskning på tvers....*, See below.

Ellis, G.: "Physics ain't what it used to be," *Nature*, vol. 438, December 2005.

Elser, J., J. D. Nagy and Y. Kuang: "Biological Stoichiometry: An Ecological Perspective on Tumor Dynamics," *Bioscience*, no. 53, 2003.

Elser, J., R. W. Sterner, E. Gorokhova et.al.: "Review: Biological Stoichiometry from Genes to Ecosystems," *Ecology Letters*, Blackwell Science, 2000.

Elser, J.: "Biological Stoichiometry: A Theoretical Framework Connecting Ecosystem Ecology, Evolution and Biochemistry for Application in Astrobiology," *International Journal of Astrobiology*, 2004.

Elzinga, A.: "Participation" in G. H. Hadorn, et al.: *Handbook of Transdisciplinary Research*, See below.

Evaluation Associates Ltd: *Interdisciplinary Research and the Research Assessment Exercise. A Report for the UK Higher Education Funding Bodies*, London 1999.

Evensen, L. S.: "Fornyelse i Tåkeheimen? Kommunikasjon på tvers av faggrenser," K. Skretting og L. Olstad (eds): *Forskning på tvers*, See below.

Fay, B. and J. D. Moon: "What Would an Adequate Philosophy of Social Science look like?" , Klemke, et al. (eds.): *Philosophy of Science*, See below.

Fedor, J.: "Special Sciences," R. Boyd, et al.(eds.): *The Philosophy of Science*, Op. cit.

Fermann, G., T. L. Knutsen (eds): *Virkelighet og Vitenskap. Perspektiver på kultur, samfunn, natur og teknologi*, Oslo, 1999.

Feynman, R. P.: "What Is Science?," *The Physics Teacher*, September 1969.

Fjelland, R.: *Innføring i vitenskapsteori*, Scandinavian University Publishing House, Oslo, 1999.

Fladmark, M. (ed): *In Search of Heritage. A Pilgrim or Tourist?*, The Robert Gordon University Heritage Library, Aberdeen, 1998.

——: (98/1) "Cultural Capital and Identity. Scotland's Democratic Intellect," M. Fladmark (ed): *In Search of Heritage......*, Op. cit.

Fler- og tverrfaglighet i miljø-og utviklingsforskning. Handlingsplan 2002–2004, RCN, Oslo 2002.

Frank, S. O.: "International Shipping on the Northern Sea Route—Russia's Perspective," C. L. Ragner (ed.): *The 21st Century,....*, See below.

Friedman, R.: "Centers and Institutes: The State of the Art," *Policy Research Centers Directory*, University of Illinois Policy Studies Organization, Urbana, 1978.

Friedman, R. M.: *The Politics of Excellence. Behind the Nobel Prize in Science*, Times Books, New York, 2001.

Faarlund, J. T.: Review of Mark C. Bakers: The Atom of the Language. The Mind's Hidden Rules of Grammar, *Norsk Lingvistisk Tidskrift*, 21, 2003.

Faarlund, J. T.: *Revolusjon i Lingvistikken. Noam Chomskys språkteori*, Det Norske Samlaget, Oslo 2005.

Fulwiler, T.: "How Well Does Writing Across the Curriculum Works?," A. Young and T. Fulwiler (eds.): *Writing Across the Disciplines. Research Into Practice*, Boynton/Cook Publishers Heinemann, Portsmouth, 1986.

Funtowicz, S., Ravetz, J.: "Values and Uncertainties" in G. H. Hadorn, et al. (eds): *Handbook of Transdiciplinary Research.......*, See below.

Gadagkar, R. "The Evolution of a Biologist in an Interdisciplinary Environment" in D. Grimm (ed): *25 Jahre Wissenschaftkolleg zu Berlin, 1981–2006.* Berlin, Akademie Verlag, 2006.

Gallagher, C., S. Greenblatt: *Practicing New Historicism*, Chicago, London, University of Chicago Press, 2000.

Garfinkel, A: "Reductionism," R. Boyd, et al. (eds.): *The Philosophy of Science,...,* Op. cit.

Geertz, C.: *The Interpretation of Culture*, Basic Books, New York, 1973.

Gibaldi, J. : *Introduction to Scholarship in Modern Languages and Literatures,* Modern Language Association, New York, 1992.

Gibbons, M. et al.: *The New Production of Knowledge*, London, Sage Publications, 1994.

Gilpin, R. and C. Wright (eds.): *Scientists and National Policy-Making*, Columbia University Press, 1964, New York.

Goonatilake, S.: *Toward a Global Science Mining Civilizational Knowledge*, Indiana University Press, Bloomington & Indianapolis, 1998.

Granberg, A. (ed): "The Problems of Interdisciplinarity: A Bibliography," *Mimeographed Report*, Research Policy Program, University of Lund, April, 1975.

Greco, J. and E. Sosa (eds): *The Blackwell Guide to Epistemology*, Blackwell Publishers, Oxford, Malden, 2001.

Greenblatt, S., G. Gunn (ed.): *Thinking Across the American Grain: Ideology, Intellect and the New Pragmatism*, University of Chicago Press, Chicago, 1992.

———: "Introduction," Greenblatt and Gunn (eds.): *Thinking Across the American Grain.*, Op. cit.

Greve, A.: "To Read a Literary Work: Human Responsiveness and the Question of Method," W. Østreng (ed.): *Consilience,* See below.

Gribbin, J.: *Almost Anyone's Guide to Science*, Yale University Press, New Haven, 1998.

———: *Deep Simplicity, Chaos, Complexity and the Emergence of Life*, Penguin Books, London, 2005.

———: *The Little Book of Science*, Barnes and Noble Inc., New York, 1999.

Gross, P., N. Levitt` and M. Lewis (eds.): *The Flight from Science and Reason,* New York Academy of Sciences, New York, 1996.

Gulbrandsen, M.: "The Role of Basic Research in Innovation," W. Østreng (ed.): *Confluence,...,* See below.

Gunn, G.: "Interdisciplinary studies," J. Gibaldi : *Introduction to Scholarship in Modern Languages and Literatures,* Modern Language Association, New York, 1992.

Hadorn, G. H. et al. (eds): *Handbook of Transdiciplinary Research*, Springer, Swiss Academies of Arts and Sciences, Bern, 2008

Hagelberg, E., J. T. Faarlund: *Introduction to PhD-Course on Evolution and Language. A Multidisciplinary Seminar in biology and Linguistics*, University of Oslo, 2004.

Hambro, C. (02/1): "Gode råd om forskningsrådet," *Chronicle* in *Aftenposten*, 21 January 2002.

———: (02/2): "Et fornyet forskningsråd," *Forskning*, RCN, vol. 10 December, 2002, no. 6.

Hansen, F., K. Simonsen: *Geografiens videnskabsteori—en introducerende discussion*, Roskilde Universitetsforlag, Roskilde, 2004.

Hansen, C.: "Truth: A Contemporary Philosophical Debate and its Bearing on Cognitive Science," W. Østreng (ed.): *Synergies........,* See below.

Harvey, D.: "On the History and Present Conditions of Geography: An Historical Materialist Manifesto," *Professional Geographer*, vol. 36, 1984.

Hawking, S. (ed): *On the Shoulders of Giants. The Great Works of Physics and Astronomy*, Running Press, Philadelphia, London, 2002.

Heffernan. M.: "Histories of Geography," S.Holloway, et al. (eds.): *Key Concepts in Geography*, See below.

Heilbron, J.: "A Regime of Disciplines. Towards a Historical Sociology of Disciplinary Knowledge," C. Camic & H. Joas (eds.): *The Dialogical Turn: New Roles for Sociology in the Post-disciplinary Age*, Lanham, Rowman & Littlefield, 2004.

Heisenberg, W.: *Physics and Philosophy. The Revolution in Modern Science*, Prometheus Books, New York, 1999.

Helle, K. (Editor in Charge) and K. Kjelstadlid, E. Lange, S. Sogner (Associate editors): *Aschehoug's Norges Historie, bind 1*, Aschehoug, Oslo, 1996.

Hemlin, S., C. M. Allwood, B. R. Martin (eds): *Creative Knowledge Environments. The Influence on Creativity in Research and Innovation*, Edward Elgar, Cheltenham, Northampton, 2004.

Herbert, D. T. and J. A. Matthews (eds.): *Unifying Geography. Common Heritage, Shared Future,* Routledge, London, New York, 2004.

Herbert, D. T., J. A. Matthews "Geography. Roots and Continuities," D. T. Herbert, J. A. Matthews (eds.): *Unifying Geography......,* Op. cit.

Hershberg, T.: "The Fragmentation of Knowledge and Practice: University, Private Sector and Public Sector Perspectives" *Issues in Integrative Studies* no. 6, 1988.

Hess, D. J.: *Science Studies. An Advanced Introduction*, New York University Press, New York, London, 1997.

Hessen, D. :"Biologien og mennesket," O. Asheim and E. Wiestad (eds.): *Filosofi & Vitenskapshistorie,* vol. 1, University of Oslo, 2003.

Hjort, J.: *The Unity of Science. A Sketch*, Gyldendal, London, Copenhagen, Christiania, 1920.

Hollinger, R.: "Introduction to the Natural and Social Sciences" in Klemke, et al. (eds.): *Philosophy of Science.,* Op. cit.

Hollaender, K., M. C. Loibl and A. Wilts: "Management," G. H. Hadorn et al. (eds): *Handbook of Transdisciplinary Research,* See below.

Holloway, S., S. Rice and G. Valentine (eds.): *Key Concepts in Geography*, Sage Publications, London, Thousand Oaks, New Dehli, 2004.

Hoffmann-Riem et at.: " Idea of the Handbook," G.H. Hadorn et al. : *Handbook of Transdiciplinary Research,* See below.

Hubbard, P., R. Kitchin, B. Bartley, D. Fuller: *Thinking Geographically. Space, Theory and Contemporary Human Geography*, Continuum, London, New York, 2002.

Hughes, J., W. Sharrock: *The Philosophy of Social Research*, Longman, London, New York, 1997.

Hønneland, G., A-K. Jørgensen: *Vårt bilde av russerne. 25 debattinnlegg om samarbeidet i nord*, Høyskole Forlaget, Oslo, 2002.

Hønneland, G.: 'Den russiske hang til matematikk' in G. Hønneland and A-K. Jørgensen: *Vårt bilde av russerne*, Op. cit.

Høyrup, J.: *Human Sciences. Reappraising the Humanities Through History and Philosophy*, State University of New York Press, New York, 2000.

Impacts of Climate Variability and Change. Pacific Northwest: A Report of the Pacific Northwest Regional Assessment Group, University of Washington, November, 1999.

INSROP Programme Report 1993–1998: Summary and Statistics of all INSROP Products and Activities, FNI, INSROP Secretariat, Lysaker, 1999.

International Northern Sea Route Programme (INSROP), *Report of the International Evaluation Committee*, Scott Polar Research Institute, University of Cambridge, Cambridge, 26 April, 1996.

Isaacson, W.: *Einstein. His Life and Universe*, Simon and Schuster, New York, 2007.

Jacoby, A.: *Senor Kon-Tiki. Boken om Thor Heyerdahl*, J.W. Cappelens Forlag, Oslo, 1965.

Jasanoff, S.:" Pluralism and Convergence in International Science Policy," *Science and Sustainability. Selected papers on IIASA's 20th anniversary*, Novographic, Vienna 1992.

—— et al.: *Handbook of Science and Technology Studies*, Thousand Oaks, Sage, 1995.

Johnston, R. B.: "Examining the validity structure of qualitative research," *Education*, 118, 1997.

Johnston, R.: "Geography and the Social Science Tradition," S. L. Holloway, et al. (eds.): *Key Concepts in Geography*, Op. cit.

Jones, M.: 'Om tverrfaglig samarbeid – kommunikasjon, verdisyn og begrepsdrøfting', *Proceedings from Forskerkonferansen om kulturlandskap*, RCN, Ås, *October 26–27, 1992*.

——:'Hva er geografi uten grenser? Om geografi og rettsgeografi', *Det Konglige Norske Videnskabers Selskabs. Forhandlinger 2001*, DKVNS, Trondheim, 2001.

Kahneman, D.: "Experiences of Collaborative Research" in *American Psychologist*, September 2003, vol. 58, no.9.

Kaje, J. H.: "Interdisciplinarity and Integrated Assessment," Mimeographed article, School of Marine Affairs, Climate Impact Group, University of Washington, Seattle, 1999.

Kennedy, P.: *The Rise and Fall of Great Powers. Economic Change and Military Conflict from 1500 to 2000*, Random House, New York, 1989.

Keohane, R. O.: *After Hegemony. Cooperation and Discord in the World Political Economy*, Princeton University Press, Princeton, 1984.

Killeen, P. R.: "The Yins and Yangs of Science," *Behavior and Philosophy*, no. 31, vol. xx-xx, Cambridge Center for Behavioral Studies, 2003.

King, G., R. O. Keohane, S. Verba: *Designing Social Inquiry. Scientific Inference in Qualitative Research,* Princeton University Press, Princeton, 1994.

Kjørup, S.: *Menneskevidenskaberne. Problemer og tradisjoner i humanioras videnskapsteori,* Roskilde Universitetsforlag, Roskilde, 1999.

Klausen, A. M. (ed.): *Den norske væremåten,* J.W. Cappelens Forlag A/S, Oslo, 1984.

Klein, J.Thompson: *Crossing Boundaries. Knowledge, Disciplinarities, and Interdisciplinarities,* University Press of Virginia, Charlotteville and London, 1996.

Klein, J. Thompson: *Interdisciplinarity. History, Theory &Practice, Wayne State University Press, Detroit, 1990.*

Klemke, E.D., et al. (eds.): *Introductory Readings in the Philosophy of Science,* Prometheus Books, New York, 1998.

——: "What is the Philosophy of Science?" in Klemke, et al. (eds.) Introductory Readings in *Philosophy of Science,*...Op. cit.

Knox, P. L. and S, A. Marston: *Places and Regions in Global Context. Human Geography,* Prentice Hall, Upper Saddle River, New Jersey, 2001.

Knudsen, T.: *The Rise and Fall of World Orders.* Manchester University Press, Manchester, New York, 1999.

Kockelman, J. (ed.): *Interdisciplinarity and Higher Education,* Pennsylvania State University Press, Pennsylvania, 1979.

Kojevnikov, A. B.: "Stalin's Great Science. The Times and Adventures of Soviet Physicists," *History of Modern Physical Science*—vol. 2, Imperial College Press, London, 2004.

Kozu, N.: "Foreword," *INSROP Programme Report 1993–1998: Summaries and Statistics of all INSROP Products and Activities,* Op.cit.

Krag, C.: "Vikingtid og rikssamling 800–1130," K. Helle (Editor in Charge), K. Kjeldstadli, E. Lange, S. Sogner (Ass. Editors): *Aschehougs Norges Historie, bind 2,* Aschehoug, Oslo, 1996.

Krohn, W.: "Learning from Cases Studies," G. H. Hadorn et al. (eds): *Handbook of Transdiciplinary Research,* Op. cit.

Kuhn, T.: *The Structure of Scientific Revolutions,* The University Press of Chicago, Chicago, 1972.

Lakatos, I. A. Musgrave: *Criticism and the Growth in Knowledge,* Proceedings of the International Colloquium in the Philosophy of Science, London, 1965.

Lambert, K. and G. Britten: "Laws and Conditional Statements," Klemke, et al. (eds.): *Philosophy of Science,* Op. cit.

Langfeldt, L.: "Nytt blikk på tverrfaglig forskning," *Forskningspolitikk,* no. 4, 2005.

Larsen, T.: "Bønder i byen – på jakt etter den norske konfigurasjonen," A. M. Klausen (ed.): *Den norske væremåten,* J.W. Cappelens Forlag A/S, Oslo, 1984.

Latour, B.: *Pandora's Hope. Essays on the Reality of Science Studies,* Harvard University Press, Cambridge, London, 1999.

——: *Science in Action,* Milton Keynes, Open University Press, 1987.

Lepenies, W.: "The Direction of the Disciplines: the Future of the Universities," *Comparative Criticism,* no. 11, 1989.

Letnes, O.: "Hva med et europeisk grunnforskningsråd?," RCN, *Forskning,* no. 1/04, vol. 12.

Lillehammer, A.: "Fra Jeger til bonde – inntil 800e.Kr," K. Helle (Editor in Carge) and K. Kjelstadlid, E. Lange, S. Sogner (Associate editors): *Aschehoug's Norges Historie...,* Op.cit.

Livingstone, D. N.: *The Geographical Tradition. Episodes in the History of a Contested Enterprise,* Blackwell, Oxford, Cambridge USA, 2005.

Lothe, J., A. Storeide (eds.): *Tidsvitner. Fortellinger fra Auschwitz og Sachsenhausen,* Gyldendal, Oslo, 2006.

Lovelock, J. E.: *Gaia,* Oxford University Press, Oxford, New York, 1979.

Lubchenco, J: "Entering the Century of the Environment: A New Social Contract for Science," *Science,* no. 279, 1998.

Løvås, S. M. and O. W. Brude: "INSROP Geographical Information System," in O.D. Brude, K.A. Moe, et al.: *Northern Sea Route Dynamic Environmental Atlas,* INSROP Working Paper no. 99, 1998.

Machlup, F.: "Are the Social Sciences Really inferior?," Klemke, et al. (eds.): *Philosophy of Science,......,* Op. cit.

Marcus, G.E.:" The Twisting and Turnings of Geography and Anthropology in Winds of Millennial Transition," I. Cook, D. Grouch, S. Naylor and J. Ryan (eds.): *Cultural Turns/Geographical Turns,* Harlow, 2000.

Marsh, D. and G. Stoker (eds.) *Theory and Method in Political Science,* Palgrave MacMillan, London, New York, 2002.

Martin, B. R.: "Research Misconduct—Does Self-Policing Work?," W. Østreng (ed): *Confluence......,* See below.

Mastermann, M.: "The Nature of a Paradigm," I. Lakatos and A. Musgrave: *Criticism and the Growth in Knowledge,* Proceedings of the International Colloquium in the Philosophy of Science, London, 1965.

Matthews, J. A. and D. Herbert (eds.): *Unifying Geography. Common Heritage, Shared Future,* Routledge, London, New York, 2004.

———: "Unity in Geography. Prospects for the Discipline," J. A. Matthews and D. T. Herbert: *Unifying Geography,* Op. cit.

May, R. N.: "Uses and Abuses of Mathematics in Biology," *Science,* vol. 303, 6 February, 2004.

McAnulla, S.: "Structure and Agency," D. Marsh and G. Stoker (eds.): *Theory and Methods in Political Science,* Palgrave Macmillan, London, New York, 2002.

Merton, J.: "The Normative Structure of Science," N. W. Storer (ed.): *The Sociology of Science. Theoretical and Empirical Investigations,* University of Chicago Press, Chicago, 1974.

Miller, H.: "The Future of Literary Theory," W. Østreng (ed.): *Consilience......,* See below.

Mische, P. M.: "Ecological Security and the Need to Reconceptualize Sovereignty," *Alternatives,* vol. XIV, no. 4, October, 1989.

Moe, K., R. Hansson: "The INSROP Environmental Assessment," O.W. Brude, K.A. Moe et al. : *Northern Sea Route Dynamic Environmental Atlas,* Op. cit.

Moe, K. A., G. N. Semanov: "Environmental Assessments," W. Østreng (ed.): *The Natural and Societal Challenges of the Northern Sea Route,* See below.

Murden, S.: "Culture in World Affairs," J. Baylis and S. Smith (eds.): *The Globaliza-tion of World Politics. An Introduction to International Relations*, Oxford University Press, New York, 2001.

Nelson, N.: "Issues in Funding and Evaluating Interdisciplinary Research," *Journal of Canadian Studies*, vol. 15, no. 3, 2000.

Northrop, F.S.C.: "Introduction," W. Heisenberg: *Physics and Philosophy*, Op. cit.

Nowotny, H., P. Scott and M. Gibbons: *Re-Thinking Science: Knowledge and the Public in an Age of Uncertainty*, Polity Press, Cambridge, 2001.

NTNU's *Program for tverrvitenskaplig forskning*, Trondheim 19. november 1997.

Nugueira, F.: "The Symmetry-Breaking Paradigm," W. Østreng (ed): *Complexity,* See below.

Nyhus, S. "Comment," G. Berge and N. Powell: "Reflections on Interdisciplinary Research...", Op.cit.

Obasi, G.O.P: "Science and the Challenges of Environmental and Developmental Issues," *Science and Sustainability. Selected Papers on IIASA's 20th Anniversary,* IIASA, Vienna, 1992

O'Hear, A.: *An Introduction to the Philosophy of Science*, Clarendon Press, Oxford, 1998.

Oppenheimer, P., H. Putnam: "Unity of Sciences as a Working Hypothesis," R. Boyd, P. Gasper, J.D. Trout (eds.): *The Philosophy of Science*, Op. cit.

O'Riordan, T.: "Beyond Environmentalism. Towards Sustainability," J. A. Mathews and D. T. Herbert (eds.): *Unifying Geography,,* Op. cit.

Pagels, H.: *The Dream of Reason. The Computer and the Rise of the Sciences of Complexity*, Simon, New York, 1988.

Pearson, K.: *The Grammar of Science,* Everyman's Library 939, J. M. Dent & Sons LTD. London, 1937.

Pecseli, H.: "Limits to Predictability and Understanding, seen from a Physicists Perspective," W. Østreng (ed.): *Convergence......,* See below.

Peet, R.: *Modern Geographical Thought,* Blackwell Publishers, Oxford, Malden, 1998.

Peterson, R.: "Why not a Separate College of Integrated Studies," M. E. Clark and S. A. Wawrytko (eds.): *Rethinking the Curriculum: Toward an Integrated Interdisciplinary College Education*, Study of Education 40, New York, 1990.

Peresypkin, V.I. : "Foreword," *Summary and Statistics of all INSROP Products and Activities,* See below.

Pfirman, S. L., J. P. Collins, et al.: "Collaborative Efforts: Promoting Interdisciplinary Scholars" in *Chronicle of Higher Education*, February 11, 2005.

Phelan, J., P. J. Rabinowitz: *A Companion to Narrative Theory*, Blackwell Publishing, Oxford, 2005.

Piaget, J. (ed.): *Main Trends in Research in the Social and Human Sciences*, UNESCO, Mouton, Paris, 1970.

Pilet, P-E.: "The Multidisciplinary Aspects of Biology: Basic and Applied Research," *Scientia,* vol. 74, no. 116, 1981.

Pippin, R. : "Philosophy among the Disciplines," D. Grimm (ed.): *25 Jahre Wissehshaftkolleg zu Berlin...,* Op. cit.

Pohl, C., L. Kerkhoff, et al.: "Integration," G. H. Hadorn et al. (eds): *Handbook of Transdisciplinary Research,* Op. cit.

Building the Europe of Knowledge, Commission of the European Community, Brussels, 2005.

Quine, W.V. *Theories and Things,* Belknap Press, Massachusetts, 1981.

Ragin, C. C.: *The Comparative Method. Moving Beyond Qualitative and Quantitative Strategies,* University of California Press, Berkeley, Los Angeles, London, 1989.

Ragner, C. L. (ed.): *The 21st Century—Turning Point for the Northern Sea Route?,* Kluwer Academic Publishers, Dordrecht, Boston, London, 2000.

Randall, L. : *Warped Passages. Unravelling the Mysteries of the Universe's Hidden Dimensions,* Harper Collins Publishers, New York, London, Toronto, Sydney, 2006

Report No. 20 (2004–2005) to the Storting: *The Will and the Way: Research in Norway,* The Royal Norwegian Ministry of Education and Research, 2005.

Report of the International Evaluation Committee on the International Northern Sea Route Programme, Scott Polar Research Institute, University of Cambridge, 26 April 1996.

Reisz, Matthew: "The perfect brainstorm," 20 March 2008: http://www.timeshigher education.co.uk/story.asp?sectioncode =26&storycode=401125 .

Richards, K.: "Geography and the Physical Science Tradition," S. L. Holloway, et al. (eds.)*: Key Concepts in Geography,* Op. cit.

Robson, B., E. Shove: "Geography and Public Policy. A Political Turn," J. A. Matthews and D. T. Herbert: *Unifying Geography,* Op. cit.

Roll-Hansen N.: *The Lysenko Effect. The Politics of Science,* Humanity Books, New York, 2005.

Roy, R.: "Interdisciplinary Science on Campus: The Elusive Dream," J. Kockelman (ed.): *Interdisciplinarity and Higher Education,* Pennsylvania State University Press, 1979.

Rosenberg, A.: " If Economics Isn't Science, What is It?," Klemke, et al. (eds.): *Philosophy of Science,* Op. cit.

Rudge, D.W.: "Introduction to Explanation and Law," Klemke, et al. (eds.): *Philosophy of Science,* Op. cit.

Rudner, R.:" The Scientist qua Scientist Makes Value Judgements," Klemke, et al. (eds.): *Philosophy of Science,* Op. cit.

Saether, R.: "What Do We Need? The Shipping Industry's Views on the Northern Sea Route's Potential and Problems," C.L. Ragner (ed.): *The 21st Century,* Op. cit.

Sandberg, B.: "Telling History/Histories: Autobiographical Writing and Testimonies of the Holocaust," W. Østreng (ed.): *Consilience,* See below.

Salter, L. and A. Hearn: *Outside the Lines: Issues in Interdisciplinary Research,* McGill-Queens University Press, Toronto, 1996.

Sandstrøm, U., M. Friberg et al.: *Tvarvetenskap—en analys,* SRC, Stockholm, 2005.

Sanz, L., M. Bordons, M.A. Zuleta: "Interdisciplinarity as a multidimensional concept: its measures in three different research areas," *Research Evaluations,* vol. 10, 2001.

Sardar, Ziauddin: *Thomas Kuhn and the Science Wars,* Postmodern Encounters, Icon Books/Totem Books USA, Cambridge, New York, 2000.

Sasakawa, Y.: "Greetings to the INSROP Symposium Tokyo '95," H. Kitagawa (ed.): *Northern Sea Rote; Future and Perspective*, Op. cit.

———: "Foreword" ,W. Østreng (ed.): *The Natural and Societal Challenges of the Northern Sea Route*, See below.

Sax, J. L.:" A Comment: Causes of Land Cover Change From a Legal Perspective," D. D. Caron, et al. (eds): *Ecological and Social Dimensions of Global Change*, Op. cit.

Schumm, S. A.: *To Interpret the Earth. Ten ways to be wrong*, Cambridge University Press, Cambridge, 1998.

Searl, J.: *Minds, Brains and Science*, Harvard University Press, Cambridge, 1997.

Selden, R., P. Widdowson, P. Brooker: *A Readers Guide to Contemporary Literary Theory*, Prentice Hall, New York, London, Sydney, 1997.

Seyfang, G. and A. Jordan: "The Johannesburg Summit and Sustainable Development: How Effective are Environmental Mega-Conferences?," O. S. Stokke and Ø. B. Thommessen (eds.): *Yearbook on International Co-operation on Environment and Development 2002/2003*, FNI/Earthscan Publications, London and Sterling.

Shapiro, I: "Announcement," *The Chronicle of Higher Education*, New Haven, 3 March 2000.

Skirbekk, G.: "Tverrvitenskaplig forskning i eitt vitskapsteoretisk perspektiv" SVT Working memo 2—1994, Centre for Scientific Theory, University of Bergen.

Skretting, K., L. Olstad (eds): *Forskning på tvers. Tverrfaglige forskningsprosjekter ved NTNU*, Trondheim, 2004.

Slaughter, A-M., A. S. Tulemello, S. Wood: International Law and International Relations Theory: A New Generation of Interdisciplinary Scholarship," *The American Journal of International Law*, vol. 92, 1998.

Smelser, J., P. B. Baltes (eds): *International Encyclopedia of the Social and Behavioral Sciences*, vol. 22, Elseviers Sciences Ltd., Oxford, 2001.

Smolin, L.: *The Trouble with Physics. The Rising of String Theory, the Fall of a Science and What Comes Next*, Allan Lane, London, 2007.

Snow, C. P.: *The Two Cultures*, Canto, Cambridge University Press, Cambridge, 1998.

Somit, A. and J. Tanenhaus: *American Political Science, a Profile of a Discipline*, Atherton, New York, 1964.

Stace, W.: "Science and the Physical World," Klemke, et al. (eds.): *Philosophy of Science*, Op. cit.

Sterner, R. W. and J. J. Elser: *Ecological Stoichiometry. The Biology of Elements from Molecules to the Biosphere*, Princeton University Press, Princeton, Oxford, 2002.

Stokke, O. S., O. Tunander (eds.): *The Barents Region. Cooperation in Arctic Europe*, Sage Publications, Oslo, 1994.

Stoltenberg, T.: "Foreword," O. S. Stokke, O. Tunander (eds.): *The Barents Region*, Op. cit.

Storer, N.W. (ed.): *The Sociology of Science. Theoretical and Empirical Investigations*, University of Chicago Press, Chicago, London, 1974.

Sørensen, K. H.: "Hvordan er vitenskap mulig?," G. Fermann, T. L. Knutsen (eds): *Virkelighet og Vitenskap*, Op.cit.

Taylor, C.: "Interpretation and the Sciences of Man," Klemke, et al. (eds.): *Philosophy of Science,* Op. cit.

The CAS Newsletter, The Center for Advanced Study, no. 1, November 2004, vol 12.

Thomassen, E.: "The Gospel of Judas, or, Is Philology a Science?," W. Østreng (ed.): *Complexity,* See below.

Travis,J.: "Science by the Masses," *Science,* 28. March, 2008, vol. 319

Trout, J.D.: "Introductory Essay on Reductionism and the Unity of Sciences," R. Boyd, P. Gasper, J.D. Trout (eds.): *The Philosophy of Science,* Op. cit.

Turner, R.: "The many Faces of American Sociology. A Discipline in Search of Identity," D. Easton, C.S. Schelling (eds.): *Divided Knowledge Across Disciplines, Across Cultures,* Sage, London.

Truelsen, J.: "Turbulence in Plasma: What is different from Turbulence in Neutral Fluids?," W. Østreng (ed.): *Convergence,* See below.

Underdal, A.: *The Politics of International Fisheries Management,* Scandinavian University Press, Oslo, 1980.

Unwin, T.: *The Place of Geography* ,Harlow, Longman, 1992.

Vatn, A.: *Institutions and the Environment,* Edward Elgar, Cheltenham, Northampton, 2005.

Vartanov, R., A. Roginko, V. Kolossov:" Russian Security Policy 1945–96: The Role of the Arctic, the Environment and the NSR," W. Østreng (ed.): *National Security and International Environmental Cooperation in the Arctic,* See below.

Waage, P.N.: *Russland er et annet sted. En kulturhistorisk bruksanvisning,* Aventura, Oslo, 1990.

Wallen, G.: "Tvarrvetenskapliga problem i ett vetenskapsteoretisk perspektiv," *Rapport no. 130,* 1 December 1981, Institutionen for Vetenskapsteori, University of Gothenburg.

Wallerstein, I. et al.: *Åpne Samfunnsvitenskapene,* Spartacus forlag, Oslo, 1997.

Walmsley, R.D.: "The Role of Multidisciplinary Research Programmes in the Management of Water Resources," *Water South Africa,* Pretoria, vol. 18, no. 3, 1992.

Weinberg, S.: *Dreams of a Final Theory. The Scientist's Search for the Ultimate Laws of Nature,* Vintage Books, New York, 1993.

Weiner, J.: *The Next Hundred Years,* Bantam Books, New York 1990.

Whitehead, A.N.: *Process and Reality,* MacMillan, New York, 1978.

Wiesmann, Urs et al.: "Enhancing Transdisciplinary Research: A Synthesis in Fifteen Propositions," Hadorn, G. H. et al (eds): *Handbook of Transdiciplinary Research,* Op. cit.

Wilber, K.: *A Theory of Everything. An Integral Vision for Business, Politics, Science, and Spirituality,* Shambala, Boston, 2000.

Williams, J.: "Regulated Turbulence in Information," W. Østreng (ed.): *Convergence,* See below.

Wilson, E. O.: *Consilience. The Unity of Knowledge, The Unity of Knowledge,* Vintage Books, New York, 1998.

——: *Sociobiology: The New Synthesis,* Harvard University Press, Boston, 1975.

Wisted, B., W. C. Mathisen: *Tverrfaglig forskning i Norden – barrierer og forskningspolitiske virkemidler,* Nordic Council, Tema Nord, Copenhagen, 1995.

Wittrock, B.: "Institutes for Advanced Study: Ideas, Histories, Rationales," *Keynote Speech* on the Occasion of the Inauguration of the Helsinki Collegium for Advanced Studies, University of Helsinki, December 2, 2002.

Young, A. and T. Fulwiler (eds.): *Writing Across the Disciplines. Research Into Practice,* Boynton/Cook Publishers Heinemann, Portsmouth, 1986.

Zeilinger, A.: "The message of the quantum," *Nature,* vol. 438, December 2005.

Ziman, J.: "What is Science?," Klemke, E.D. et al. (eds.): *Introductory Readings in the Philosophy of Science.....,* Op. cit.

——: *Prometheus Bound: Science in a Dynamic Steady State,* Cambridge University Press, Cambridge, 1999.

——: *Real Sciences. What it is, and what it means,* Cambridge University Press, Cambridge, 2000.

Østreng, W.: *The Soviet Union in Arctic Waters. Security Implications for the Northern Flank of NATO,* Occasional Paper No. 36, The Law of the Sea Institute, University of Hawaii, Honolulu, 1987.

——: "Polar Science and Politics Close Twins or Opposite Poles in International Cooperation," S. Andresen and W. Østreng (eds.): *International Resource Management: The Role of Science and Politics,* Belhaven Press, London 1989.

——: "Political and Military Relations among the Ice states: The Conceptual Basis for State Behaviour," F. Griffith (ed.): *Arctic Alternatives: Civility or Militarism in the Circumpolar North,* Toronto, Science for Peace/Samuel Stevens, 1992.

——: (ed) (99/1) *National Security and International Environmental Cooperation— the Case of the Northern Sea Route,* Kluwer Acadmic Publishers, Dordrecht, Boston, London, 1999.

——: (ed): (99/2) The Natural and Societal Challenges of the Northern Sea Route. A Reference Work, Kluwer Academic Publishers, Dordrecht, Boston, London, 1999.

——: "National Security and the Evolving Issues of Arctic Environmental Cooperation" in W. Østreng (ed.): *National Security and International Environmental Cooperation in the Arctic,* Op. cit.

——: "National Security and the Evolving Issues of Arctic Environmental Cooperation," W. Østreng (ed.): *National Security and International Environmental Cooperation in the Arctic,* Op. cit.

——: "The Multiple Realities of the Northern Sea Route: Geographical Hot and Cool Spots of Navigation," W. Østreng (ed.): *The Natural and Societal Challenges of the Northern Sea Route,*Op. cit.

——: "International Use of the Northern Sea Route: What is the Problem?," W. Østreng (ed): *National Security and International Environmental Cooperation in the Arctic.* Op. cit.

——: " The INSROP Sweat. What was it all about and how was it handled?," W. Østreng (ed.): *The Natural and Societal Challenges of the Northern Sea Route,* Op. cit.

——: "What do we know about the Northern Sea Route? INSROP's Theoretical and Applied Research Design: Operational Features," C. L. Ragner (ed): *The 21st Century,* Op. cit.

———: (ed.) *Synergies. CAS on Interdisciplinary communication 2003/2004*, Centre for Advanced Study at the Norwegian Academy of Science and Letters, Oslo, 2004.

———: "The Gribbin Syndrome and the Entities of Knowledge Integration," W. Østreng (ed.): *Synergies. Interdisciplinary Communications 2003/2004*, Centre for Advanced Study at the Norwegian Academy of Science and Letters, Oslo, 2005.

———: (ed.) *Convergence. Interdisciplinary Communications 2004/2005*, Centre for Advanced Study at the Norwegian Academy of Science and Letters, Oslo, 2005.

———: (ed.) *Consilience. Interdisciplinary Communication 2005/2006*, Centre for Advanced Study at the Norwegian Academy of Science and Letters, Oslo, January 2007.

———: (ed.): *Complexity. Interdisciplinary Communications 2006/2007*, Centre for Advanced Study, Oslo, 2007.

———: "Crossing Scientific Boundaries by Way of Disciplines," W. Østreng (ed): *Complexity,* Op. cit.

———: "The International Northern Sea Route Programme (INSROP): Applicable Lessons Learned" in *Polar Record*, Cambridge University Press, no. 42 (220), 2006.

———: (ed.) *Confluence. Interdisciplinary Communications 2006/2007*, Center for Advanced Study at the Norwegian Academy of Science and Letters, Oslo, 2009.

———: "Science versus Scholarship or Scholarship as Science," W. Østreng (ed): *Confluence.....,*Op. cit.

Index

Abbott, Kenneth W., 121–22
Abrikosov, Alexei, 17, 68
academic interdisciplinarity, 25–29,
    33–34, 90–91, 171–74, 183–84; and
    circuits of credibility, 260–62; and
    methodology, 281; and
    specialization, 94; and synergetic
    discourse, 108
acupuncture, 10
adaptability, 28
adversarial collaboration, 99–100, 104,
    129, 210–11, 232
Agenda 21. *See Rio Declaration of 1992*
Agricultural Research Council of
    Norway, 133n106
alternative medicine, 36n49
animal rights, 73
Annan, Kofi, 254
anthropology, 36n49, 175; and INSROP,
    195
Anton, Ted, 103
archaeology, 101
architecture, and interdisciplinarity,
    112–13
Arctic Climate Impact Assessment
    Study (ACIA), 253
Arctic Council, 74, 225, 250
*Arctic Marine Shipping Assessment
    (AMSA)*, 225, 226, 235, 241, 250

Aristarchos of Samos, 8
Aristotle, 57
art, 81
astrology, 10
astronomy, 10
atom, 48
atomic bomb, 62, 251
attitudes, scientific, 91, 96, 128;
    aggregative approach, 94, 129, 237;
    and level of interdisciplinarity, 127–28
autobiography, 173–74
autonomization, 92, 130n6

Barents Euro-Arctic Region (BEAR),
    199–200, 239
Barnes, Barry, 93
Becher, Tony, 57
Bechtel, William, 77, 120, 146
Beethoven, Ludwig van, 13
biology, 101; and multidisciplinarity,
    175. *See also* stoichiometry
biosphere, 70
Blumberg, Baruch, 27, 115
Blyth, Mark, 145
Boolean algebra, 101–2
border interdisciplinarity. *See* academic
    interdisciplinarity
borderland interdependence. *See*
    academic interdisciplinarity

301

Breinigsville, PA USA
13 December 2009
229155BV00002B/2/P